Annals of Mathematics Studies
Number 194

Fourier Restriction for Hypersurfaces in Three Dimensions and Newton Polyhedra

Isroil A. Ikromov
Detlef Müller

PRINCETON UNIVERSITY PRESS
PRINCETON AND OXFORD
2016

Published by Princeton University Press, 41 William Street,
Princeton, New Jersey 08540

In the United Kingdom: Princeton University Press, 6 Oxford Street,
Woodstock, Oxfordshire OX20 1TW

press.princeton.edu

Library of Congress Cataloging-in-Publication Data

Names: Ikromov, Isroil A., 1961– author. |Müller, Detlef, 1954– author.
Title: Fourier restriction for hypersurfaces in three dimensions and
Newton polyhedra / Isroil A. Ikromov and Detlef Müller.
Description: Princeton : Princeton University Press, [2016] | Series: Annals of mathematics studies ; number 194 | Includes bibliographical references and index.
Identifiers: LCCN 2015041649 | ISBN 9780691170541
(hardcover : alk. paper) | ISBN 9780691170558 (pbk. : alk. paper)
Subjects: LCSH: Hypersurfaces. | Polyhedra. | Surfaces, Algebraic. | Fourier analysis.
Classification: LCC QA571 .I37 2016 | DDC 516.3/52—dc23 LC record available at http://lccn.loc.gov/2015041649

British Library Cataloging-in-Publication Data is available

This book has been composed in Times

Printed on acid-free paper. ∞

Typeset by S R Nova Pvt Ltd, Bangalore, India

Printed in the United States of America

10 9 8 7 6 5 4 3 2 1

To Eli Stein

Contents

Fourier Restriction for Hypersurfaces in Three Dimensions and Newton Polyhedra

Chapter One

Introduction

Let S be a smooth hypersurface in $\mathbb{R}^3$ with Riemannian surface measure $d\sigma$. We shall assume that S is of *finite type*, that is, that every tangent plane has finite order of contact with S. Consider the compactly supported measure $d\mu := \rho d\sigma$ on S, where $0 \le \rho \in C_0^\infty(S)$. The central problem that we shall investigate in this monograph is the determination of the range of exponents p for which a Fourier restriction estimate

$$\left(\int_S |\widehat{f}|^2\, d\mu\right)^{1/2} \le C_p \|f\|_{L^p(\mathbb{R}^3)}, \qquad f \in \mathcal{S}(\mathbb{R}^3), \tag{1.1}$$

holds true.

This problem is a special case of the more general Fourier restriction problem, which asks for the exact range of exponents p and q for which an L^p-L^q Fourier restriction estimate

$$\left(\int_S |\widehat{f}|^q\, d\mu\right)^{1/q} \le C_p \|f\|_{L^p(\mathbb{R}^n)}, \qquad f \in \mathcal{S}(\mathbb{R}^n), \tag{1.2}$$

holds true and which can be formulated for much wider classes of subvarieties S in arbitrary dimension n and suitable measures $d\mu$ supported on S. In fact, as observed by G. Mockenhaupt [M00] (see also the more recent work by I. Łaba and M. Pramanik [LB09]), it makes sense in much wider settings, even for measures $d\mu$ supported on "thin" subsets S of $\mathbb{R}^n$, such as Salem subsets of the real line.

The Fourier restriction problem presents one important instance of a wide circle of related problems, such as the boundedness properties of Bochner Riesz means, dimensional properties of Kakeya type sets, smoothing effects of averaging over time intervals for solutions to the wave equation (or more general dispersive equations), or the study of maximal averages along hypersurfaces. The common question underlying all these problems asks for the understanding of the interplay between the Fourier transform and properties of thin sets in Euclidean space, for instance geometric properties of subvarieties. Some of these aspects have been outlined in the survey article [M14], from which parts of this introduction have been taken.

The idea of Fourier restriction goes back to E. M. Stein, and a first instance of this concept is the determination of the sharp range of L^p-L^q Fourier restriction estimates for the circle in the plane through work by C. Fefferman and E. M. Stein [F70] and A. Zygmund [Z74], who obtained the endpoint estimates (see also L. Hörmander [H73] and L. Carleson and P. Sjölin [CS72] for estimates on more general related oscillatory integral operators). For subvarieties of higher dimension, the first fundamental result was obtained (in various steps) for Euclidean spheres

S^{n-1} by E. M. Stein and P. A. Tomas [To75], who proved that an L^p-L^2 Fourier restriction estimate holds true for S^{n-1}, $n \geq 3$, if and only if $p' \geq 2(2/(n-1)+1)$, where p' denotes the exponent conjugate to p, that is, $1/p + 1/p' = 1$ (cf. [S93] for the history of this result). A crucial property of Euclidean spheres which is essential for this result is the non-vanishing of the Gaussian curvature on these spheres, and indeed an analogous result holds true for every smooth hypersurface S with nonvanishing Gaussian curvature (see [Gl81]).

Fourier restriction estimates have turned out to have numerous applications to other fields. For instance, their great importance to the study of dispersive partial differential equations became evident through R. Strichartz' article [Str77], and in the PDE-literature dual versions which invoke also Plancherel's theorem are often called Strichartz estimates.

The question as to which L^p-L^q Fourier restriction estimates hold true for Euclidean spheres is still widely open. It is conjectured that estimate (1.2) holds true for $S = S^{n-1}$ if and only if $p' > 2n/(n-1)$ and $p' \geq q(2/(n-1)+1)$, and there has been a lot of deep work on this and related problems by numerous mathematicians, including J. Bourgain, T. Wolff, A. Moyua, A. Vargas, L. Vega, and T. Tao (see, e.g., [Bou91], [Bou95], [W95], [MVV96], [TVV98], [W00], [TV00], [T03], and [T04] for a few of the relevant articles, but this list is far from being complete). There has been a lot of work also on conic hypersurfaces and some on even more general classes of hypersurfaces with vanishing Gaussian curvature, for instance in Barcelo [Ba85], [Ba86], in Tao, Vargas, and Vega [TVV98], in Wolff [W01], and in Tao and Vargas [TV00], and more recently by A. Vargas and S. Lee [LV10] and S. Buschenhenke [Bu12]. Again, these citations give only a sample of what has been published on this subject.

Recent work by J. Bourgain and L. Guth [BG11], making use also of multilinear estimates from work by J. Bennett, A. Carbery, and T. Tao [BCT06], has led to further important progress. Nevertheless, this and related problems continue to represent one of the major challenges in Euclidean harmonic analysis, bearing various deep connections with other important open problems, such as the Bochner-Riesz conjecture, the Kakeya conjecture and C. Sogge's local smoothing conjecture for solutions to the wave equation. We refer to Stein's book [S93] for more information on and additional references to these topics and their history until 1993, and to more recent related essays by Tao, for instance in [T04].

As explained before, we shall restrict ourselves to the study of the Stein-Tomas-type estimates (1.1). For convex hypersurfaces of finite line type, a good understanding of this type of restriction estimates is available, even in arbitrary dimension (we refer to the article [I99] by A. Iosevich, which is based on work by J. Bruna, A. Nagel and S. Wainger [BNW88], providing sharp estimates for the Fourier transform of the surface measure on convex hypersurfaces). However, our emphasis will be to allow for very general classes of hypersurfaces $S \subset \mathbb{R}^3$, not necessarily convex, whose Gaussian curvature may vanish on small, or even large subsets.

Given such a hypersurface S, one may ask in terms of which quantities one should describe the range of ps for which (1.1) holds true. It turns out that an extremely useful concept to answer this question is the notion of

Newton polyhedron. The importance of this concept to various problems in analysis and related fields has been revealed by V.I. Arnol'd and his school, in particular through groundbreaking work by A. N. Varchenko [V76] and subsequent work by V. N. Karpushkin [K84] on estimates for oscillatory integrals, and came up again in the seminal article [PS97] by D. H. Phong and E. M. Stein on oscillatory integral operators.

Indeed, there is a close connection between estimates for oscillatory integrals and L^p-L^2 Fourier restriction estimates, which had become evident already through the aforementioned work by Stein and Tomas. The underlying principles had been formalized in a subsequent article by A. Greenleaf [Gl81]. For the case of hypersurfaces, Greenleaf's classical restriction estimate reads as follows:

THEOREM 1.1 (Greenleaf). *Assume that* $|\widehat{d\mu}(\xi)| \lesssim |\xi|^{-1/h}$. *Then the restriction estimate* (1.1) *holds true for every* $p \geq 1$ *such that* $p' \geq 2(h+1)$.

Observe next that in order to establish the restriction estimate (1.1), we may localize the estimate to a sufficiently small neighborhood of a given point x^0 on S. Notice also that if estimate (1.1) holds for the hypersurface S, then it is valid also for every affine-linear image of S, possibly with a different constant if the Jacobian of this map is not one. By applying a suitable Euclidean motion of $\mathbb{R}^3$ we may and shall therefore assume in the sequel that $x^0 = (0, 0, 0)$ and that S is the graph

$$S = S_\phi = \{(x_1, x_2, \phi(x_1, x_2)) : (x_1, x_2) \in \Omega\}$$

of a smooth function ϕ defined on a sufficiently small neighborhood Ω of the origin, such that $\phi(0, 0) = 0$ and $\nabla\phi(0, 0) = 0$.

Then we may write $\widehat{d\mu}(\xi)$ as an oscillatory integral,

$$\widehat{d\mu}(\xi) = J(\xi) := \int_\Omega e^{-i(\xi_3\phi(x_1,x_2)+\xi_1x_1+\xi_2x_2)}\eta(x)\,dx_1dx_2, \qquad \xi \in \mathbb{R}^3,$$

where $\eta \in C_0^\infty(\Omega)$. Since $\nabla\phi(0, 0) = 0$, the complete phase in this oscillatory integral will have no critical point on the support of η unless $|\xi_1| + |\xi_2| \ll |\xi_3|$, provided Ω is chosen sufficiently small. Integrations by parts then show that $\widehat{\mu}(\xi) = O(|\xi|^{-N})$ as $|\xi| \to \infty$, for every $N \in \mathbb{N}$, unless $|\xi_1| + |\xi_2| \ll |\xi_3|$.

We may thus focus on the latter case. In this case, by writing $\lambda = -\xi_3$ and $\xi_j = -s_j\lambda$, $j = 1, 2$, we are reduced to estimating two-dimensional oscillatory integrals of the form

$$I(\lambda; s) := \int e^{i\lambda(\phi(x_1,x_2)+s_1x_1+s_2x_2)}\eta(x_1, x_2)\,dx_1dx_2,$$

where we may assume without loss of generality that $\lambda \gg 1$ and that $s = (s_1, s_2) \in \mathbb{R}^2$ is sufficiently small, provided that η is supported in a sufficiently small neighborhood of the origin. The complete phase function is thus a small, linear perturbation of the function ϕ.

If $s = 0$, then the function $I(\lambda; 0)$ is given by an oscillatory integral of the form $\int e^{i\lambda\phi(x)}\eta(x)\,dx$, and it is well known ([BG69], [At70]) that for any analytic phase function ϕ defined on a neighborhood of the origin in $\mathbb{R}^n$ satisfying $\phi(0) = 0$, such an integral admits an asymptotic expansion as $\lambda \to \infty$ of the form

$$\sum_{k=0}^{\infty}\sum_{j=0}^{n-1} a_{j,k}(\eta)\lambda^{-r_k}\log(\lambda)^j, \tag{1.3}$$

provided the support of η is sufficiently small. Here, the r_k form an increasing sequence of rational numbers consisting of a finite number of arithmetic progressions, which depends only on the zero set of ϕ, and the $a_{j,k}(\eta)$ are distributions with respect to the cutoff function η. The proof is based on Hironaka's theorem on the resolution of singularities.

We are interested in the case $n = 2$. If we denote the leading exponent r_0 in (1.3) by $r_0 = 1/h$, then we find that the following estimate holds true:

$$|I(\lambda; 0)| \le C\lambda^{-1/h}\log(\lambda)^{\nu}, \quad \lambda \gg 1, \tag{1.4}$$

where ν may be 0, or 1. Assuming that this estimate is stable under sufficiently small analytic perturbations of ϕ, then we find in particular that $I(\lambda; s)$ satisfies the same estimate (1.4) for $|s|$ sufficiently small, so that we obtain the following *uniform estimate* for $\widehat{d\mu}$,

$$|\widehat{d\mu}(\xi)| \le C(1 + |\xi|)^{-1/h}\log(2 + |\xi|)^{\nu}, \quad \xi \in \mathbb{R}^3, \tag{1.5}$$

provided the support of ρ is sufficiently small. Greenleaf's theorem then shows that the Fourier restriction estimate (1.1) holds true for $p' \ge 2(h+1)$, if $\nu = 0$, and at least for $p' > 2(h+1)$, if $\nu = 1$, where $1/h$ denotes the decay rate of the oscillatory integral $I(\lambda; 0)$, hence ultimately of $\widehat{d\mu}$. Moreover, for instance for hypersurfaces with nonvanishing Gaussian curvature, this yields the sharp restriction result mentioned before.

However, as we shall see, for large classes of hypersurfaces, the relation between the decay rate of the Fourier transform of $d\mu$ and the range of $p's$ for which (1.1) holds true will not be so close anymore.

Nevertheless, uniform decay estimates of the form (1.5) will still play an important role.

The first major question that arises is thus the following one: given a smooth phase function ϕ, how can one determine the sharp decay rate $1/h$ and the exponent ν in the estimate (1.4) for the oscillatory integral $I(\lambda; 0) = \int e^{i\lambda\phi(x)}\eta(x)\,dx$? This question has been answered by Varchenko for analytic ϕ in [V76], where he identified h as the so-called height of the Newton polyhedron associated to ϕ in "adapted" coordinates and also gave a corresponding interpretation of the exponent ν. Subsequently, Karpushkin [K84] showed that the estimates given by Varchenko are stable under small analytic perturbations of the phase function ϕ, which in particular leads to uniform estimates of the form (1.5). More recently, in [IM11b], we proved, by a quite different method, that Karpushkin's result remains valid even for smooth, finite-type functions ϕ, at least for linear perturbations, which is sufficient in order to establish uniform estimates of the form (1.5).

In order to present these results in more detail, let us review some basic notations and results concerning Newton polyhedra (see [V76], [IM11a]).

1.1 NEWTON POLYHEDRA ASSOCIATED WITH ϕ, ADAPTED COORDINATES, AND UNIFORM ESTIMATES FOR OSCILLATORY INTEGRALS WITH PHASE ϕ

We shall build on the results and technics developed in [IM11a] and [IKM10], which will be our main sources, also for references to earlier and related work. Let us first recall some basic notions from [IM11a], which essentially go back to Arnol'd (cf. [Arn73], [AGV88]) and his school, most notably Varchenko [V76].

If ϕ is given as before, consider the associated Taylor series

$$\phi(x_1, x_2) \sim \sum_{\alpha_1,\alpha_2=0}^{\infty} c_{\alpha_1,\alpha_2} x_1^{\alpha_1} x_2^{\alpha_2}$$

of ϕ centered at the origin. The set

$$\mathcal{T}(\phi) := \left\{(\alpha_1, \alpha_2) \in \mathbb{N}^2 : c_{\alpha_1,\alpha_2} = \frac{1}{\alpha_1!\alpha_2!}\partial_1^{\alpha_1}\partial_2^{\alpha_2}\phi(0,0) \neq 0\right\}$$

will be called the *Taylor support* of ϕ at $(0, 0)$. We shall always assume that the function ϕ is of *finite type* at every point, that is, that the associated graph S of ϕ is of finite type. Since we are also assuming that $\phi(0, 0) = 0$ and $\nabla\phi(0, 0) = 0$, the finite-type assumption at the origin just means that

$$\mathcal{T}(\phi) \neq \emptyset.$$

The *Newton polyhedron* $\mathcal{N}(\phi)$ of ϕ at the origin is defined to be the convex hull of the union of all the quadrants $(\alpha_1, \alpha_2) + \mathbb{R}_+^2$ in $\mathbb{R}^2$, with $(\alpha_1, \alpha_2) \in \mathcal{T}(\phi)$. The associated *Newton diagram* $\mathcal{N}_d(\phi)$ of ϕ in the sense of Varchenko [V76] is the union of all compact faces of the Newton polyhedron; here, by a *face,* we shall mean an edge or a vertex.

We shall use coordinates (t_1, t_2) for points in the plane containing the Newton polyhedron, in order to distinguish this plane from the (x_1, x_2)-plane.

The *Newton distance* in the sense of Varchenko, or shorter *distance*, $d = d(\phi)$ between the Newton polyhedron and the origin is given by the coordinate d of the point (d, d) at which the bisectrix $t_1 = t_2$ intersects the boundary of the Newton polyhedron. (See Figure 1.1.)

The principal face $\pi(\phi)$ of the Newton polyhedron of ϕ is the face of minimal dimension containing the point (d, d). Deviating from the notation in [V76], we shall call the series

$$\phi_{\text{pr}}(x_1, x_2) := \sum_{(\alpha_1,\alpha_2)\in\pi(\phi)} c_{\alpha_1,\alpha_2} x_1^{\alpha_1} x_2^{\alpha_2}$$

the *principal part* of ϕ. In case that $\pi(\phi)$ is compact, ϕ_{pr} is a mixed homogeneous polynomial; otherwise, we shall consider ϕ_{pr} as a formal power series.

Note that the distance between the Newton polyhedron and the origin depends on the chosen local coordinate system in which ϕ is expressed. By a *local coordinate system (at the origin)* we shall mean a smooth coordinate system defined near the origin which preserves 0. The *height* of the smooth function ϕ is defined by

$$h(\phi) := \sup\{d_y\},$$

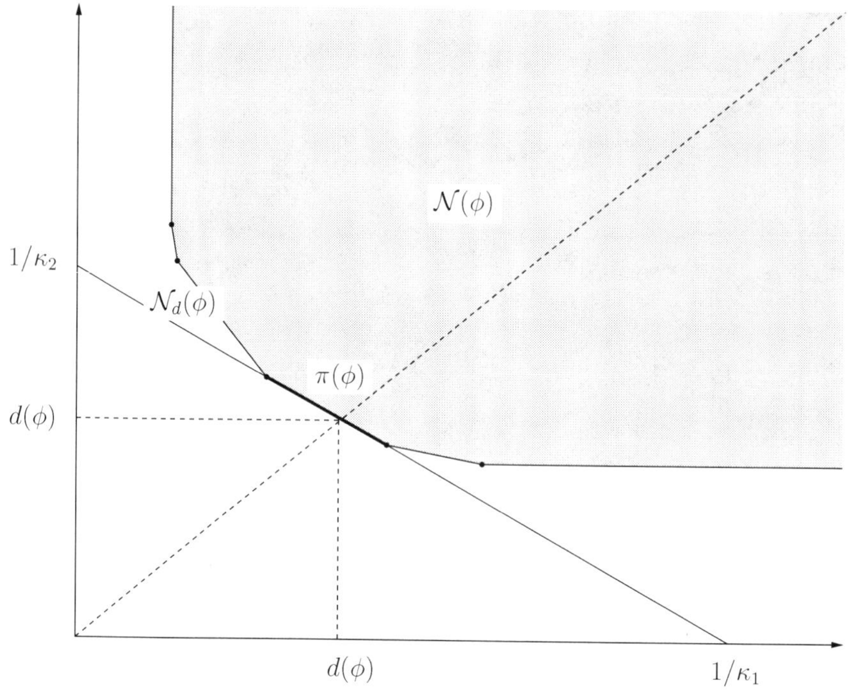

Figure 1.1 Newton polyhedron

where the supremum is taken over all local coordinate systems $y = (y_1, y_2)$ at the origin and where d_y is the distance between the Newton polyhedron and the origin in the coordinates y.

A given coordinate system x is said to be *adapted* to ϕ if $h(\phi) = d_x$. In [IM11a] we proved that one can always find an adapted local coordinate system in two dimensions, thus generalizing the fundamental work by Varchenko [V76] who worked in the setting of real-analytic functions ϕ (see also [PSS99]).

Notice that if the principal face of the Newton polyhedron $\mathcal{N}(\phi)$ is a compact edge, then it lies on a unique *principal line*

$$L := \{(t_1, t_2) \in \mathbb{R}^2 : \kappa_1 t_1 + \kappa_2 t_2 = 1\},$$

with $\kappa_1, \kappa_2 > 0$. By permuting the coordinates x_1 and x_2, if necessary, we shall always assume that $\kappa_1 \le \kappa_2$. The weight $\kappa = (\kappa_1, \kappa_2)$ will be called the *principal weight* associated with ϕ. It induces dilations $\delta_r(x_1, x_2) := (r^{\kappa_1} x_1, r^{\kappa_2} x_2),\ r > 0$, on $\mathbb{R}^2$, so that the principal part ϕ_{pr} of ϕ is κ-homogeneous of degree one with respect to these dilations, that is, $\phi_{\text{pr}}(\delta_r(x_1, x_2)) = r\phi_{\text{pr}}(x_1, x_2)$ for every $r > 0$, and we find that

$$d = \frac{1}{\kappa_1 + \kappa_2} = \frac{1}{|\kappa|}.$$

It can then easily be shown (cf. Proposition 2.2 in [IM11a]) that $\phi_{\rm pr}$ can be factored as

$$\phi_{\rm pr}(x_1, x_2) = c x_1^{\nu_1} x_2^{\nu_2} \prod_{l=1}^{M} (x_2^q - \lambda_l x_1^p)^{n_l}, \tag{1.6}$$

with $M \geq 1$, distinct nontrivial "roots" $\lambda_l \in \mathbb{C} \setminus \{0\}$ of multiplicities $n_l \in \mathbb{N} \setminus \{0\}$, and trivial roots of multiplicities $\nu_1, \nu_2 \in \mathbb{N}$ at the coordinate axes. Here, p and q are positive integers without common divisor, and $\kappa_2/\kappa_1 = p/q$.

More generally, assume that $\kappa = (\kappa_1, \kappa_2)$ is any weight with $0 < \kappa_1 \leq \kappa_2$ such that the line $L_\kappa := \{(t_1, t_2) \in \mathbb{R}^2 : \kappa_1 t_1 + \kappa_2 t_2 = 1\}$ is a supporting line to the Newton polyhedron $\mathcal{N}(\phi)$ of ϕ (recall that a *supporting line* to a convex set K in the plane is a line such that K is contained in one of the two closed half planes into which the line divides the plane and such that this line intersects the boundary of K). Then $L_\kappa \cap \mathcal{N}(\phi)$ is a face of $\mathcal{N}(\phi)$, i.e., either a compact edge or a vertex, and the *κ-principal part* of ϕ

$$\phi_\kappa(x_1, x_2) := \sum_{(\alpha_1, \alpha_2) \in L_\kappa} c_{\alpha_1, \alpha_2} x_1^{\alpha_1} x_2^{\alpha_2}$$

is a nontrivial polynomial which is κ-homogeneous of degree 1 with respect to the dilations associated to this weight as before, which can be factored in a similar way as in (1.6). By definition, we then have

$$\phi(x_1, x_2) = \phi_\kappa(x_1, x_2) + \text{ terms of higher } \kappa\text{-degree}.$$

Adaptedness of a given coordinate system can be verified by means of the following proposition (see [IM11a]):

If P is any given polynomial that is κ-homogeneous of degree one (such as $P = \phi_{\rm pr}$), then we denote by

$$n(P) := \operatorname{ord}_{S^1} P \tag{1.7}$$

the maximal order of vanishing of P along the unit circle S^1. Observe that by homogeneity, the Taylor support $\mathcal{T}(P)$ of P is contained in the face $L_\kappa \cap \mathcal{N}(P)$ of $\mathcal{N}(P)$. We therefore define the *homogeneous distance* of P by $d_h(P) := 1/(\kappa_1 + \kappa_2) = 1/|\kappa|$. Notice that $(d_h(P), d_h(P))$ is just the point of intersection of the line L_κ with the bisectrix $t_1 = t_2$, and that $d_h(P) = d(P)$ if and only if $L_\kappa \cap \mathcal{N}(P)$ intersects the bisectrix. We remark that the height of P can then easily be computed by means of the formula

$$h(P) = \max\{n(P), d_h(P)\} \tag{1.8}$$

(see Corollary 3.4 in [IM11a]). Moreover, in [IM11a] (Corollary 4.3 and Corollary 5.3), we also proved the following characterization of adaptedness of a given coordinate system.

Proposition 1.2. *The coordinates x are adapted to ϕ if and only if one of the following conditions is satisfied:*

(a) *The principal face $\pi(\phi)$ of the Newton polyhedron is a compact edge, and $n(\phi_{\rm pr}) \leq d(\phi)$.*

(b) $\pi(\phi)$ *is a vertex.*
(c) $\pi(\phi)$ *is an unbounded edge.*

These conditions had already been introduced by Varchenko, who has shown that they are sufficient for adaptedness when ϕ is analytic.

We also note that in case (a) we have $h(\phi) = h(\phi_{\rm pr}) = d_h(\phi_{\rm pr})$. Moreover, it can be shown that we are in case (a) whenever $\pi(\phi)$ is a compact edge and $\kappa_2/\kappa_1 \notin \mathbb{N}$; in this case we even have $n(\phi_{\rm pr}) < d(\phi)$ (cf. [IM11a], Corollary 2.3).

1.1.1 Construction of adapted coordinates

In the case where the coordinates (x_1, x_2) are not adapted to ϕ, the previous results show that the principal face $\pi(\phi)$ must be a compact edge, that $m := \kappa_2/\kappa_1 \in \mathbb{N}$, and that $n(\phi_{\rm pr}) > d(\phi)$. One easily verifies that this implies that $p = m$ and $q = 1$ in (1.6), and that there is at least one nontrivial, real root $x_2 = \lambda_l x_1^m$ of $\phi_{\rm pr}$ of multiplicity $n_l = n(\phi_{\rm pr}) > d(\phi)$. Indeed, one can show that this root is unique. Putting $b_1 := \lambda_l$, we shall denote the corresponding root $x_2 = b_1 x_1^m$ of $\phi_{\rm pr}$ as its *principal root.*

Changing coordinates

$$y_1 := x_1, \; y_2 := x_2 - b_1 x_1^m,$$

we arrive at a "better" coordinate system $y = (y_1, y_2)$. Indeed, this change of coordinates will transform $\phi_{\rm pr}$ into a function $\widetilde{\phi_{\rm pr}}$, where the principal face of $\widetilde{\phi_{\rm pr}}$ will be a horizontal half line at level $t_2 = n(\phi_{\rm pr})$, so that $d(\widetilde{\phi_{\rm pr}}) > d(\phi)$, and correspondingly one finds that $d(\tilde{\phi}) > d(\phi)$ if $\tilde{\phi}$ expresses ϕ in the coordinates y (cf. [IM11a]).

In particular, if the new coordinates y are still not adapted, then the principal face of $\mathcal{N}(\tilde{\phi})$ will again be a compact edge, associated to a weight $\tilde{\kappa} = (\tilde{\kappa}_1, \tilde{\kappa}_2)$ such that $\tilde{m} := \tilde{\kappa}_2/\tilde{\kappa}_1$ is again an integer and $\tilde{m} > m \geq 1$.

Somewhat oversimplifying, by iterating this procedure, we essentially arrive at Varchenko's algorithm for the construction of an adapted coordinate system (cf. [IM11a] for details).

In conclusion, one can show (compare Theorem 5.1 in [IM11a]) that there exists a smooth real-valued function ψ (which we may choose as the *principal root jet* of ϕ) of the form

$$\psi(x_1) = b_1 x_1^m + O(x_1^{m+1}), \tag{1.9}$$

with $b_1 \in \mathbb{R} \setminus \{0\}$, defined on a neighborhood of the origin such that an adapted coordinate system (y_1, y_2) for ϕ is given locally near the origin by means of the (in general nonlinear) shear

$$y_1 := x_1, \; y_2 := x_2 - \psi(x_1). \tag{1.10}$$

In these adapted coordinates, ϕ is given by

$$\phi^a(y) := \phi(y_1, y_2 + \psi(y_1)). \tag{1.11}$$

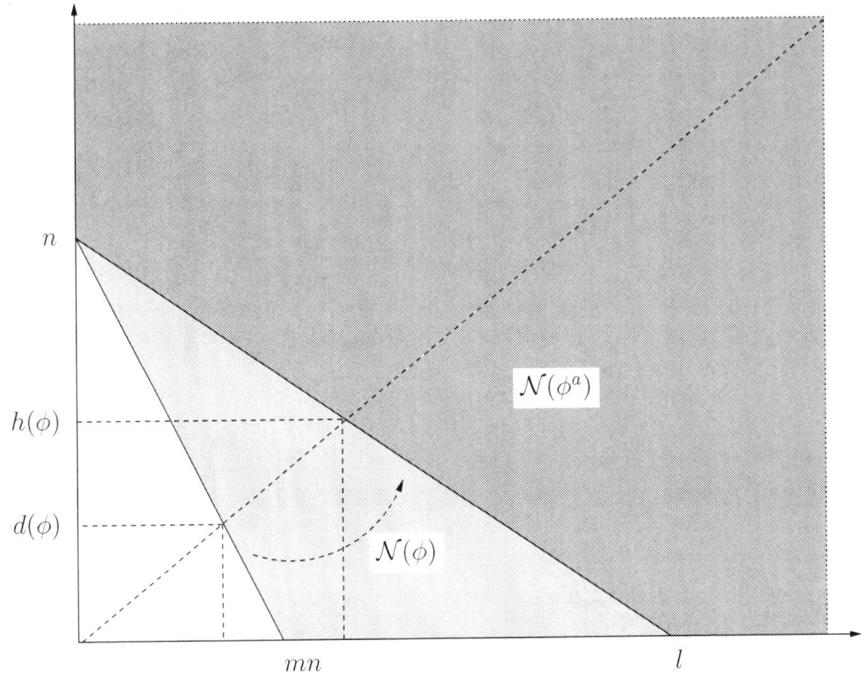

Figure 1.2 $\phi(x_1, x_2) := (x_2 - x_1^m)^n + x_1^\ell \quad (\ell > mn)$

Example 1.3.

$$\phi(x_1, x_2) := (x_2 - x_1^m)^n + x_1^\ell.$$

Assume that $\ell > mn$. Then the coordinates are not adapted. Indeed, $\phi_{\rm pr}(x_1, x_2) = (x_2 - x_1^m)^n$, $d(\phi) = 1/(1/n + 1/(mn)) = mn/(m+1)$ and $n(\phi_{\rm pr}) = n > d(\phi)$. Adapted coordinates are given by $y_1 := x_1$, $y_2 := x_2 - x_1^m$, in which ϕ is expressed by $\phi^a(y) = y_2^n + y_1^\ell$. (See Figure 1.2.)

Remark 1.4. *An alternative proof of Varchenko's theorem on the existence of adapted coordinates for analytic functions ϕ of two variables has been given by Phong, Sturm, and Stein in* [PSS99], *by means of Puiseux series expansions of the roots of ϕ.*

We are now in the position to identify the exponents h and ν in (1.4) and (1.5) in terms of Newton polyhedra associated to ϕ:

If there exists an adapted local coordinate system y near the origin such that the principal face $\pi(\phi^a)$ of ϕ, when expressed by the function ϕ^a in the new coordinates, is a vertex, and if $h(\phi) \geq 2$, then we put $\nu(\phi) := 1$; otherwise, we put $\nu(\phi) := 0$. We remark [IM11b] that the first condition is equivalent to the following one: *If y is any adapted local coordinate system at the origin, then either $\pi(\phi^a)$ is a vertex or a compact edge, and $n(\phi^a_{\rm pr}) = d(\phi^a)$.*

Varchenko [V76] has shown for analytic ϕ that the leading exponent in (1.3) is given by $r_0 = 1/h(\phi)$, and $\nu(\phi)$ is the maximal j for which $a_{j,0}(\eta) \neq 0$. Correspondingly, in [IM11b] we prove, by means of a quite different method, that estimate (1.5) holds true with $h = h(\phi)$ and $\nu = \nu(\phi)$, that is, that the following estimate holds true for ϕ smooth and of finite type:

$$|\widehat{d\mu}(\xi)| \leq C(1+|\xi|)^{-1/h(\phi)} \log(2+|\xi|)^{\nu(\phi)}, \quad \xi \in \mathbb{R}^3. \tag{1.12}$$

The special case where $\xi = (0, 0, \xi_3)$ is normal to S at the origin is due to Greenblatt [Gb09].

One can also show that this estimate is sharp in the exponents even when ϕ is not analytic, except for the case where the principal face $\pi(\phi^a)$ is an unbounded edge (see [IM11b], [M14]).

1.2 FOURIER RESTRICTION IN THE PRESENCE OF A LINEAR COORDINATE SYSTEM THAT IS ADAPTED TO ϕ

Coming back to the restriction estimate (1.1) for our hypersurface S in $\mathbb{R}^3$, we begin with the case where there exists a linear coordinate system that is adapted to the function ϕ. For this case, the following complete answer was given in [IM11b].

Theorem 1.5. *Let $S \subset \mathbb{R}^3$ be a smooth hypersurface of finite type, and fix a point $x^0 \in S$. After applying a suitable Euclidean motion of $\mathbb{R}^3$, let us assume that $x^0 = 0$ and that near x^0 we may view S as the graph S_ϕ of a smooth function ϕ of finite type satisfying $\phi(0,0) = 0$ and $\nabla\phi(0,0) = 0$.*

Assume that, after applying a suitable linear change of coordinates, the coordinates (x_1, x_2) are adapted to ϕ. We then define the critical exponent p_c by

$$p_c' := 2h(\phi) + 2,$$

where p' denotes the exponent conjugate to p, that is, $1/p + 1/p' = 1$.

Then there exists a neighborhood $U \subset S$ of the point x^0 such that for every non-negative density $\rho \in C_0^\infty(U)$, the Fourier restriction estimate (1.1) *holds true for every p such that*

$$1 \leq p \leq p_c. \tag{1.13}$$

Moreover, if $\rho(x^0) \neq 0$, then the condition (1.13) *on p is also necessary for the validity of* (1.1).

Earlier results for particular classes of hypersurfaces in $\mathbb{R}^3$ (which can be seen to satisfy the assumptions of this theorem) are, for instance, in the work by E. Ferreyra and M. Urciuolo [FU04], [FU08] and [FU09], who studied certain classes of quasi-homogeneous hypersurfaces, for which they were able to prove L^p-L^q-restriction estimates when $p < \frac{4}{3}$. For further progress in the study of these classes of hypersurfaces, we refer to the work by S. Buschenhenke, A. Vargas and the second named author [BMV15]. We also like to mention work by A. Magyar [M09] on L^p-L^2 Fourier restriction estimates for some classes of analytic hypersurfaces, which preceeded [IM11b] .

As shown in [IM11b], the necessity of condition (1.13) follows easily by means of Knapp type examples (a related discussion of Knapp type arguments is given in Section 1.4). It is here where we need to assume that there is a linear coordinate system which is adapted to ϕ. The sufficiency of condition (1.13) is immediate from Greenleaf's Theorem 1.1 in combination with (1.12), in the case where $\nu(\phi) = 0$. Notice that this is true, no matter whether or not there exists a linear coordinate system that is adapted to ϕ.

If $\nu(\phi) = 1$, then we just miss the endpoint $p = p_c$, which ultimately can be dealt with by means of Littlewood-Paley theory (for more details, we refer the reader to [IM11b] and also the survey article [M14]). An analogous argument based on Littlewood-Paley theory will appear in Chapter 3.

In view of Theorem 1.5, from now on we shall always make the following assumption, unless stated explicitly otherwise.

Assumption 1.6 (NLA). *There is no linear coordinate system that is adapted to* ϕ.

Our main goal will be to understand which Fourier restriction estimates of the form (1.1) will hold under this assumption.

1.3 FOURIER RESTRICTION WHEN NO LINEAR COORDINATE SYSTEM IS ADAPTED TO ϕ—THE ANALYTIC CASE

Under the preceding assumption, but not yet assuming that ϕ is analytic, let us have another look at the first step of Varchenko's algorithm. If here $m = \kappa_2/\kappa_1 = 1$, then this leads to a linear change of coordinates of the form $y_1 = x_1$, $y_2 = x_2 - b_1 x_1$, which will transform ϕ into a function $\tilde{\phi}$ for which, by our assumption, the coordinates (y_1, y_2) are still not adapted. Replacing ϕ by $\tilde{\phi}$, it is also immediate that estimate (1.1) will hold for the graph of ϕ if and only if it holds for the graph of $\tilde{\phi}$. Replacing ϕ by $\tilde{\phi}$, we may and shall therefore always assume that our original coordinate system (x_1, x_2) is chosen so that

$$m = \frac{\kappa_2}{\kappa_1} \in \mathbb{N} \qquad \text{and } m \geq 2. \tag{1.14}$$

The next proposition will show that such a linear coordinate system is linearly adapted to ϕ in the following sense.

In analogy with Varchenko's notion of height, we can introduce the notion of *linear height* of ϕ, which measures the upper limit of all Newton distances of ϕ in linear coordinate systems:

$$h_{\text{lin}}(\phi) := \sup\{d(\phi \circ T) : T \in GL(2, \mathbb{R})\}.$$

Note that

$$d(\phi) \leq h_{\text{lin}}(\phi) \leq h(\phi).$$

We also say that a linear coordinate system $y = (y_1, y_2)$ is *linearly adapted* to ϕ if $d_y = h_{\text{lin}}(\phi)$. Clearly, if there is a linear coordinate system that is adapted to ϕ, it is in particular linearly adapted to ϕ. The following proposition, whose proof will be given in Chapter 9 (Appendix A), gives a characterization of linearly adapted coordinates under Assumption 1.6 (NLA).

Proposition 1.7. *If ϕ satisfies Assumption (NLA) and if $\phi = \phi(x)$, then the following are equivalent:*

(a) *The coordinates x are linearly adapted to ϕ.*
(b) *If the principal face $\pi(\phi)$ is contained in the line*

$$L = \{(t_1, t_2) \in \mathbb{R}^2 : \kappa_1 t_1 + \kappa_2 t_2 = 1\},$$

then either $\kappa_2/\kappa_1 \geq 2$ or $\kappa_1/\kappa_2 \geq 2$.

Moreover, in all linearly adapted coordinates x for which $\kappa_2/\kappa_1 > 1$, the principal face of the Newton polyhedron is the same, so that in particular the numbers $m := \kappa_2/\kappa_1$ and d_x do not depend on the choice of the linearly adapted coordinate system $x = (x_1, x_2)$.

This result shows in particular that linearly adapted coordinates always do exist under Assumption (NLA), since either the original coordinates for ϕ are already linearly adapted or we arrive at such coordinates after applying the first step in Varchenko's algorithm (when $\kappa_2/\kappa_1 = 1$ in the original coordinates).

Let us then look at the Newton polyhedron $\mathcal{N}(\phi^a)$ of ϕ^a, which expresses ϕ in the adapted coordinates (y_1, y_2) of (1.11), and denote the vertices of the Newton polyhedron $\mathcal{N}(\phi^a)$ by (A_l, B_l), $l = 0, \ldots, n$, where we assume that they are ordered so that $A_{l-1} < A_l$, $l = 1, \ldots, n$, with associated compact edges given by the intervals $\gamma_l := [(A_{l-1}, B_{l-1}), (A_l, B_l)]$, $l = 1, \ldots, n$. The unbounded horizontal edge with left endpoint (A_n, B_n) will be denoted by γ_{n+1}. To each of these edges γ_l, we associate the weight $\kappa^l = (\kappa_1^l, \kappa_2^l)$ so that γ_l is contained in the line

$$L_l := \{(t_1, t_2) \in \mathbb{R}^2 : \kappa_1^l t_1 + \kappa_2^l t_2 = 1\}.$$

For $l = n + 1$, we have $\kappa_1^{n+1} := 0$, $\kappa_2^{n+1} = 1/B_n$. We denote by

$$a_l := \frac{\kappa_2^l}{\kappa_1^l}, \quad l = 1, \ldots, n,$$

the reciprocal of the (modulus of the) slope of the line L_l. For $l = n + 1$, we formally set $a_{n+1} := \infty$. (See Figure 1.3.)

If $l \leq n$, then the κ^l-principal part ϕ_{κ^l} of ϕ corresponding to the supporting line L_l can easily be shown to be of the form

$$\phi_{\kappa^l}(y) = c_l \, y_1^{A_{l-1}} y_2^{B_l} \prod_\alpha \left(y_2 - c_l^\alpha \, y_1^{a_l}\right)^{N_\alpha} \tag{1.15}$$

(cf. Proposition 2.2 in [IM11a]; also [IKM10]).

Remark 1.8. *When ϕ is analytic, then this expression is linked to the Puiseux series expansion of roots of ϕ^a as follows* [PS97] (*compare also* [IM11a]): *We may then factor*

$$\phi^a(y_1, y_2) = U(y_1, y_2) y_1^{\nu_1} y_2^{\nu_2} \prod_r (y_2 - r(y_1)),$$

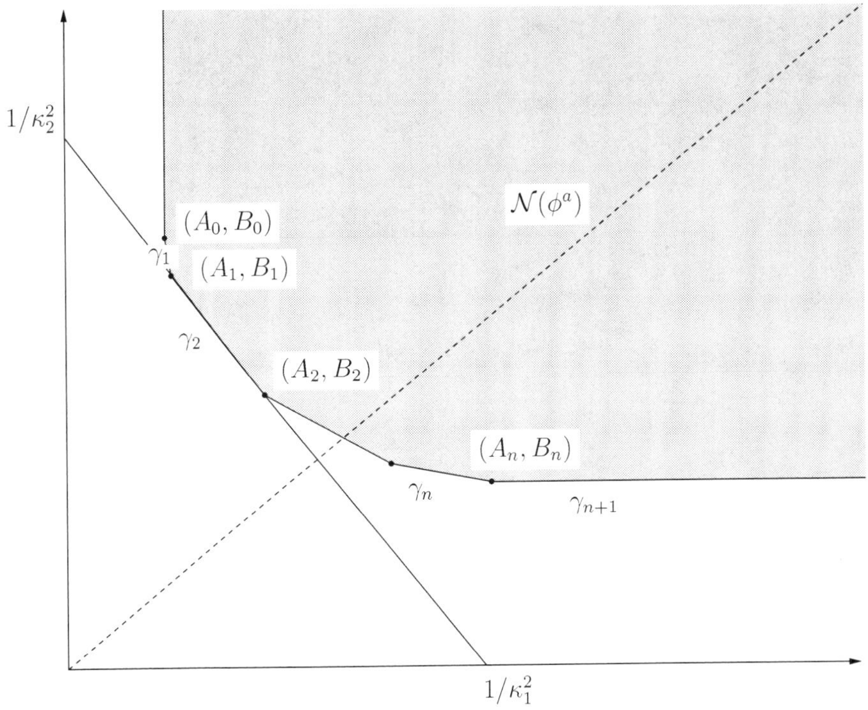

Figure 1.3 Edges and weights

where the product is indexed by the set of all nontrivial roots $r = r(y_1)$ *of* ϕ^a *(which may also be empty) and where* U *is analytic, with* $U(0,0) \neq 0$. *Moreover, these roots can be expressed in a small neighborhood of* 0 *as Puiseux series*

$$r(y_1) = c_{l_1}^{\alpha_1} y_1^{a_{l_1}} + c_{l_1 l_2}^{\alpha_1\alpha_2} y_1^{a_{l_1 l_2}^{\alpha_1}} + \cdots + c_{l_1\cdots l_p}^{\alpha_1\cdots\alpha_p} y_1^{a_{l_1\cdots l_p}^{\alpha_1\cdots\alpha_{p-1}}} + \cdots,$$

where

$$c_{l_1\cdots l_p}^{\alpha_1\cdots\alpha_{p-1}\beta} \neq c_{l_1\cdots l_p}^{\alpha_1\cdots\alpha_{p-1}\gamma} \quad \textit{for} \quad \beta \neq \gamma,$$

$$a_{l_1\cdots l_p}^{\alpha_1\cdots\alpha_{p-1}} > a_{l_1\cdots l_{p-1}}^{\alpha_1\cdots\alpha_{p-2}},$$

with strictly positive exponents $a_{l_1\cdots l_p}^{\alpha_1\cdots\alpha_{p-1}} > 0$ *(which are all multiples of a fixed positive rational number) and nonzero complex coefficients* $c_{l_1\cdots l_p}^{\alpha_1\cdots\alpha_p} \neq 0$ *and where we have kept enough terms to distinguish between all the nonidentical roots of* ϕ^a. *The leading exponents in these series are the numbers*

$$a_1 < a_2 < \cdots < a_n.$$

One can therefore group the roots into the clusters of roots $[l], l = 1, \ldots, n$, *where the lth cluster* $[l]$ *consists of all roots with leading exponent* a_l.

Correspondingly, we can decompose

$$\phi^a(y_1, y_2) = U(y_1, y_2) y_1^{\nu_1} y_2^{\nu_2} \prod_{l=1}^{n} \Phi_{[l]}(y_1, y_2),$$

where

$$\Phi_{[l]}(y_1, y_2) := \prod_{r \in [l]} (y_2 - r(y_1)).$$

More generally, by the cluster $\begin{bmatrix} \alpha_1 & \cdots & \alpha_p \\ l_1 & \ldots & l_p \end{bmatrix}$, *we shall designate all the roots* $r(y_1)$, *counted with their multiplicities, that satisfy*

$$r(y_1) - \left(c_{l_1}^{\alpha_1} y_1^{a_{l_1}} + c_{l_1 l_2}^{\alpha_1 \alpha_2} y_1^{a_{l_1 l_2}^{\alpha_1}} + \cdots + c_{l_1 \cdots l_p}^{\alpha_1 \cdots \alpha_p} y_1^{a_{l_1 \cdots l_p}^{\alpha_1 \cdots \alpha_{p-1}}} \right) = O(y_1^b)$$

for some exponent $b > a_{l_1 \cdots l_p}^{\alpha_1 \cdots \alpha_{p-1}}$. *Then the cluster* $[l] = [l_1]$ *will split into the clusters* $\begin{bmatrix} \alpha_1 \\ l_1 \end{bmatrix}$, *these cluster into the finer "subclusters"* $\begin{bmatrix} \alpha_1 & \alpha_2 \\ l_1 & l_2 \end{bmatrix}$, *and so on.*

Observe also the following: If $\delta_s^l(y_1, y_2) = (s^{\kappa_1^l} y_1, s^{\kappa_2^l} y_2)$, $s > 0$, *denote the dilations associated to the weight* κ^l, *and if* $r \in [l_1]$ *is a root in the cluster* $[l_1]$, *then one easily checks that for* $y = (y_1, y_2)$ *in a bounded set we have* $\delta_s^l y_2 = s^{\kappa_2^l} y_2$ *and* $r(\delta_s^l y_1) = s^{a_{l_1} \kappa_1^l} c_{l_1}^{\alpha_1} y_1^{a_{l_1}} + O(s^{a_{l_1} \kappa_1^l + \varepsilon})$ *as* $s \to 0$, *for some* $\varepsilon > 0$. *Consequently,*

$$\delta_s^l y_2 - r(\delta_s^l y_1) = \begin{cases} -s^{a_{l_1} \kappa_1^l} c_{l_1}^{\alpha_1} y_1^{a_{l_1}} + O(s^{a_{l_1} \kappa_1^l + \varepsilon}), & \text{if } l_1 < l, \\ s^{\kappa_2^l} (y_2 - c_l^{\alpha_l} y_1^{a_l}) + O(s^{\kappa_2^l + \varepsilon}), & \text{if } l_1 = l, \\ s^{\kappa_2^l} y_2 + O(s^{\kappa_2^l + \varepsilon}), & \text{if } l_1 > l. \end{cases}$$

This shows that the κ^l*-principal part of* ϕ^a *is given by*

$$\phi_{\kappa^l}^a = C_l y_1^{\nu_1 + \sum_{l_1 < l} \|[l_1]\| a_{l_1}} y_2^{\nu_2 + \sum_{l_1 > l} \|[l_1]\|} \prod_{\alpha_1} (y_2 - c_l^{\alpha_1} y_1^{a_l})^{N_{l,\alpha_1}}, \tag{1.16}$$

where N_{l,α_1} *denotes the number of roots in the cluster* $[l]$ *with leading term* $c_l^{\alpha_1} y_1^{a_l}$, *and where by* $|M|$ *we denote the cardinality of a set* M.

A look at the Newton polyhedron reveals that the exponents of y_1 *and* y_2 *in* (1.16) *can be expressed in terms of the vertices* (A_j, B_j) *of the Newton polyhedron:*

$$\nu_1 + \sum_{l_1 < l} \|[l_1]\| a_{l_1} = A_{l-1}, \qquad \nu_2 + \sum_{l_1 > l} \|[l_1]\| = B_l.$$

Notice also that

$$\prod_{\alpha_1} (y_2 - c_l^{\alpha_1} y_1^{a_l})^{N_{l,\alpha_1}} = (\Phi_{[l]})_{\kappa^l}.$$

Comparing this with (1.15), *the close relation between the Newton polyhedron of* ϕ^a *and the Puiseux series expansion of roots becomes evident, and accordingly we say that the edge* $\gamma_l := [(A_{l-1}, B_{l-1}), (A_l, B_l)]$ *is associated to the cluster of roots* $[l]$.

Consider next the line parallel to the bisectrix

$$\Delta^{(m)} := \{(t, t+m+1) : t \in \mathbb{R}\}.$$

For any edge $\gamma_l \subset L_l := \{(t_1, t_2) \in \mathbb{R}^2 : \kappa_1^l t_1 + \kappa_2^l t_2 = 1\}$, define h_l by

$$\Delta^{(m)} \cap L_l = \{(h_l - m, h_l + 1)\},$$

that is,

$$h_l = \frac{1 + m\kappa_1^l - \kappa_2^l}{\kappa_1^l + \kappa_2^l}, \tag{1.17}$$

and define the *restriction height*, or for short, *r-height*, of ϕ by

$$h^r(\phi) := \max(d, \max_{\{l=1,\dots,n+1 : a_l > m\}} h_l).$$

Here, $d = d_x$ denotes again the Newton distance $d(\phi)$ of ϕ with respect to our original, linearly adapted coordinates $x = (x_1, x_2)$. Recall also that by Proposition 1.7 the numbers m and d_x are well defined, that is, they do not depend on the chosen linearly adapted coordinate system x. For a more invariant description of $h^r(\phi)$, we refer the reader to Proposition 1.17.

Remarks 1.9. (a) *For L in place of L_l and κ in place of κ^l, one has $m = \kappa_2/\kappa_1$ and $d = 1/(\kappa_1 + \kappa_2)$, so that one gets d in place of h_l in* (1.17).

(b) *Since $m < a_l$, we have $h_l < 1/(\kappa_1^l + \kappa_2^l)$, hence $h^r(\phi) < h(\phi)$. On the other hand, since the line $\Delta^{(m)}$ lies above the bisectrix, it is obvious that $h^r(\phi) + 1 \geq h(\phi)$, so that*

$$h(\phi) - 1 \leq h^r(\phi) < h(\phi). \tag{1.18}$$

(See Figure 1.4.)

It is easy to see from Remark 1.9(a) that the r-height admits the following *geometric interpretation.*

By following Varchenko's algorithm (cf. Subsection 8.2 of [IKM10]), one realizes that the Newton polyhedron of ϕ^a intersects the line L of the Newton polyhedron of ϕ in a compact face, either in a single vertex or a compact edge. That is, the intersection contains at least one and at most two vertices of ϕ^a, and we choose (A_{l_0-1}, B_{l_0-1}) as the one with smallest second coordinate. Then l_0 is the smallest index l such that γ_l has a slope smaller than the slope of L, that is, $a_{l_0-1} \leq m < a_{l_0}$.

We may thus consider the *augmented Newton polyhedron* $\mathcal{N}^r(\phi^a)$ of ϕ^a, which is the convex hull of the union of $\mathcal{N}(\phi^a)$ with the half line $L^+ \subset L$ with right endpoint (A_{l_0-1}, B_{l_0-1}). Then $h^r(\phi) + 1$ is the second coordinate of the point at which the line $\Delta^{(m)}$ intersects the boundary of $\mathcal{N}^r(\phi^a)$.

We remind the reader that all notions that we have introduced so far (with the exception of those discussed in Remark 1.8) make perfect sense for arbitrary smooth functions ϕ of finite type, in particular, for analytic ϕ. For real analytic hypersurfaces, it turns out that we now have all necessary notions at hand in order to formulate the central result of this monograph. The extension to more general classes of smooth, finite-type hypersurfaces will require further notions and will be discussed in the next section (compare Theorem 1.14).

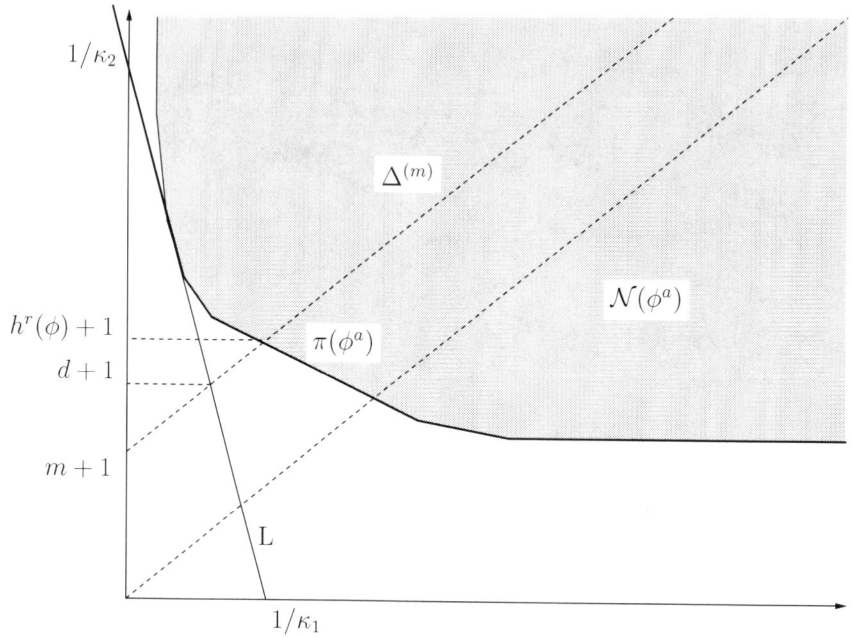

Figure 1.4 r-height

Theorem 1.10. *Let $S \subset \mathbb{R}^3$ be a real analytic hypersurface of finite type, and fix a point $x^0 \in S$. After applying a suitable Euclidean motion of $\mathbb{R}^3$, let us assume that $x^0 = 0$ and that near x^0 we may view S as the graph S_ϕ of a real analytic function ϕ satisfying $\phi(0,0) = 0$ and $\nabla\phi(0,0) = 0$.*

Assume that there is no linear coordinate system adapted to ϕ. Then there exists a neighborhood $U \subset S$ of x^0 such that for every nonnegative density $\rho \in C_0^\infty(U)$, the Fourier restriction estimate (1.1) *holds true for every $p \geq 1$ such that $p' \geq p'_c := 2h^r(\phi) + 2$.*

Remarks 1.11. (a) *An application of Greenleaf's result would imply, at best, that the condition $p' \geq 2h(\phi) + 2$ is sufficient for* (1.1) *to hold, which is a strictly stronger condition than $p' \geq p'_c$.*

(b) *In a preprint, which regretfully has remained unpublished and which has been brought to our attention by A. Seeger after we had found our results,* H. Schulz [Sc90] *had already observed this kind of phenomenon for particular examples of surfaces of revolution.*

Example 1.12.

$$\phi(x_1, x_2) := (x_2 - x_1^m)^n, \qquad n, m \geq 2.$$

The coordinates (x_1, x_2) are not adapted. Adapted coordinates are $y_1 := x_1$, $y_2 := x_2 - x_1^m$, in which ϕ is given by

$$\phi^a(y_1, y_2) = y_2^n.$$

Here

$$\kappa_1 = \frac{1}{mn}, \quad \kappa_2 = \frac{1}{n},$$

$$d := d(\phi) = \frac{1}{\kappa_1 + \kappa_2} = \frac{nm}{m+1},$$

and

$$p_c' = \begin{cases} 2d+2, & \text{if } n \le m+1, \\ 2n, & \text{if } n > m+1. \end{cases}$$

On the other hand, $h := h(\phi) = n$, so that $2h + 2 = 2n + 2 > p_c'$.

1.4 SMOOTH HYPERSURFACES OF FINITE TYPE, CONDITION (R), AND THE GENERAL RESTRICTION THEOREM

Theorem 1.10 can be extended to smooth, finite-type functions ϕ under an additional Condition (R), which, however, is always satisfied when ϕ is real analytic. To state this more general result and in order to prepare a more invariant description of the notion of r-height, we need to introduce more notation. Again, we shall assume that the coordinates (x_1, x_2) are linearly adapted to ϕ.

1.4.1 Fractional shears and condition (R)

We need a few more definitions. Consider the half lines $\mathbb{R}_\pm := \{x_1 \in \mathbb{R} : \pm x_1 > 0\}$, and denote by $H^\pm := \mathbb{R}_\pm \times \mathbb{R}$ the corresponding right (respectively, left) half plane.

We say that a function $f = f(x_1)$ defined in $U \cap \mathbb{R}_+$ (respectively, $U \cap \mathbb{R}_-$), where U is an open neighborhood of the origin, is *fractionally smooth* if there exist a smooth function g on U and a positive integer q such that $f(x_1) = g(|x_1|^{1/q})$ for $x_1 \in U \cap \mathbb{R}_+$ (respectively, $x_1 \in U \cap \mathbb{R}_-$). Moreover, we shall say that a fractionally smooth function f is *flat* if $f(x_1) = O(|x_1|^N)$ as $x_1 \to 0$, for every $N \in \mathbb{N}$. Notice that this notion of flatness describes only the behavior at the origin. Observe also that a fractionally smooth function that is flat is even smooth.

Two fractionally smooth functions f and g defined on a neighborhood of the origin will be called *equivalent*, and we shall write $f \sim g$, if $f - g$ is flat. Finally, a *fractional shear* in $H^\pm$ will be a change of coordinates of the form

$$y_1 := x_1, \; y_2 := x_2 - f(x_1),$$

where f is real valued and fractionally smooth but not flat. If we express the smooth function ϕ on, say, the half plane H^+ as a function of $y = (y_1, y_2)$, the resulting function

$$\phi^f(y) := \phi(y_1, y_2 + f(y_1))$$

will in general no longer be smooth at the origin, but fractionally smooth in the sense described next.

For such functions, there are straightforward generalizations of the notions of Newton polyhedron, and so forth. Namely, following [IKM10] and assuming without loss of generality that we are in H^+ where $x_1 > 0$, let Φ be any smooth function of the variables $x_1^{1/q}$ and x_2 near the origin; that is, there exists a smooth function $\Phi^{[q]}$ near the origin such that $\Phi(x) = \Phi^{[q]}(x_1^{1/q}, x_2)$ (more generally, one could assume that Φ is a smooth function of the variables $x_1^{1/q}$ and $x_2^{1/p}$, where p and q are positive integers, but we won't need this generality here and shall, therefore, always assume that $p = 1$). Such functions Φ will also be called *fractionally smooth.* If the formal Taylor series of $\Phi^{[q]}$ is given by

$$\Phi^{[q]}(x_1, x_2) \sim \sum_{\alpha_1, \alpha_2 = 0}^{\infty} c_{\alpha_1, \alpha_2} x_1^{\alpha_1} x_2^{\alpha_2},$$

then Φ has the formal Puiseux series expansion

$$\Phi(x_1, x_2) \sim \sum_{\alpha_1, \alpha_2 = 0}^{\infty} c_{\alpha_1, \alpha_2} x_1^{\alpha_1/q} x_2^{\alpha_2}.$$

We therefore define the *Taylor-Puiseux support*, or *Taylor support*, of Φ by

$$\mathcal{T}(\Phi) := \{(\tfrac{\alpha_1}{q}, \alpha_2) \in \mathbb{N}_q^2 : c_{\alpha_1, \alpha_2} \neq 0\},$$

where $\mathbb{N}_q^2 := (\frac{1}{q}\mathbb{N}) \times \mathbb{N}$. The *Newton-Puiseux polyhedron* (or *Newton polyhedron*) $\mathcal{N}(\Phi)$ of Φ at the origin is then defined to be the convex hull of the union of all the quadrants $(\alpha_1/q, \alpha_2) + \mathbb{R}_+^2$ in $\mathbb{R}^2$, with $(\alpha_1/q, \alpha_2) \in \mathcal{T}(\Phi)$, and other notions, such as the notion of principal face, Newton distance, or homogenous distance, are defined in analogy with our previous definitions for smooth functions ϕ.

Coming back to our fractional shear f, assume that $f(x_1)$ has the formal Puiseux series expansion (say for $x_1 > 0$)

$$f(x_1) \sim \sum_{j \geq 0} c_j x_1^{m_j}, \tag{1.19}$$

with nonzero coefficients c_j and exponents m_j, which are growing with j and are all multiples of $1/q$. We then isolate the leading exponent m_0 and choose the weight κ^f so that $\kappa_2^f/\kappa_1^f = m_0$ and such that the line

$$L^f := \{(t_1, t_2) \in \mathbb{R}^2 : \kappa_1^f t_1 + \kappa_2^f t_2 = 1\}$$

is a supporting line to $\mathcal{N}(\phi^f)$.

In analogy with $h^r(\phi)$, by replacing the exponent m by m_0 and the line L by L^f, we can then define the *r-height $h^f(\phi)$ associated with f* by putting

$$h^f(\phi) = \max(d^f, \max_{\{l : a_l > m_0\}} h_l^f), \tag{1.20}$$

where (d^f, d^f) is the point of intersection of the line L^f with the bisectrix and where h_l^f is associated to the edge γ_l of $\mathcal{N}(\phi^f)$ by the analogue of formula (1.17), that is,

$$h_l^f = \frac{1 + m_0\kappa_1^l - \kappa_2^l}{\kappa_1^l + \kappa_2^l}, \tag{1.21}$$

if γ_l is again contained in the line L_l defined by the weight κ^l.

In a similar way as for the notion of r-height, we can reinterpret $h^f(\phi)$ *geometrically* as follows. We define the *augmented Newton polyhedron* $\mathcal{N}^f(\phi^f)$ as the convex hull of the union of $\mathcal{N}(\phi^f)$ with the half line $(L^f)^+ \subset L^f$, whose right endpoint is the vertex of $\mathcal{N}(\phi^f) \cap L^f$ with the smallest second coordinate. Then $h^f(\phi) + 1$ is the second coordinate of the point at which the line $\Delta^{(m_0)}$ intersects the boundary of $\mathcal{N}^f(\phi^f)$.

Finally, let us say that a fractionally smooth function $f(x_1)$ *agrees with the principal root jet* $\psi(x_1)$ *up to terms of higher order* if the following holds: if ψ is not a polynomial, then we require that $f \sim \psi$, and if ψ is a polynomial of degree D, then we require that the leading exponent in the formal Puiseux series expansion of $f - \psi$ is strictly bigger than D.

Such functions f will indeed arise in Section 6.5 in the course of an algorithm that will allow us to analyze the fine splitting of certain roots near the principal root jet. It is this fine splitting that will lead to terms of higher order that have to be added to ψ.

We can now formulate the extra "root" condition that we need when ϕ is nonanalytic.

Condition (R). For every fractionally smooth, real function $f(x_1)$ that agrees with the principal root jet $\psi(x_1)$ up to terms of higher order, the following holds true:

If $B \in \mathbb{N}$ is maximal such that $\mathcal{N}(\phi^f) \subset \{(t_1, t_2) : t_2 \geq B\}$, and if $B \geq 1$, then ϕ factors as $\phi(x_1, x_2) = (x_2 - \tilde{f}(x_1))^B \tilde{\phi}(x_1, x_2)$, where $\tilde{f} \sim f$ and where $\tilde{\phi}$ is fractionally smooth.

Examples 1.13. (a) If $\varphi(x_1)$ is flat and nontrivial and if $m \geq 2$ and $B \geq 2$, then the function $\phi_1(x_1, x_2) := (x_2 - x_1^m - \varphi(x_1))^B$ does satisfy Condition (R), whereas (R) fails for $\phi_2(x_1, x_2) := (x_2 - x_1^m)^B + \varphi(x_1)$. In these examples, we have $\psi(x_1) = x_1^m$.

(b) $\phi_3(x_1, x_2) := (x_2 - x_1^2 - \varphi_1(x_1))^5\big((x_2 - 3x_1^2)^2 + \varphi_2(x_1)\big)$ does satisfy Condition (R) for arbitrary flat functions $\varphi_1(x_1)$ and $\varphi_2(x_1)$. Indeed, in this example the principal face $\pi(\phi_3)$ is the interval $[(0, 7), (14, 0)]$, and so one easily finds that $d = 1/(\frac{1}{7} + \frac{1}{14}) = \frac{14}{3}$ and $\psi(x_1) = x_1^2$. Notice that $\psi(x_1)$ is a root of multiplicity $5 > \frac{14}{3}$ of $(\phi_3)_{\mathrm{pr}}$, and hence the principal root.

Real analytic functions ϕ are easily seen to satisfy Condition (R). Indeed, the definition of B implies that

$$\phi^f(y_1, y_2) = y_2^B h(y_1, y_2) + \varphi(y_1, y_2),$$

where $\varphi(y_1, y_2)$ is flat in y_1 and where the mapping $y_1 \mapsto h(y_1, 0)$ is of finite type. In particular, $g(x_1) := \phi(x_1, f(x_1)) = \phi^f(y_1, 0) = \varphi(y_1, 0)$ is flat. On the other hand, if the function ϕ is analytic, then we may factor it as

$$\phi(x_1, x_2) = U(x_1, x_2)x_1^{\nu_1}x_2^{\nu_2}\prod_{j=1}^{J}(x_2 - r_j(x_1))^{n_j},$$

where the $r_j = r_j(x_1)$ denote the distinct nontrivial roots of ϕ and n_j their multiplicities, and where U is analytic near the origin such that $U(0,0) \neq 0$. Moreover, the roots r_j admit Puiseux series expansions near the origin (compare Remark 1.8). But then

$$g(x_1) = U(x_1, f(x_1))x_1^{\nu_1} f(x_1)^{\nu_2}\prod_{j=1}^{J}(f(x_1) - r_j(x_1))^{n_j},$$

and since g is flat, this shows that necessarily $f - r_k$ is flat for some $k \in \{1, \dots, J\}$, that is, $f \sim r_k$. Then r_k must be a real root of ϕ, and the identity

$$\phi^f(y_1, y_2) = U(y_1, y_2 + f(y_1))y_1^{\nu_1}\left(y_2 + f(y_1)\right)^{\nu_2}\prod_{j=1}^{J}\left(y_2 + f(y_1) - r_j(y_1)\right)^{n_j}$$

shows that $B = n_k$. By choosing $\tilde{f} := r_k$, we thus find that indeed $\phi(x_1, x_2) = (x_2 - \tilde{f}(x_1))^B\tilde{\phi}(x_1, x_2)$, with $\tilde{\phi}(x_1, x_2) := U(x_1, x_2)x_1^{\nu_1}x_2^{\nu_2}\prod_{j\neq k}(x_2 - r_j(x_1))^{n_j}$.

1.4.2 The general restriction theorem and sharpness of its conditions

We can now state our main result.

Theorem 1.14. *Let $S \subset \mathbb{R}^3$ be a smooth hypersurface of finite type, and fix a point $x^0 \in S$. After applying a suitable Euclidean motion of $\mathbb{R}^3$, let us assume that $x^0 = 0$ and that near x^0 we may view S as the graph S_ϕ of a smooth function ϕ of finite type satisfying $\phi(0,0) = 0$ and $\nabla\phi(0,0) = 0$.*

Assume that the coordinates (x_1, x_2) are linearly adapted to ϕ but not adapted and that Condition (R) is satisfied.

Then there exists a neighborhood $U \subset S$ of x^0 such that for every nonnegative density $\rho \in C_0^\infty(U)$, the Fourier restriction estimate (1.1) *holds true for every $p \geq 1$ such that $p' \geq p'_c := 2h^r(\phi) + 2$.*

The main body of this monograph will be devoted to the proof of this theorem. A question that remains open at this stage is whether Condition (R) is really needed in this theorem, or whether it can be removed completely, respectively replaced by a weaker condition.

Before returning to the proof, we shall show by means of Knapp type examples that the conditions in this theorem are sharp in the following sense:

Theorem 1.15. *Let ϕ be smooth of finite type, and assume that the Fourier restriction estimate* (1.1) *holds true in a neighborhood of x^0. Then, if $\rho(x^0) \neq 0$, necessarily $p' \geq p'_c$.*

Since the proof will also help to illuminate the notion of r-height, we shall give it right away. In fact, we shall prove the following more general result (notice that we are making no assumption on adaptedness of ϕ here).

Proposition 1.16. *Assume that the coordinates* $x = (x_1, x_2)$ *are linearly adapted to* ϕ *and that the restriction estimate* (1.1) *holds true in a neighborhood of* $x^0 = 0$, *where* $\rho(x^0) \neq 0$. *Consider any fractional shear, say on* H^+, *given by*

$$y_1 := x_1, \; y_2 := x_2 - f(x_1),$$

where f *is real valued and fractionally smooth but not flat. Let* $\phi^f(y) = \phi(y_1, y_2 + f(y_1))$ *be the function expressing* ϕ *in the coordinates* $y = (y_1, y_2)$. *Then, necessarily,*

$$p' \geq 2h^f(\phi) + 2.$$

Theorem 1.15 will follow by choosing for f the principal root jet ψ.

Proof. The proof will be based on suitable Knapp-type arguments.

Let us use the same notation for the Newton polyhedron of ϕ^f as we did for ϕ^a in Section 1.3, that is, the vertices of the Newton polyhedron $\mathcal{N}(\phi^f)$ will be denoted by (A_l, B_l), $l = 0, \dots, n$, where we assume that they are ordered so that $A_{l-1} < A_l$, $l = 1, \dots, n$, with associated compact edges $\gamma_l := [(A_{l-1}, B_{l-1}), (A_l, B_l)]$, $l = 1, \dots, n$, contained in the supporting lines L_l to $\mathcal{N}(\phi^f)$ and associated with the weights κ^l. The unbounded horizontal edge with left endpoint (A_n, B_n) will be denoted by γ_{n+1}. For $l = n + 1$, we have $\kappa_1^{n+1} := 0$, $\kappa_2^{n+1} = 1/B_n$. Again, we put $a_l := \kappa_2^l/\kappa_1^l$, and $a_{n+1} := \infty$.

Because of (1.20), we have to prove the following estimates:

$$p' \geq 2d^f + 2; \tag{1.22}$$

$$p' \geq 2h_l^f + 2 \quad \text{for every } l \text{ such that } a_l > m_0, \tag{1.23}$$

where, according to (1.21),

$$h_l^f = \frac{1 + m_0\kappa_1^l - \kappa_2^l}{\kappa_1^l + \kappa_2^l}.$$

Consider first any nonhorizontal edge γ_l of $\mathcal{N}(\phi^f)$ with $a_l > m_0$, and denote by D_ε^f the region

$$D_\varepsilon^f := \{y \in \mathbb{R}^2 : |y_1| \leq \varepsilon^{\kappa_1^l}, |y_2| \leq \varepsilon^{\kappa_2^l}\}, \qquad \varepsilon > 0,$$

in the coordinates y. In the original coordinates x, it corresponds to

$$D_\varepsilon := \{x \in \mathbb{R}^2 : |x_1| \leq \varepsilon^{\kappa_1^l}, |x_2 - f(x_1)| \leq \varepsilon^{\kappa_2^l}\}.$$

Assume that ε is sufficiently small. Since

$$\phi^f(\varepsilon^{\kappa_1^l} y_1, \varepsilon^{\kappa_2^l} y_2) = \varepsilon(\phi_{\kappa^l}^f(y_1, y_2) + O(\varepsilon^\delta)),$$

for some $\delta > 0$, where $\phi_{\kappa^l}^f$ denotes the κ^l-principal part of ϕ^f, we have that $|\phi^f(y)| \leq C\varepsilon$ for every $y \in D_\varepsilon^f$, that is,

$$|\phi(x)| \leq C\varepsilon \qquad \text{for every } x \in D_\varepsilon. \tag{1.24}$$

Moreover, for $x \in D_\varepsilon$, because $|f(x_1)| \lesssim |x_1|^{m_0}$ and $m_0 \leq a_l = \kappa_2^l/\kappa_1^l$, we have

$$|x_2| \leq \varepsilon^{\kappa_2^l} + |f(x_1)| \lesssim \varepsilon^{\kappa_2^l} + \varepsilon^{m_0\kappa_1^l} \lesssim \varepsilon^{m_0\kappa_1^l}.$$

We may thus assume that D_ε is contained in the box where $|x_1| \le 2\varepsilon^{\kappa_1^l}$, $|x_2| \le 2\varepsilon^{m_0\kappa_1^l}$. Choose a Schwartz function φ_ε such that

$$\widehat{\varphi_\varepsilon}(x_1, x_2, x_3) = \chi_0\Big(\frac{x_1}{\varepsilon^{\kappa_1^l}}\Big)\chi_0\Big(\frac{x_2}{\varepsilon^{m_0\kappa_1^l}}\Big)\chi_0\Big(\frac{x_3}{C\varepsilon}\Big),$$

where χ_0 is a smooth cutoff function supported in $[-2, 2]$ and identically 1 on $[-1, 1]$.

Then by (1.24) we see that $\widehat{\varphi_\varepsilon}(x_1, x_2, \phi(x_1, x_2)) \ge 1$ on D_ε; hence, if $\rho(0) \ne 0$, then

$$\Big(\int_S |\widehat{\varphi_\varepsilon}|^2 \rho\, d\sigma\Big)^{1/2} \ge C_1 |D_\varepsilon|^{1/2} = C_1 \varepsilon^{(\kappa_1^l + \kappa_2^l)/2},$$

where $C_1 > 0$ is a positive constant. Since $\|\varphi_\varepsilon\|_p \simeq \varepsilon^{((1+m_0)\kappa_1^l+1)/p'}$, we find that the restriction estimate (1.1) can hold only if

$$p' \ge 2\frac{(1+m_0)\kappa_1^l + 1}{\kappa_1^l + \kappa_2^l} = 2h_l^f + 2.$$

The case $l = n+1$, where γ_l is the horizontal edge for which $h_l^f = B_n - 1$ (with $B_n = 1/\kappa_l^2$), requires a minor modification of this argument. Observe that, by Taylor expansion, in this case ϕ^f can be written as

$$\phi^f(y_1, y_2) = y_2^{B_n} h(y_1, y_2) + \sum_{j=0}^{B_n - 1} y_2^j g_j(y_1), \tag{1.25}$$

where the functions g_j are flat and h is fractionally smooth and continuous at the origin. Choose a small $\delta > 0$, and define

$$D_\varepsilon^f := \{y \in \mathbb{R}^2 : |y_1| \le \varepsilon^\delta, |y_2| \le \varepsilon^{\kappa_2^l}\}, \quad \varepsilon > 0.$$

Then (1.25) shows that again $|\phi^f(y)| \le C\varepsilon$ for every $y \in D_\varepsilon^f$, so that (1.24) holds true again. Moreover, for $x \in D_\varepsilon$, we now find that

$$|x_2| \le \varepsilon^{\kappa_2^l} + |f(x_1)| \lesssim \varepsilon^{\kappa_2^l} + \varepsilon^{m_0\delta} \lesssim \varepsilon^{m_0\delta}$$

for δ sufficiently small. Choosing

$$\widehat{\varphi_\varepsilon}(x_1, x_2, x_3) = \chi_0\Big(\frac{x_1}{\varepsilon^\delta}\Big)\chi_0\Big(\frac{x_2}{\varepsilon^{m_0\delta}}\Big)\chi_0\Big(\frac{x_3}{C\varepsilon}\Big),$$

and arguing as before, we find that here (1.1) implies that

$$p' \ge 2\frac{(1+m_0)\delta + 1}{\delta + \kappa_2^l} \quad \text{for every } \delta > 0;$$

hence, $p' \ge 2B_n = 2h_{l_{n+1}}^f + 2$. This finishes the proof of (1.23).

Notice finally that the argument for the nonhorizontal edges still works if we replace the line L_l by the line L^f and the weight κ^l by the weight κ^f associated with that line. Since here $m_0\kappa_1^f = \kappa_2^f$, this leads to condition (1.22). Q.E.D.

1.5 AN INVARIANT DESCRIPTION OF THE NOTION OF r-HEIGHT

Finally, we can also give a more invariant description of the notion of r-height, which conceptually resembles more closely Varchenko's definition of the notion of height, only that we restrict the admissible changes of coordinates to the class of fractional shears in the half planes H^+ and H^-.

In this section, we do not work under Assumption (NLA), but we may and shall again assume that our initial coordinates (x_1, x_2) are linearly adapted to ϕ. Then we set

$$\tilde{h}^r(\phi) := \sup_f h^f(\phi),$$

where the supremum is taken over all nonflat fractionally smooth, real functions $f(x_1)$ of $x_1 > 0$ (corresponding to a fractional shear in H^+) or of $x_1 < 0$ (corresponding to a fractional shear in H^-). Then, obviously,

$$h^r(\phi) \le \tilde{h}^r(\phi),$$

but in fact there is equality.

Proposition 1.17. *Assume that the coordinates (x_1, x_2) are linearly adapted to ϕ, where ϕ is smooth and of finite type and satisfies $\phi(0,0) = 0$, $\nabla\phi(0,0) = 0$.*

(a) *If the coordinates (x_1, x_2) are not adapted to ϕ, then for every nonflat fractionally smooth, real function $f(x_1)$ and the corresponding fractional shear in H^+ (respectively, H^-), we have $h^f(\phi) \le h^r(\phi)$. Consequently, $h^r(\phi) = \tilde{h}^r(\phi)$.*
(b) *If the coordinates (x_1, x_2) are adapted to ϕ, then $\tilde{h}^r(\phi) = d(\phi) = h(\phi)$.*

In particular, the critical exponent for the restriction estimate (1.1) *is in all cases given by $p_c' := 2\tilde{h}^r(\phi) + 2$.*

Let us content ourselves at this stage with a short, but admittedly indirect, proof of part (b) and of part (a) under the assumption that ϕ is analytic. Our arguments will again be based on Proposition 1.16. Since these arguments will rely on the validity of Theorems 1.5 and 1.10, which is somewhat unsatisfactory, we shall give a direct, but lengthier, proof in Chapter 9, which will in addition not require analyticity of ϕ.

On the proof of Proposition 1.17. Recall that we assume that the original coordinates (x_1, x_2) are linearly adapted to ϕ.

In order to prove (a) for analytic ϕ, assume furthermore that the coordinates (x_1, x_2) are not adapted to ϕ, and let $f(x_1)$ be any nonflat fractionally smooth, real-valued function of x_1, with corresponding fractional shear, say in H^+. We have to show that

$$h^f(\phi) \le h^r(\phi). \tag{1.26}$$

According to Theorem 1.10 the restriction estimate (1.1) holds true for $p = p_c$, where $p_c' = 2h^r(\phi) + 2$. Moreover, choosing ρ so that $\rho(x^0) \neq 0$, then

Proposition 1.16 implies that $p' \geq 2h^f(\phi) + 2$. Combining these estimates, we obtain (1.26).

In order to prove (b), we assume that the coordinates (x_1, x_2) are adapted to ϕ, so that $d(\phi) = h(\phi)$. We have to prove that

$$\tilde{h}^r(\phi) = d(\phi). \tag{1.27}$$

Let us first observe that Theorem 1.5 and Proposition 1.16 imply, in a similar way as in the proof of (a), that $2h(\phi) + 2 \geq 2h^f(\phi) + 2$; hence $d(\phi) \geq h^f(\phi)$. We thus see that

$$\tilde{h}^r(\phi) \leq d(\phi).$$

On the other hand, when the principal face $\pi(\phi)$ is compact, then we can choose a supporting line

$$L = \{(t_1, t_2) \in \mathbb{R}^2 : \kappa_1 t_1 + \kappa_2 t_2 = 1\}$$

to the Newton polyhedron of ϕ containing $\pi(\phi)$ and such that $0 < \kappa_1 \leq \kappa_2$. We then put $f(x_1) := x_1^{m_0}$, where $m_0 := \kappa_2/\kappa_1$. Then $d(\phi) = 1/(\kappa_1 + \kappa_2) = d^f \leq h^f(\phi) \leq \tilde{h}^r(\phi)$, and we obtain (1.27).

Assume finally that $\pi(\phi)$ is an unbounded horizontal half line, with left endpoint (A, B), where $A < B$. We then choose $f_n(x_1) := x_1^n$, $n \in \mathbb{N}$. Then it is easy to see that for n sufficiently large, the line L^{f_n} will pass through the point (A, B), and thus $\lim_{n\to\infty} h^{f_n}(\phi) = B = d(\phi)$. Therefore, $\tilde{h}^r(\phi) \geq d(\phi)$, which shows that (1.27) is also valid in this case. Q.E.D.

1.6 ORGANIZATION OF THE MONOGRAPH AND STRATEGY OF PROOF

In Chapter 2 we shall begin to prepare the proof of Theorem 1.14 by compiling various auxiliary results. This will include variants of van der Corput type estimates for one-dimensional oscillatory integrals and related sublevel estimates through "integrals of sublevel type," which will be used all over the place. In some situations, we shall also need more specific information, in particular on Airy-type oscillatory integrals, as well as on some special classes of integrals of sublevel type, which will also be provided. We shall also derive a straightforward variant of a beautiful real interpolation method that has been devised by Bak and Seeger in [BS11] and that will allow us in some cases to replace the more classical complex interpolation methods in the proof of Stein-Tomas-type Fourier restriction estimates by substantially shorter arguments. Nevertheless, complex interpolation methods will play a major role in many other situations, and a crucial tool in our application of Stein's interpolation theorem for analytic families of operators will be provided by certain uniform estimates for oscillatory sums, respectively, double sums (cf. Lemmas 2.7 and 2.9). Last, we shall derive normal forms for phase functions ϕ of linear height < 2 for which no linear coordinate system adapted to ϕ does exist. These normal forms will provide the basis for our discussion of the case $h_{\text{lin}}(\phi) < 2$ in Chapter 4.

As a first step in our proof of Theorem 1.14, in which we shall always assume that the coordinates x are linearly adapted, but not adapted to ϕ, we shall show in Chapter 3 that one may reduce the desired Fourier restriction estimate to a piece S_ψ of the surface S lying above a small, "horn-shaped" neighborhood D_ψ of the principal root jet ψ, on which $|x_2 - \psi(x_1)| \leq \varepsilon x_1^m$. Here, $\varepsilon > 0$ can be chosen as small as we wish. This step works in all cases, no matter which value $h_{\text{lin}}(\phi)$ takes. The proof will give us the opportunity to introduce some of the basic tools which will be applied frequently, such as dyadic domain decompositions, rescaling arguments based on the dilations associated to a given edge of the Newton polyhedron (in this chapter given by the principal face $\pi(\phi)$ of the Newton polyhedron of ϕ), in combination with Greenleaf's restriction Theorem 1.1 and Littlewood-Paley theory, which will allow us to sum the estimates that we obtain for the dyadic pieces.

From here on, following our approach in [IKM10] and [IM11b], it will be natural to distinguish between the cases where $h_{\text{lin}}(\phi) < 2$ and where $h_{\text{lin}}(\phi) \geq 2$, since, in contrast to the first case, in the latter case a reduction to estimates for one-dimensional oscillatory integrals will be possible in many situations, which in return can be performed by means of the van der Corput–type Lemma 2.1.

Chapters 4 and 5 will be devoted to the case where $h_{\text{lin}}(\phi) < 2$. The starting point for our discussion will be the normal forms for ϕ provided by Proposition 2.11. Some of the main tools will again consist of various kinds of dyadic domain decompositions in combination with Littlewood-Paley theory and rescaling arguments. In addition, we shall have to localize frequencies to dyadic intervals in each component, which then also leads us to distinguish a variety of different cases, depending on the relative sizes of these dyadic intervals as well as of another parameter related to the Littlewood-Paley decomposition. It turns out that the particular case where $m = 2$ in (1.9) and (1.14) will require, in some situations (these are listed in Proposition 4.2), a substantially more refined analysis than the case $m \geq 3$. Indeed, in some cases, namely, those described in Proposition 4.2(a) and (b), our arguments from Chapter 4 will almost give the complete answer, except that we miss the endpoint $p = p_c$. In order to capture also the corresponding endpoint estimates, we shall devise rather intricate complex interpolation arguments in Section 5.3. These will be prototypical for many more arguments of this type that we shall devise in later chapters.

Even more of a problem will be presented by the cases described in Proposition 4.2(c). In these situations, we not only miss the endpoint estimate in our discussion in Chapter 4, but it turns out that we even have to close a large gap in the L^p range that we need to cover. In order to overcome this problem, we shall perform a further dyadic decomposition in frequency space with respect to the distance to a certain "Airy cone." This refined Airy-type analysis will be developed in Sections 5.1 and 5.2. Again, in order to capture also the endpoint $p = p_c$, we need to apply a complex interpolation argument. Useful tools in these complex interpolation arguments will be Lemmas 2.7 and 2.9 on oscillatory sums and double sums.

In Chapter 6 we shall turn to the case where $h_{\text{lin}}(\phi) \geq 2$. In a first step, following some ideas from the article [PS97] by Phong and Stein (compare also [IKM10] and [IM11b]), we shall perform a decomposition of the remaining piece S_ψ of the surface S, which will be adapted in some sense to the "root structure" of the function ϕ within the domain D_ψ. When speaking of roots, we will, in fact, always have the case of analytic ϕ in mind; for nonanalytic ϕ, these statements may no longer make strict sense, but the ideas from the analytic case may still serve as a very useful guideline.

In order to understand the root structure within our narrow horn-shaped neighborhood D_ψ of the principal root jet ψ, it is natural to look at the Newton polyhedron $\mathcal{N}(\phi^a)$ of ϕ when expressed in the adapted coordinates (y_1, y_2) given by (1.10), (1.11). More precisely, we shall associate to every edge γ_l of $\mathcal{N}(\phi^a)$ lying above the bisectrix a domain D_l, which will be homogeneous in the adapted coordinates (y_1, y_2) under the natural dilations $(y_1, y_2) \mapsto (r^{\kappa_1^l} y_1, r^{\kappa_2^l} y_2)$, $r > 0$, defined by the weight κ^l that we had associated to the edge γ_l. We shall then partition the domain D_ψ into these domains D_l, intermediate domains E_l, and a residue domain D_{pr} and consider the corresponding decomposition of the surface S. The remaining domain D_{pr}, which contains the principal root jet $x_2 = \psi(x_1)$, will in some sense be associated with the principal face $\pi(\phi^a)$ of the Newton polyhedron of ϕ^a and, hence, homogeneous in the coordinates (y_1, y_2). Each domain E_l can be viewed as a "transition" domain between two different types of homogeneity (in adapted coordinates).

In the domains D_l we can again apply our dyadic decomposition techniques in combination with rescaling arguments, making use of the dilations associated with the weight κ^l, but serious new problems do arise, caused by the nonlinear change from the coordinates (x_1, x_2) to the adapted coordinates (y_1, y_2). Following again [PS97], the discussion of the transition domains E_l will be based on bidyadic domain decompositions in the coordinates (y_1, y_2).

In our discussion of the domains D_l, we shall have to distinguish three cases, Cases 1–3, depending on the behavior of the κ^l-principal part $\phi^a_{\kappa^l}$ of ϕ^a near a given point $v \neq 0$. The first case will be easy to handle, and the same is true even for the residue domain D_{pr}. However, in the other two cases, a large difference between the domains D_l and the domain D_{pr} will appear. The reason for this is that for the edges γ_l lying above the bisectrix, we shall be able to prove a favorable control on the multiplicities of the roots of $\phi^a_{\kappa^l}$ (more precisely of $\partial_2\phi^a_{\kappa^l}$), but this breaks down on D_{pr}.

We shall, therefore, start to have a closer look at the domain D_{pr} in Section 6.4. In order to handle Case 3, which is the case where $\phi^a_{\kappa^{l_{\text{pr}}}}$ has a critical point at v, in Section 6.5 we shall devise a further decomposition of the domain D_{pr} into various subdomains of "type" $D_{(l)}$ and $E_{(l)}$, where each domain $D_{(l)}$ will be homogeneous in suitable "modified adapted" coordinates, and the domains $E_{(l)}$ can again be viewed as "transition domains." This domain decomposition algorithm, roughly speaking, reflects the "fine splitting" of roots of $\partial_2\phi^a$. The new transition domains $E_{(l)}$ can be treated in a similar way as the domains E_l before, and in the end we shall be left with domains of type $D_{(l)}$. Now, under the assumption that $h_{\text{lin}}(\phi) \geq 5$, it turns out that these remaining domains can be handled by means of a fibration of the given piece of surface into a family of curves, in combination with Drury's Fourier restriction theorem for curves with nonvanishing torsion [Dru85]. However, that method breaks down when $h_{\text{lin}}(\phi) < 5$, so that in the subsequent two chapters we shall devise an alternative approach for dealing with these remaining domains $D_{(l)}$. That approach will work equally well whenever $h_{\rm lin}(\phi) \geq 2$.

In Chapter 7 we shall mostly consider the domains of type $D_{(1)}$, which are in some sense "closest" to the principal root jet, since it will turn out that the other domains $D_{(l)}$ with $l \geq 2$ are easier to handle (compare Section 7.10). Within the domains of type $D_{(1)}$, we shall have to deal with functions ϕ which, in suitable "modified" adapted coordinates, look like $\tilde{\phi}^a(y_1, y_2) = y_2^B\, b_B(y_1, y_2) + y_1^n \alpha(y_1) + \sum_{j=1}^{B-1} y_2^j\, b_j(y_1)$, where $B \geq 2$ and b_B is nonvanishing. In a first step, by means of some lower bounds on the r-height, we shall be able to establish favorable restriction estimates in most situations, with the exception of certain cases where $m = 2$ and $B = 3$ or $B = 4$. Along the way, in some cases we shall have to apply interpolation arguments in order to capture the endpoint estimates for $p = p_c$. Sometimes this can be achieved by means of the aforementioned variant of the Fourier restriction theorem by Bak and Seeger, whose assumptions, when satisfied, are easily checked. However, in most of these cases we shall have to apply complex interpolation, in a similar way as we did before in Section 5.3.

Eventually we can thus reduce considerations to certain cases where $m = 2$ and $B = 3$ or $B = 4$. The most difficult situations will occur when frequencies $\xi = (\xi_1, \xi_2, \xi_3)$ are localized to domains on which all components ξ_i of ξ are comparable in size. These cases, which will be discussed in Chapter 8, turn out to be the most challenging ones among all, the case $B = 3$ being the worst. Again, we shall have to apply a refined Airy-type analysis in combination with complex interpolation arguments as in Chapter 5, but further methods are needed— for instance, rescaling arguments going back to Duistermaat [Dui74]— in order to control the dependence of certain classes of oscillatory integrals on some small parameters, and the technical complexity will become quite demanding.

The monograph will conclude with the proof of Proposition 1.7 on the characterization of linearly adapted coordinates in Appendix A and a direct proof of Proposition 1.17 on an invariant description of the notion of r-height in Appendix B of Chapter 9.

Conventions: Throughout this monograph, we shall use the "variable constant" notation, that is, many constants appearing in the course of our arguments, often denoted by C, will typically have different values on different lines. Moreover, we shall use symbols such as $\sim$, $\lesssim$, or $\ll$ in order to avoid writing down constants. By $A \sim B$ we mean that there are constants $0 < C_1 \leq C_2$ such that $C_1 A \leq B \leq C_2 A$, and these constants will not depend on the relevant parameters arising in the context in which the quantities A and B appear. Similarly, by $A \lesssim B$ we mean that there is a (possibly large) constant $C_1 > 0$ such that $A \leq C_1 B$, and by $A \ll B$ we mean that there is a sufficiently small constant $c_1 > 0$ such that $A \leq c_1 B$, and again these constants do not depend on the relevant parameters.

By χ_0 and χ_1 we shall always denote smooth cutoff functions with compact support on $\mathbb{R}^n$, where χ_0 will be supported in a neighborhood of the origin and usually be identically 1 near the origin, whereas $\chi_1 = \chi_1(x)$ will be support away from the origin, sometimes in each of its coordinates x_j, that is, $|x_j| \sim 1$ for $j = 1, \ldots, n$, for every x in the support of χ_1. These cutoff functions may also vary from line to line and may, in some instances, where several of such functions of different variables appear within the same formula, even designate different functions.

Also, if we speak of the *slope* of a line such as a supporting line to a Newton polyhedron, then we shall actually mean the modulus of the slope. Finally, when speaking of domain decompositions, we shall not always keep to the mathematical convention of a domain being on open connected set but shall occasionally use the word *domain* in a more colloquial way.

Finally, by $\mathbb{N}^\times$, $\mathbb{Q}^\times$, $\mathbb{R}^\times$, and so on, we shall denote the set of nonzero elements in $\mathbb{N}$, $\mathbb{Q}$, $\mathbb{R}$, and so on.

ACKNOWLEDGMENTS

We thank the referees for numerous suggestions which greatly helped to improve the exposition. Moreover, we are indepted to Eugen Zimmermann for providing some of the graphics for this monograph. Finally, we gratefully acknowledge the support for this work by the Deutsche Forschungsgemeinschaft (DFG). We also wish to acknowledge permission to reprint excerpts from the following previously published materials:

"Uniform Estimates for the Fourier Transform of Surface Carried Measures in $\mathbb{R}^3$ and an Application to Fourier Restriction" by Isroil A. Ikromov and Detlef Müller from *Journal of Fourier Analysis and Applications*, December 2011, Volume 17, Issue 6, pp. 1292–1332 (16 July 2011) © Springer Science+ Business Media, LLC 2011. Published with permission of Springer Science+ Business Media.

"On Adapted Coordinate Systems" by Isroil A. Ikromov and Detlef Müller. First published in *Transactions of the American Mathematical Society* in 363 (2011), 2821–2848, published by the American Mathematical Society. Reprinted by permission of the American Mathematical Society.

"Estimates for maximal functions associated to hypersurfaces in $\mathbb{R}^3$ and related problems of harmonic analysis"; I.A. Ikromov, M. Kempe and D. Müller, *Acta Mathematica* 204 (2010), 151–271. Reprinted by permission of Acta Mathematica.

Advances in Analysis: The Legacy of Elias M. Stein edited by Charles Fefferman, Alexandru Ionescu, D.H. Phong, and Stephen Wainger, Princeton University Press, 2014. Reprinted by permission of Princeton University Press and Stephen Wainger, Charles Fefferman, D.H. Phong, and Alexandru Ionescu.

Chapter Two

Auxiliary Results

In the course of the proof of the main Theorem 1.14, various auxiliary results will be useful. These will be compiled in this chapter.

First, we shall recall some classical van der Corput–type estimates for one-dimensional oscillatory integrals $\int_I e^{i\lambda f(s)} g(s)\, ds$ and related integrals $\int_I G(\lambda f(s))\, ds$, which will be frequently needed. Integrals of the latter type (and related integrals) will be called *integrals of sublevel type*, because if we choose for G the characteristic function of the interval $[-1, 1]$, then the integral computes the measure of the sublevel set where $|f(s)| \le 1/\lambda$ in I.

In some situations, we shall need more precise information on the asymptotic behavior of oscillatory integrals. In particular, we shall frequently make use of the method of stationary phase (in higher dimension), which covers the case of phase functions with nondegenerate critical points (see, e.g., [H90],[So93] or [S93] for references). We shall also encounter phase functions with degenerate critical points, in particular Airy-type integrals, in the presence of perturbation terms, for which precise asymptotic information will be crucial. Therefore, we shall analyze a model class of such Airy-type integrals (and integrals with even higher order degeneracies) in Section 2.2.

A rather specific class of integrals of sublevel type will be studied in Section 2.3. These will become quite important to various complex interpolation arguments of Chapters 7 and 8 in order to prove certain endpoint results of Theorem 1.14 when $2 \le h_{\text{lin}}(\phi) < 5$.

Next, as mentioned before, in some situations we may work with a variant of the real interpolation method devised by J. G. Bak and A. Seeger [BS11] in place of a complex interpolation argument in order to establish certain endpoint estimates. This variant of the results by Bak and Seeger will be stated in Section 2.4. Whenever it is possible to apply this method, it will allow for substantially simplified proofs, as for instance in Chapters 5 and 7.

Yet another important tool for our complex interpolation arguments in Chapters 7 and 8 will be provided by the uniform estimates for oscillatory sums and doublesums of Section 2.5. Such sums will arise from dyadic and bidyadic frequency domain decompositions (also with respect to the distance to certain "Airy cones") in Chapters 5, 7, and 8.

Finally, for our treatment of the case where $h_{\text{lin}}(\phi) < 2$, we shall make use of normal forms for phase functions ϕ of linear height < 2 for which no linear coordinate system adapted to ϕ does exist. These will be derived in the last Section 2.6. When ϕ has only isolated critical points, they will actually correspond to singularities of type A_n and D_n in Arnol'ds classification of singularities.

2.1 VAN DER CORPUT–TYPE ESTIMATES

We shall often make use of van der Corput–type estimates. These include the classical van der Corput lemma [vdC21] (see also [S93]) as well as variants of it, going back to J. E. Björk (see [D77]) and G. I. Arhipov [AKC79], and also related classical sublevel estimates, which originated in work by van der Corput too [vdC21] (see also [AKC79], [CaCW99], and [G09]).

Lemma 2.1. *Let $M \geq 2$ ($M \in \mathbb{N}$), and let f be a real-valued function of class C^M defined on an interval $I \subset \mathbb{R}$. Assume that either*

(i) *$|f^{(M)}(s)| \geq 1$ on I, or that*
(ii) *f is of polynomial type $M \geq 2$, that is, there are positive constants c_1, $c_2 > 0$ such that*

$$c_1 \leq \sum_{j=1}^{M} |f^{(j)}(s)| \leq c_2 \quad \text{for every } s \in I,$$

and I is compact.

Then the following hold true: For every $\lambda \in \mathbb{R}$,

(a)

$$\left| \int_I e^{i\lambda f(s)} g(s)\, ds \right| \leq C(\|g\|_{L^\infty(I)} + \|g'\|_{L^1(I)})\,(1 + |\lambda|)^{-1/M},$$

where the constant C depends only on M in case (i), and on M, c_1, c_2 and I in case (ii).

(b) *If $G \in L^1(I)$ is a nonnegative function which is majorized by a function $H \in L^1(I)$ such that $\hat{H} \in L^1(\mathbb{R})$, then*

$$\int_I G(\lambda f(s))\, ds \leq C|\lambda|^{-1/M},$$

where the constant C depends only on M and $\|H\|_1 + \|\hat{H}\|_1$ in case (i), and on M, c_1, c_2, I and $\|H\|_1 + \|\hat{H}\|_1$ in case (ii).

Proof. For (a), we refer to [vdC21], [S93], [D77], and [AKC79]. Moreover, it is well known (see [vdC21]) that (b) is an immediate consequence of (a). Indeed, by means of the Fourier inversion formula and Fubini's theorem, we may estimate

$$\int_I G(\lambda f(s))\, ds \leq \frac{1}{2\pi}\left| \int \hat{H}(\xi) \int_I e^{i\xi\lambda f(s)}\, ds\, d\xi \right| \leq C|\lambda|^{-1/M} \int_{\mathbb{R}} |\hat{H}(\xi)||\xi|^{-1/M}\, d\xi.$$

Q.E.D.

We remark that the conditions on the function G in (b) are satisfied in particular if $G = |\varphi|$, where φ is of Schwartz class.

2.2 AIRY-TYPE INTEGRALS

In various situations, we shall need not only estimates but more precise information on the asymptotic behavior of certain oscillatory integrals of Airy type. The next lemma and its proof will be prototypical for the situations that will arise. We shall actually only make use of the case $B = 3$ of Airy-type integrals, but since the method of proof works in the same way for arbitrary $B \geq 3$, we shall state it more generally. We shall sketch a proof, since the method of proof will later also be applied to similar situations that are not completely covered by the lemma and also since we shall need somewhat more specific results than those we found in the literature (compare, for instance, Lemma 1 in [R69] or [Dui74]).

Lemma 2.2. *Let $B \geq 3$ be an integer, and let*

$$J(\lambda, u, s) := \int_{\mathbb{R}} e^{i\lambda(b(t,s)t^B - ut - \sum_{j=2}^{B-1} b_j(u)t^j)}\, a(t, s)\, dt, \quad \lambda \geq 1, u \in \mathbb{R}, |u| \lesssim 1,$$

where a, b are smooth, real-valued functions of (t, s) on an open neighborhood of $I \times K$, where I is a compact neighborhood of the origin in $\mathbb{R}$ and K is a compact subset of $\mathbb{R}^m$. The functions b_j are assumed to be real valued and also smooth. Assume also that $b(t, s) \neq 0$ on $I \times K$, that $|t| \leq \varepsilon$ on the support of a, and that

$$|b_j(u)| \leq C|u|, \qquad j = 2, \ldots, B-1.$$

If $\varepsilon > 0$ is chosen sufficiently small and λ sufficiently large, then the following hold true:

(a) *If $\lambda^{(B-1)/B}|u| \lesssim 1$, then*

$$J(\lambda, u, s) = \lambda^{-1/B}\, g(\lambda^{(B-1)/B} u, \lambda, s),$$

where $g(v, \lambda, s)$ is a smooth function of (v, λ, s) whose derivates of any order are uniformly bounded on its natural domain.

(b) *If $\lambda^{(B-1)/B}|u| \gg 1$, let us assume first that u and b have the same sign, and that B is odd. Then*

$$\begin{aligned} J(\lambda, u, s) = {} & \lambda^{-1/2}|u|^{-(B-2)/(2B-2)}\, \chi_0\left(\frac{u}{\varepsilon}\right) \\ & \times \Big(a_+(|u|^{1/(B-1)}, s)\, e^{i\lambda|u|^{B/(B-1)} q_+(|u|^{1/(B-1)}, s)} \\ & + a_-(|u|^{1/(B-1)}, s)\, e^{i\lambda|u|^{B/(B-1)} q_-(|u|^{1/(B-1)}, s)}\Big) \\ & + (\lambda|u|)^{-1} E(\lambda|u|^{B/(B-1)}, |u|^{1/(B-1)}, s), \end{aligned}$$

where $a_\pm, q_\pm$ are smooth functions and where E is smooth and satisfies estimates

$$|\partial_\mu^\alpha \partial_v^\beta \partial_s^\gamma E(\mu, v, s)| \leq C_{N,\alpha,\beta,\gamma}|v|^{-\beta}|\mu|^{-N}, \qquad \forall N, \alpha, \beta, \gamma \in \mathbb{N}.$$

Moreover, when $|u|$ is sufficiently small, then

$$q_{\pm}(v,s) = \mp \operatorname{sgn} b(0,s)|b(0,s)|^{1/(B-1)}\rho(v,s),$$

where ρ is smooth and $\rho(0,s) = (B-1)\cdot B^{-B/(B-1)}$.

Finally, if u and b have opposite signs, then the same formula remains valid, even with $a_+ \equiv 0$, $a_- \equiv 0$. And, if B is even, we do have a similar result, but without the presence of the term containing a_-.

Proof. In Case (a), scaling in t by the factor $\lambda^{-1/B}$ allows us to rewrite

$$J(\lambda,u,s) = \lambda^{-1/B}\int e^{i\left(b(\lambda^{-1/B}t,s)t^B - \lambda^{(B-1)/B}ut - \sum_{j=2}^{B-1}\lambda^{(B-j)/B}b_j(u)t^j\right)}\,a(\lambda^{-1/B}t,s)\,dt.$$

Choose a smooth cutoff function χ_0 on $\mathbb{R}$ that is identically 1 on $[-1,1]$, and $M \gg 1$, and decompose

$$\lambda^{1/B}J(\lambda,u,s) = G_0(\lambda^{(B-1)/B}u,\lambda,s) + G_\infty(\lambda^{(B-1)/B}u,\lambda,s),$$

where, for $|v| \lesssim 1$,

$$G_0(v,\lambda,s) := \int e^{i(b(\lambda^{-1/B}t,s)t^B - vt - \sum_{j=2}^{B-1}\lambda^{(B-j)/B}b_j(\lambda^{(1-B)/B}v)t^j)}\,\chi_0\left(\frac{t}{M}\right)a(\lambda^{-1/B}t,s)\,dt,$$

$$G_\infty(v,\lambda,s) := \int e^{i(b(\lambda^{-1/B}t,s)t^B - vt - \sum_{j=2}^{B-1}\lambda^{(B-j)/B}b_j(\lambda^{(1-B)/B}v)t^j)}$$

$$\times\left(1-\chi_0\left(\frac{t}{M}\right)\right)a(\lambda^{-1/B}t,s)\,dt.$$

Notice that for $j \geq 2$,

$$|\lambda^{(B-j)/B}b_j(\lambda^{(1-B)/B}v)| \leq C\lambda^{(B-j)/B}\lambda^{(1-B)/B}|v| \lesssim \lambda^{-1/B}.$$

It is then easy to see that G_0 is a smooth function of (v,λ,s) whose derivates of any order are uniformly bounded on its natural domain, and the same can easily be verified for G_∞ by means of iterated integrations by parts. This proves (a).

In order to prove (b), consider first the case where $|u| \geq \varepsilon$. If $\Phi = \Phi(t)$ denotes the complete phase in the oscillatory integral defining $J(\lambda,u,s)$, recalling that $|t| \leq \varepsilon$, we easily see that

$$|\Phi'(t)| \geq C\lambda|u|,$$

provided we choose ε sufficiently small. Integrations by parts then show that we can represent $J(\lambda,u,s)$ by the third term $(\lambda|u|)^{-1}E(\lambda|u|^{B/(B-1)},|u|^{1/(B-1)},\lambda,s)$.

Let us therefore assume that $|u| < \varepsilon$. We shall also assume that $u > 0$; the case $u < 0$ can be treated in a similar way. Here, we scale t by the factor $u^{1/(B-1)}$, and rewrite

$$J(\lambda,u,s) = u^{1/(B-1)}\int e^{i\lambda u^{B/(B-1)}(b(u^{1/(B-1)}t,s)t^B - t - \sum_{j=2}^{B-1}u^{-(B-j)/(B-1)}b_j(u)t^j)}$$

$$\times\, a(u^{1/(B-1)}t,s)\,dt.$$

Again, we decompose this as

$$J(\lambda,u,s) = J_0(\lambda,u^{1/(B-1)},s) + J_\infty(\lambda,u^{1/(B-1)},s),$$

where, with $v := u^{1/(B-1)}$,

$$J_0(\lambda, v, s) := v \int e^{i\lambda v^B (b(vt,s)t^B - t - \sum_{j=2}^{B-1} v^{-(B-j)} b_j(v^{B-1})t^j)} \chi_0\left(\frac{t}{M}\right) a(vt, s)\, dt,$$

$$J_\infty(\lambda, v, s) := v \int e^{i\lambda v^B (b(vt,s)t^B - t - \sum_{j=2}^{B-1} v^{-(B-j)} b_j(v^{B-1})t^j)} \left(1 - \chi_0\left(\frac{t}{M}\right)\right) a(vt, s)\, dt.$$

Observe that

$$|v^{-(B-j)} b_j(v^{B-1})| \le C v^{j-1} \lesssim \varepsilon^{1/(B-1)}, \qquad j = 2, \dots, B-1.$$

Assume that ε is sufficiently small. If B is odd, then, in the first integral J_0, the phase has exactly two nondegenerate critical points $t_\pm(v, s) \sim \pm 1$ if $b > 0$, and thus the method of stationary phase shows that

$$\begin{aligned} J_0(\lambda, v, s) &= v(\lambda v^B)^{-1/2} a_+(v, s) e^{i\lambda v^B q_+(v,s)} \\ &\quad + v(\lambda v^B)^{-1/2} a_-(v, s) e^{i\lambda v^B q_-(v,s)} + v E_1(\lambda v^B, v, s), \end{aligned}$$

where $a_\pm$ are smooth functions and where E_1 is smooth and rapidly decaying with respect to the first variable. If $b < 0$, then there are no critical points, and we get the term E_1 only. Moreover,

$$q_\pm(v, s) = b(vt_\pm(v, s), s) t_\pm(v, s)^B - t_\pm(v, s) + O(v).$$

Note that if $v = 0$, then $t_\pm(0, s) = \pm(Bb(0, s))^{-1/(B-1)}$, so that

$$q_\pm(0, s) = \mp \frac{B-1}{B^{B/(B-1)} b(0, s)^{1/(B-1)}} \neq 0,$$

which proves the statement about $q_\pm$. A similar discussion applies when B is even. In this case, there is only one critical point, namely, $t_+(v, s)$.

In the second integral J_∞, we may apply integrations by parts in order to rewrite it as

$$\begin{aligned} J_\infty(\lambda, v, s) := v(\lambda v^B)^{-N} \int & e^{i\lambda v^B (b(vt,s)t^B - t - \sum_{j=2}^{B-1} v^{-(B-j)} b_j(v^{B-1})t^j)} \\ & \times a_N(t, v, s)\, dt, \qquad N \in \mathbb{N}, \end{aligned}$$

where a_N is supported where $|t| \ge M$ and $|a_N(t, v, s)| \le C_N |t|^{-2N}$. Similarly, if we take derivatives with respect to s, we produce additional powers of t in the integrand, which, however, can be compensated by integrations by parts. Analogous considerations apply to derivatives with respect to v (where we produce negative powers of v) and with respect to λv^B. Altogether, we find that

$$J_\infty(\lambda, v, s) = \frac{1}{\lambda v^{B-1}} E_2(\lambda v^B, v, s),$$

where E_2 is smooth and

$$|\partial_\mu^\alpha \partial_v^\beta \partial_s^\gamma E_2(\mu, v, s)| \le C_{N,\alpha,\beta,\gamma} |v|^{-\beta} |\mu|^{-N}, \qquad \forall N, \alpha, \beta, \gamma \in \mathbb{N}.$$

Summing up all terms, and putting $E := E_1 + E_2$, we obtain the statements in (b). Q.E.D.

The following remark, which will become relevant, for instance, in Chapter 5, can be verified easily by well-known versions of the method of stationary phase for oscillatory integrals whose amplitude depends also on the parameter λ as symbols of order 0 (see, e.g., [So93]).

Remark 2.3. *We may even allow in Lemma 2.2 that the function $a(t,s)$ depends on λ too, that is, $a = a(t,s;\lambda)$, in such a way that it is a symbol of order 0 with respect to λ, uniformly in the other parameters; that is,*

$$\left|\left(\frac{\partial}{\partial\lambda}\right)^{\alpha}\left(\frac{\partial}{\partial t}\right)^{\beta_1}\left(\frac{\partial}{\partial s}\right)^{\beta_2} a(t,s;\lambda)\right| \le C_{\alpha,\beta}(1+\lambda)^{-\alpha}$$

for all $\alpha, \beta_1, \beta_2 \in \mathbb{N}$. Then the same conclusions hold true, only with $a_\pm$ and E depending additionally on λ as symbols of order 0 in a uniform way.

2.3 INTEGRAL ESTIMATES OF VAN DER CORPUT TYPE

Lemma 2.4. *Let $b = b(y_1, y_2)$ be a C^2-function on $\mathbb{R}\times[-1,1]$ such that $b(0,0)\neq 0$, $\|b(y_1,\cdot)\|_{C^2([-1,1])}\le c_1$ for every $y_1\in\mathbb{R}$ and*

$$|b(y_1,y_2)-b(0,0)|\le\varepsilon, \quad \textit{and} \quad c_2|y_2|^{3-j}\le\left|\partial_{y_2}^j\left(b(y_1,y_2)y_2^3\right)\right|, \quad j=1,2, \tag{2.1}$$

for every $(y_1,y_2)\in\mathbb{R}\times[-1,1]$, where $0<c_1\le c_2$. Furthermore, let $Q=Q(y_2)$ be a smooth function on $[-1,1]$ such that $\|Q\|_{C^2([-1,1])}\le c_1$, and let A, B, T be real numbers so that $\max\{|A|,|B|\}\ge L$, $T\ge L$ *and*

$$|A|\le T^3, \quad |B|\le T^2. \tag{2.2}$$

Moreover, let $r_i = r_i(y_1)$, $i=1,2$, be measurable functions on $\mathbb{R}$ such that $|r_i(y_1)|\le c(1+|y_1|)$. For $\epsilon\ge 0$ and $N\ge 2$, we put

$$I_\epsilon(A,B,T) := \int_{-T}^{T}\int_{\infty}^{\infty}\Big(1+\Big|A-\Big(B+Q\Big(\frac{y_2}{T}\Big)\Big)y_2 - b\Big(y_1,\frac{y_2}{T}\Big)y_2^3 + r_1(y_1)+\frac{y_2}{T}r_2(y_1)\Big|\Big)^{-N}(1+|y_1|)^{-N}|y_2|^\epsilon\,dy_1\,dy_2.$$

Then, for N sufficiently large, there are constants $C>0$ and $\varepsilon_0>0$, which depend only on the constants c, c_1, and c_2, such that for all functions b and q and all A, B, T with the assumed properties and all $\varepsilon\le\varepsilon_0$ and $L\ge C$, we have

$$|I_\epsilon(A,B,T)|\le C\max\{|A|^{1/3},|B|^{1/2}\}^{\epsilon-1/2}.$$

Proof. It will be convenient for the proof to call a constant C admissible if it depends only on the constants $\varepsilon, \epsilon, b(0,0)$ and c, c_1, c_2 from the statement of the lemma. All constants C appearing within the proof will be admissible but may change from line to line.

We begin by the observation that the second assumption in (2.1) implies also that

$$c_2|y_2|^{3-j}\le\left|\partial_{y_2}^j\left(b(y_1,\tau y_2)y_2^3\right)\right|, \quad j=1,2,\ |y_2|<\tau^{-1}, \tag{2.3}$$

for every $\tau > 0$. Indeed, if we fix y_1 and put $\psi_\tau(y_2) := b(y_1, \tau y_2)y_2^3$, then $\psi_\tau(y_2) = \psi_1(\tau y_2)\tau^{-3}$, so that $\psi_\tau^{(j)}(y_2) = \psi_1^{(j)}(\tau y_2)\tau^{j-3}$. This immediately leads to (2.3).

Let us first assume that $|A| \geq L \geq 1$. Without loss of generality, we may assume that $A, B \geq 0$ (if necessary, we may change the signs of r or b, q as well as of y_2). We may then choose $\alpha, \beta \geq 0$ so that $A = \alpha^3$, $B = \beta^2$.

Next, by convolving $(1 + |\cdot|)^{-N}$ with a suitable smooth bump function, we may choose a smooth, nonnegative function ρ on $\mathbb{R}$ that is integrable and such that its Fourier transform is also integrable and so that $(1 + |x|)^{-N} \leq \rho(x) \leq 2(1 + |x|)^{-N}$ and put

$$J_\epsilon(\alpha, \beta, T) := \int_{-T}^{T} \int_{\infty}^{\infty} \rho\left(\alpha^3 - \left(\beta^2 + Q\left(\frac{y_2}{T}\right)\right) y_2 - b\left(y_1, \frac{y_2}{T}\right) y_2^3 + r_1(y_1) + \frac{y_2}{T} r_2(y_1)\right) (1 + |y_1|)^{-N} |y_2|^\epsilon \, dy_1 \, dy_2.$$

It then suffices to prove that

$$|J_\epsilon(\alpha, \beta, T)| \leq C \max\{\alpha, \beta\}^{\epsilon - 1/2} \tag{2.4}$$

whenever $L^{1/3} \leq \alpha \leq T$, $0 \leq \beta \leq T$ (recall (2.2)).

To this end, performing the change of variables $y_2 = \alpha s$, we rewrite

$$J_\epsilon(\alpha, \beta, T) = \alpha^{1+\epsilon} \int_{\infty}^{\infty} \int_{-T/\alpha}^{T/\alpha} \rho\left(\alpha^3\left[1 - \left(\gamma + \frac{1}{\alpha^2} Q\left(\frac{\alpha}{T} s\right)\right) s - b\left(y_1, \frac{\alpha}{T} s\right) s^3\right] + r_1(y_1) + s\frac{\alpha}{T} r_2(y_1)\right) \left(1 + |y_1|\right)^{-N} |s|^\epsilon \, ds \, dy_1, \tag{2.5}$$

where

$$\gamma := \left(\frac{\beta}{\alpha}\right)^2.$$

Case 1. $\gamma \geq 1$, that is, $\beta \geq \alpha$. Changing variables $s = \gamma^{1/2} t$, we rewrite

$$J_\epsilon(\alpha, \beta, T) = \beta^{1+\epsilon} \int_{\infty}^{\infty} \int_{-T/\beta}^{T/\beta} \rho\left(\alpha^3 - \beta^3\left(\left(1 + \frac{1}{\beta^2} Q\left(\frac{\beta}{T} t\right)\right) t - b\left(y_1, \frac{\beta}{T} t\right) t^3\right) + r_1(y_1) + t\frac{\beta}{T} r_2(y_1)\right) (1 + |y_1|)^{-N} |t|^\epsilon \, dt \, dy_1.$$

Now, there are admissible constants $C_1, C_2 \geq 1$ so that if $|t| \geq C_1$, then

$$\left|\alpha^3 - \beta^3\left(\left(1 + \frac{1}{\beta^2} Q\left(\frac{\beta}{T} t\right)\right) t - b\left(y_1, \frac{\beta}{T} t\right) t^3\right)\right| \geq C_2 \beta^3 |t|^3.$$

Notice also that $|r_1(y_1) + t\frac{\beta}{T} r_2(y_1)| \leq 2c(1 + |y_1|)$. Integrating separately in y_1 over the sets where $1 + |y_1| \leq C_2\beta^3|t|^3/(4c)$ and where $1 + |y_1| > C_2\beta^3|t|^3/(4c)$, one easily finds that the contribution $J_I(\alpha, \beta, T)$ by the region where $|t| \geq C_1$ to the integral $J_\epsilon(\alpha, \beta, T)$ can be estimated by

$$|J_I(\alpha, \beta, T)| \leq C\beta^{1+\epsilon}\left((\beta^3)^{-N} + (\beta^3)^{1-N}\right) \leq 2C\beta^{4+\epsilon-3N}.$$

Assume next that $|t| < C_1$. We then choose $\chi \in C_0^\infty(\mathbb{R})$ such that $\chi(t) = 1$ when $|t| \leq C_1$ and $\chi(t) = 0$ when $|t| \geq 2C_1$, with corresponding control of the derivatives of χ. The contribution by the region where $|t| < C_1$ to the integral $J_\epsilon(\alpha, \beta, T)$ can then be estimated by

$$J_{II}(\alpha, \beta, T) := \beta^{1+\epsilon} \iint \rho\left(\alpha^3 - \beta^3\phi_{y_1}(t) + r_1(y_1)\right)(1 + |y_1|)^{-N}\chi(t)|t|^\epsilon\, dt\, dy_1,$$

where we have set

$$\phi_{y_1}(t) := \left(1 + \frac{1}{\beta^2}Q\left(\frac{\beta}{T}t\right) - \frac{r_2(y_1)}{\beta^2 T}\right)t - b\left(y_1, \frac{\beta}{T}t\right)t^3, \quad |t| \leq C_1.$$

Recall here that $T/\beta \geq 1$ (for $1 \leq |t| \leq 2C_1$, we may extend the function b in a suitable way, if necessary).

By Fourier inversion, this can be estimated by

$$\begin{aligned}|J_{II}(\alpha, \beta, T)| \leq C\beta^{1+\epsilon}\Big| \iint \int e^{-i\xi\beta^3\phi_{y_1}(t)}\chi(t)|t|^\epsilon\, dt\, e^{i\xi(\alpha^3 + r_1(y_1))} \\ \times\left(1 + |y_1|\right)^{-N}\hat{\rho}(\xi)\, d\xi\, dy_1\Big| \\ \leq C\beta^{1+\epsilon}\iint \left|\int e^{-i\xi\beta^3\phi_{y_1}(t)}\chi(t)|t|^\epsilon\, dt\right|\left(1 + |y_1|\right)^{-N}|\hat{\rho}(\xi)|\, d\xi\, dy_1.\end{aligned}$$

Now, if $|r_2(y_1)/(\beta^2 T)| \geq \frac{1}{2}$, then $c(1 + |y_1|) \geq \beta^2 T/2 \geq 1$, so trivially the integration in y_1 yields that

$$|J_{II}(\alpha, \beta, T)| \leq C\beta^{1+\epsilon}(\beta^2 T)^{1-N} \leq C\beta^{3+\epsilon-2N}$$

for every $N \in \mathbb{N}$.

So, assume that $|r_2(y_1)/(\beta^2 T)| < \frac{1}{2}$. Then the phase $\phi_{y_1}(t)$ has no degenerate critical point on the support of $\chi(t)$ if we assume ε to be sufficiently small, since then b is a small perturbation of the constant function $b(0, 0)$ in the sense of (2.1). It is then easily verified that our assumptions (in particular, (2.3)) imply that $\phi_{y_1}(t)$ satisfies the hypotheses of the van der Corput–type Lemma 2.1(ii), with $M = 2$ and constants $c_1, c_2 > 0$, which are admissible provided we choose L sufficiently large. Therefore, the lemma shows that the inner integral with respect to t is bounded by $C(1 + |\xi|\beta^3)^{-1/2}$, which implies that

$$|J_{II}(\alpha, \beta, T)| \leq C\beta^{1+\epsilon}(\beta^3)^{-1/2} = C\beta^{\epsilon-1/2}.$$

Case 2. $\gamma < 1$, that is, $\beta < \alpha$. Then there are admissible constants $C_3, C_4 \geq 1$ so that if $|s| \geq C_3$, then $\alpha^3|1 - (\gamma + \frac{1}{\alpha^2}Q(\frac{\alpha}{T}s))\,s - b(y_1, \frac{\alpha}{T}s)\,s^3| \geq C_4\alpha^3|s|^3$ in (2.5). Arguing in a similar way as in the first case, this implies that the contribution $J_{III}(\alpha, \beta, T)$ by the region where $|s| \geq C_3$ to the integral $J_\epsilon(\alpha, \beta, T)$ in (2.5) can be estimated by

$$|J_{III}(\alpha, \beta, T)| \leq C\alpha^{4+\epsilon-3N}.$$

Similarly, there are admissible constants $C_5, C_6 > 0$ so that if $|s| \le C_5$, then $|1 - (\gamma + \frac{1}{\alpha^2} Q(\frac{\alpha}{T} s))\, s - b(y_1, \frac{\alpha}{T} s)\, s^3| \ge C_6$ in (2.5), and this implies that the contribution $J_{IV}(\alpha, \beta, T)$ by the region where $|s| \le C_5$ to the integral $J_\epsilon(\alpha, \beta, T)$ can be estimated by

$$|J_{IV}(\alpha, \beta, T)| \le C\alpha^{4+\epsilon-3N}.$$

Finally, on the set where $C_5 < |s| < C_3$, the phase $\phi_{y_1}(s) := (\frac{r_2(y_1)}{\alpha^2 T} - (\gamma + \frac{1}{\alpha^2} Q(\frac{\alpha}{T} s)))s - b(y_1, \frac{\alpha}{T} s)s^3$ has again no degenerate critical point, and we conclude in a similar way as in the first case that the contribution $J_V(\alpha, \beta, T)$ by this region can be estimated by

$$|J_V(\alpha, \beta, T)| \le C\alpha^{1+\epsilon}(\alpha^3)^{-1/2} = C\alpha^{\epsilon-1/2}.$$

Combining all these estimates, we arrive at (2.4), which concludes the proof of the lemma when $|A| \ge L$.

Finally, when $|A| \le L$ and $|B| \ge L$, then we may indeed assume without loss of generality that $\alpha = 0$. Arguments very similar to those that we have applied before then show that $|J_\epsilon(\alpha, \beta, T)| \le C\beta^{\epsilon-1/2}$, which concludes the proof of the lemma. Q.E.D.

Lemma 2.5. *Let $b = b(y)$ be a C^2-function on $[-1, 1]$ such that $b(0) \ne 0$ and $\|b\|_{C^2([-1,1])} \le c_1$. Furthermore, let A and B be real numbers, and let $\delta_0 \in]0, 1[$. For $T \ge L$ and $\delta > 0$ such that $\delta < 1$ and $\delta T \le \delta_0$, we put*

$$I(A, B) := \int_{\mathbb{R}} \left|\rho\big(A + By + b(\delta y)y^3\big)\right| \chi_0\left(\frac{y}{T}\right) dy,$$

where $\rho \in \mathcal{S}(\mathbb{R})$ denotes a fixed Schwartz function and $\chi_0 \in C_0^\infty(\mathbb{R})$, a nonnegative bump function supported in the interval $[-1, 1]$. Then, for δ_0 sufficiently small and L sufficiently large, we have

$$|I(A, B)| \le C\big(1 + \max\{|A|^{1/3}, |B|^{1/2}\}\big)^{-1/2}, \tag{2.6}$$

where the constant C depends only on c_1, δ_0, ρ, and χ_0 but not on A, B, T, and δ.

Proof. We may dominate the function $|\rho|$ by a nonnegative Schwartz function. Let us therefore assume without loss of generality that $\rho \ge 0$.

Let us first assume that $\max\{|A|, |B|\} \ge L$. If $|A| \le T^3/\delta_0^3$ and $|B| \le T^2/\delta_0^2$, estimate (2.6) follows easily from the previous Lemma 2.4 (replace $\chi_0(\cdot/T)$ by the characteristic function of the interval $[-1/\delta, 1/\delta]$, choose $r_1 \equiv 0$ and $r_2 \equiv 0$, and put $T := 1/\delta$ in the previous lemma).

Next, if $|B| > T^2/\delta_0^2$, using Fourier inversion, we may estimate

$$|I(A, B)| \le C \int \left| \int e^{is\phi(y)} \chi_0\left(\frac{y}{T}\right) dy \right| |\hat{\rho}(s)|\, ds, \tag{2.7}$$

where $\phi(y) := By + b(\delta y)y^3$. And, by means of integration by parts, we obtain that

$$\left| \int e^{is\phi(y)} \chi_0\left(\frac{y}{T}\right) dy \right| \le C|s|^{-1} \int_{-T}^{T} \left(\left|\frac{\phi''(y)}{\phi'(y)^2}\right| + \left|\frac{1}{T\phi'(y)}\right| \right) dy.$$

Since for $|y| \le T$ we have

$$|\phi'(y)| = \left|B + y^2\left(3b(\delta y) + \frac{\delta y}{3}b'(\delta y)\right)\right| \ge \frac{|B|}{2} \quad \text{and} \quad |\phi''(y)| \le CT,$$

if we choose δ_0 sufficiently small, we see that $\left|\int e^{is\phi(y)}\chi_0\left(\frac{y}{T}\right)dy\right| \le C/|sB|$ for δ_0 sufficiently small. On the other hand, trivially we have $\left|\int e^{is\phi(y)}\chi_0\left(\frac{y}{T}\right)dy\right| \le CT$, and taking the geometric mean of these estimates and using $T < \delta_0|B|^{1/2}$, we find that

$$\left|\int e^{is\phi(y)}\chi_0\left(\frac{y}{T}\right)dy\right| \le C|s|^{-1/2}|B|^{-1/4}.$$

In combination with (2.7) this leads to the estimate $|I(A,B)| \le C\big(1+|B|\big)^{-1/4}$.

Next, if $|A| > T^3/\delta_0^3$ and $|B| \le |A|^{2/3}$, then we may estimate

$$\begin{aligned}|A + By + b(\delta y)y^3| &\ge |A| - |B|T - \|b\|_\infty T^3\\ &\ge |A| - |A|^{2/3}\delta_0|A|^{1/3} - \|b\|_\infty\delta_0|A| \ge \frac{|A|}{2},\end{aligned}$$

provided δ_0 is sufficiently small. This implies that $|I(A,B)| \le C|A|^{-N}T \le C|A|^{-N+1/3}$ for every $N \in \mathbb{N}$, which is stronger than what we need for (2.6).

We have thus confirmed estimate (2.6) when $\max\{|A|, |B|\} \ge L$.

Finally, if $\max\{|A|, |B|\} < L$, then the rapid decay of ρ easily implies that $|I(A,B)| \le C$. Q.E.D.

2.4 FOURIER RESTRICTION VIA REAL INTERPOLATION

In order to establish the restriction estimate (1.1) at the endpoint $p = p_c$, we shall have to apply interpolation arguments in several situations. Indeed, our approach will be based on various kinds of dyadic frequency domain decompositions, and it turns out that the estimates for the operators corresponding to the dyadic frequency domains arising in this way do not sum at the endpoint.

Classically, one uses complex interpolation methods, based on Stein's interpolation theorem for analytic families of operators (Theorem 4.1 in [SW71]), as in the seminal work by Stein and Tomas on the Euclidean sphere (see [To75]).

An alternative "classical" approach, which has turned out to be quite useful in order to obtain even mixed L^p-estimates in time and space for solutions to dispersive partial differential equations, is based on the Hardy-Littlewood-Sobolev inequality. This approach has been used in work by J. Ginibre and G. Velo [GV92] and further developed, for instance, by M. Keel and T. Tao [KT98]. However, it does not seem clear at all how to adapt the underlying method to our problem, so that we shall not pursue this question here, even though it might be interesting to investigate which mixed L^p-norm estimates do hold in our context.

Yet another, alternative approach, based on real interpolation, has been devised more recently by J. G. Bak and A. Seeger [BS11], which yields an even stronger endpoint estimate in terms of Lorentz spaces (see Theorem 1.1 in [BS11]) and whose assumptions often can be verified quite easily.

Regretfully, in the majority of situations that we shall encounter in the course of the proof of Theorem 1.14, it does not seem possible to make use of this result (respectively, the method of proof), so the last chapters of this monograph will be devoted to the study of various rather intricate analytic families of operators constructed by summing various pieces of dyadic decompositions that have been performed earlier, to which we shall apply Stein's interpolation theorem.

Nevertheless, in a few cases, we shall be able to make use of the following variant of Theorem 1.1 in [BS11], whose proof follows directly by means of the methods developed in [BS11]. What prevents us from applying Theorem 1.1 in [BS11] directly is that we shall encounter complex measures, whereas in the work by Bak and Seeger it is important to start with a positive measure.

In the sequel, if μ is any bounded, complex Borel measure on $\mathbb{R}^d$, then we shall often denote by T_μ the convolution operator

$$T_\mu : \varphi \mapsto \varphi * \hat{\mu}.$$

Proposition 2.6. *Let μ be a positive Borel measure on $\mathbb{R}^d$ of total mass $\|\mu\|_1 \le 1$, and let $p_0 \in [1, 2[$. Assume that μ can be decomposed into a finite sum*

$$\mu = \mu^b + \sum_{i \in I} \mu^i$$

of bounded complex Borel measures μ^b and μ^i, $i \in I$, such that the following hold true. There is a constant $A \ge 0$ such that:

(a) *The operator T_{μ^b} is bounded from $L^{p_0}(\mathbb{R}^d)$ to $L^{p_0'}(\mathbb{R}^d)$, with*

$$\|T_{\mu^b}\|_{p_0 \to p_0'} \le A. \tag{2.8}$$

(b) *Each of the measures μ^i decomposes as*

$$\mu^i = \sum_{j=1}^{K_i} \mu^i_j = \sum_{j=1}^{K_i} \mu * \phi^i_j,$$

where $K_i \in \mathbb{N} \cup \{\infty\}$, and where the ϕ^i_j are integrable functions such that

$$\|\phi^i_j\|_1 \le 1. \tag{2.9}$$

Assume also that there are constants $a_i > 0, b_i > 0$ such that for all i and j,

$$\|\mu^i_j\|_\infty \le A 2^{j a_i}; \tag{2.10}$$

$$\|\widehat{\mu^i_j}\|_\infty \le A 2^{-j b_i}; \tag{2.11}$$

$$p_0 = 2\frac{a_i + b_i}{2a_i + b_i}; \tag{2.12}$$

that is, if $\theta a_i - (1-\theta) b_i = 0$, then $\frac{1}{p_0} = \frac{\theta}{2} + (1-\theta)$.

Then there is a constant C that depends only on d and any given compact interval in $]0, \infty[$ containing the a_i and b_i such that for every i,

$$\|T_{\mu^i} f\|_{L^{p_0'}} \le C A \|f\|_{L^{p_0}}, \tag{2.13}$$

and consequently

$$\int |\hat{f}|^2 \, d\mu \leq CA\|f\|^2_{L^{p_0}(\mathbb{R}^d)}. \tag{2.14}$$

Proof. We shall denote by $L^{p,s}$ the Lorentz space of type (p, s). By essentially following the proof of Proposition 2.1 in [BS11], we define the interpolation parameter $\theta := 2/p_0'$. Observe that by (2.12) we have $\theta = b_i/(a_i + b_i)$; hence $(1-\theta)(-b_i) + \theta a_i = 0$ for every i. Thus, the two inequalities (2.10) and (2.11) allow us to apply an interpolation trick due to Bourgain [Bou85] and to conclude that each of the operators T_{μ^i} is of restricted weak-type (p_0, p_0'), with operator norm $\leq CA$. Moreover, if J is any compact subinterval of $]0, \infty[$, then for $a_i, b_i \in J$ we may chose the constant C so that it depends only on J. In combination with (2.8), this implies that also T_μ is of restricted weak-type (p_0, p_0'), with operator norm $\leq CA$, where C may be different from the previous constant but with similar properties. By applying Tomas' R^*R-argument for the restriction operator R, we get

$$\int |\hat{f}|^2 \, d\mu \leq CA\|f\|^2_{L^{p_0,1}}.$$

In combination with Plancherel's theorem and (2.10) and (2.9) we can next use this estimate as in [BS11] to control

$$\|T_{\mu^i_j} f\|_2^2 = \|f * \widehat{\mu^i_j}\|_2^2 = c\|\check{f}\mu^i_j\|_2^2 \leq A2^{ja_i}\|\phi^i_j\|_1 \, CA\|f\|^2_{L^{p_0,1}} \leq CA^2 2^{ja_i}\|f\|^2_{L^{p_0,1}}.$$

Here, we have denoted by $\check{f}$ the inverse Fourier transform of f. It is in this estimate that we make use of the positivity of the measure μ in an essential way. The remaining part of the argument in [BS11] does not require positivity of the underlying measure, so that it applies to each of the complex measures μ^i as well, and we may conclude that for any $s \in [0, \infty]$,

$$\|T_{\mu^i} f\|_{L^{p_0',s}} \leq CA\|f\|_{L^{p_0,s}}$$

(compare Proposition 2.1, (2.2), in [BS11]). Choosing $s = 2$, so that $p_0 \leq 2 \leq p_0'$, by the nesting properties of the scale of Lorentz spaces this implies in particular that

$$\|T_{\mu^i} f\|_{L^{p_0',p_0'}} \leq CA\|f\|_{L^{p_0,p_0}}$$

and, hence, (2.13). The same type of estimate then holds also for the operator T_μ, and Tomas' argument then implies (2.14). Q.E.D.

2.5 UNIFORM ESTIMATES FOR FAMILIES OF OSCILLATORY SUMS

The following simple lemma, which may also be of independent interest, will become crucial in several of our complex interpolation arguments, in order to control certain sums corresponding to the dyadic decompositions mention before in the previous section.

Lemma 2.7. *Let $Q = \prod_{j=1}^{n}[-R_k, R_k] \subset \mathbb{R}^n$ be a compact cuboid, with $R_k > 0$, $k = 1, \dots, n$, and let H be a C^1-function on an open neighborhood of Q. Moreover, let $\alpha, \beta^1, \dots, \beta^n \in \mathbb{R}^\times$ be given. For any given real numbers $a_1, \dots, a_n \in \mathbb{R}^\times$ and $M \in \mathbb{N}$, we put*

$$F(t) := \sum_{l=0}^{M} 2^{i\alpha l t}(H\chi_Q)\big(2^{\beta^1 l}a_1, \dots, 2^{\beta^n l}a_n\big). \tag{2.15}$$

Then there is a constant C depending on Q and the numbers α and β^k but not on H, the a_k, M, and t, such that

$$|F(t)| \le C\frac{\|H\|_{C^1(Q)}}{|2^{i\alpha t} - 1|}, \qquad \text{for all } t \in \mathbb{R}, a_1, \dots, a_2 \in \mathbb{R}^\times \text{ and } M \in \mathbb{N}. \tag{2.16}$$

Proof. For $y = (y_1, \dots, y_n)$ in an open neighborhood of Q, Taylor's integral formula allows us to write $H(y) = H(0) + \sum_{k=1}^{n} y_k H_k(y)$, with continuous functions H_k whose C^0-norms on Q are controlled by the $C^1(Q)$-norm of H. Accordingly, we shall decompose $F(t) = F_0(t) + \sum_k F_k(t)$, where

$$F_0(t) := H(0)\sum_{l=0}^{M} 2^{i\alpha l t}\chi_Q\big(2^{\beta^1 l}a_1, \dots, 2^{\beta^n l}a_n\big),$$

$$F_k(t) := \sum_{l=0}^{M} 2^{i\alpha l t}(y_k H_k \chi_Q)\big(2^{\beta^1 l}a_1, \dots, 2^{\beta^n l}a_n\big), \qquad k = 1, \dots, n.$$

It will thus suffice to establish estimates of the form (2.16) for each of these functions F_0 and F_k, $k = 1, \dots, n$. We begin with F_0.

Observe that in the sum defining $F_0(t)$, we are effectively summing a geometric series over the integers contained in an interval, $l \in \{M_1, M_1 + 1, \dots, M_2 - 1, M_2\}$, where $M_1, M_2 \in \mathbb{N}$ depend on M, the a_ks, and the β^ks, and therefore

$$F_0(t) = H(0)\frac{2^{i\alpha(M_2+1)t} - 2^{i\alpha M_1 t}}{2^{i\alpha t} - 1}.$$

This implies an estimate of the form (2.16) for $F_0(t)$. Next, if $k \ge 1$, then trivially

$$|F_k(t)| \le C' \sum_{\{l : 2^{\beta^k l}|a_k| \le R_k\}} 2^{\beta^k l}|a_k| \le C R_k,$$

by summing a geometric series. Again this implies an estimate of the form (2.16).
Q.E.D.

Remark 2.8. *The estimate in* (2.16) *can be sharpened as follows:*
Assume that there are constants $\epsilon \in]0, 1]$ and C_k, $k = 1, \dots, n$, such that

$$\int_0^1 \left|\frac{\partial H}{\partial y_k}(sy)\right| ds \le C_k |y_k|^{\epsilon - 1}, \qquad \text{for all } y \in Q. \tag{2.17}$$

Then, under the hypotheses of Lemma 2.7, there is a constant C depending on Q, the numbers α and β^k and ϵ, but not on H, the a_k, M, and t, such that

$$|F(t)| \le C\frac{|H(0)| + \sum_k C_k}{|2^{i\alpha t} - 1|}, \qquad \text{for all } t \in \mathbb{R}, a_1, \dots, a_2 \in \mathbb{R}^\times \text{ and } M \in \mathbb{N}.$$

Indeed, Taylor's integral formula and (2.17) imply that $|y_k H_k(y)| \le C_k |y_k|^\epsilon$, which is sufficient for us to reach a conclusion in a similar way as before.

We shall also need the following analogue of Lemma 2.7 for oscillatory double sums. Its proof follows similar ideas but is technically more involved.

LEMMA 2.9. *Let $Q = \prod_{j=1}^n [-R_k, R_k] \subset \mathbb{R}^n$ be a compact cuboid, with $R_k > 0$, $k = 1, \dots, n$, and let H be a C^2-function on an open neighborhood of Q. Moreover, let $\alpha_1, \alpha_2 \in \mathbb{Q}^\times$ and $\beta_1^k, \beta_2^k \in \mathbb{Q}$ such that the vectors (α_1, α_2) and (β_1^k, β_2^k) are linearly independent, for every $k = 1, \dots, n$, that is,*

$$\alpha_1 \beta_2^k - \alpha_2 \beta_1^k \neq 0, \qquad k = 1, \dots, n. \tag{2.18}$$

For any given real numbers $a_1, \dots, a_n \in \mathbb{R}^\times$ and $M_1, M_2 \in \mathbb{N}$, we then put

$$F(t) := \sum_{m_1=0}^{M_1} \sum_{m_2=0}^{M_2} 2^{i(\alpha_1 m_1 + \alpha_2 m_2)t} (H\chi_Q)\big(2^{(\beta_1^1 m_1 + \beta_2^1 m_2)} a_1, \dots, 2^{(\beta_1^n m_1 + \beta_2^n m_2)} a_n\big). \tag{2.19}$$

Then there is a constant C depending on Q and the numbers α_i and β_i^k, but not on H, the a_k, M_1, M_2, and t, and a number $N \in \mathbb{N}^\times$ depending on the β_i^k such that

$$|F(t)| \le C \frac{\|H\|_{C^2(Q)}}{|\rho(t)|}, \qquad \textit{for all } t \in \mathbb{R}, a_1, \dots, a_2 \in \mathbb{R}^\times \textit{ and } M_1, M_2 \in \mathbb{N}, \tag{2.20}$$

where $\rho(t) := \prod_{\nu=1}^N \tilde{\rho}(\nu t)\tilde{\rho}(-\nu t)$, with

$$\tilde{\rho}(t) := (2^{i\alpha_1 t} - 1)(2^{i\alpha_2 t} - 1) \prod_{k=1}^n (2^{i(\alpha_1 \beta_2^k - \alpha_2 \beta_1^k)t} - 1).$$

REMARK 2.10. *For $\zeta \in \mathbb{C}$ and $0 < \theta < 1$, let us put*

$$\tilde{\gamma}(\zeta) := (2^{\alpha_1 \zeta} - 1)(2^{\alpha_2 \zeta} - 1) \prod_{k=1}^n (2^{(\alpha_1 \beta_2^k - \alpha_2 \beta_1^k)\zeta} - 1)$$

and

$$\gamma_\theta(\zeta) := \prod_{1 \le \nu \le N} \frac{\tilde{\gamma}(\nu(\zeta - 1))\tilde{\gamma}(-\nu(\zeta - 1))}{\tilde{\gamma}(\nu(\theta - 1))\tilde{\gamma}(-\nu(\theta - 1))}$$

(notice that for $\nu = 1, \dots, N$, we have $\tilde{\gamma}(\pm\nu(\theta - 1)) \neq 0$). Then γ_θ is a well-defined entire analytic function such that $\gamma_\theta(\theta) = 1$. Moreover, for ζ in the complex strip $\Sigma := \{\zeta \in \mathbb{C} : 0 \le \operatorname{Re} \zeta \le 1\}$, this function is uniformly bounded and $\gamma_\theta(1 + it) = c_\theta \rho(t)$, so that

$$\Big|\gamma(1 + it)F(t)\Big| \le C \qquad \textit{for all } t \in \mathbb{R}, a_1, \dots, a_2 \in \mathbb{R}^\times \textit{ and } M_1, M_2 \in \mathbb{N}$$

if $F(t)$ is defined as in (2.19).

Proof. The basic idea of the proof becomes most transparent under the additional assumption that also the vectors (β_1^k, β_2^k), $k = 1, \dots, n$, are pairwise linearly independent, that is,

$$\beta_1^l \beta_2^k - \beta_1^k \beta_2^l \neq 0, \qquad \text{for all } l \neq k. \tag{2.21}$$

We shall, therefore, begin with this case and later indicate the modifications needed for the general case.

For $y = (y_1, \dots, y_n)$ in an open neighborhood of Q, Taylor's integral formula allows us to write $H(y) = H(0) + \sum_{k=1}^{n} y_k H_k(y)$, with C^1-functions H_k whose C^1-norms on Q are controlled by the $C^2(Q)$-norm of H.

Similarly, by putting $h_k(y_k) := H_k(0, \dots, 0, y_k, 0, \dots, 0)$, we may decompose $H_k(y) = h_k(y_k) + \sum_{\{l:l\neq k\}} y_l H_{kl}(y)$, with continuous functions H_{kl} whose $C(Q)$-norms are controlled by the $C^2(Q)$-norm of H. This allows to write

$$H(y) = H(0) + \sum_k y_k h_k(y_k) + \sum_{l\neq k} y_k y_l H_{kl}(y).$$

Accordingly, we shall decompose $F(t) = F_0(t) + \sum_k F_k(t) + \sum_{l\neq k} F_{kl}(t)$, where

$$F_0(t) := H(0) \sum_{m_1=0}^{M_1} \sum_{m_2=0}^{M_2} 2^{i(\alpha_1 m_1+\alpha_2 m_2)t} \chi_Q\big(2^{(\beta_1^1 m_1+\beta_2^1 m_2)} a_1, \dots, 2^{(\beta_1^n m_1+\beta_2^n m_2)} a_n\big),$$

$$F_k(t) := \sum_{m_1=0}^{M_1} \sum_{m_2=0}^{M_2} 2^{i(\alpha_1 m_1+\alpha_2 m_2)t} (y_k h_k)(2^{(\beta_1^k m_1+\beta_2^k m_2)} a_k)$$
$$\times \chi_Q\big(2^{(\beta_1^1 m_1+\beta_2^1 m_2)} a_1, \dots, 2^{(\beta_1^n m_1+\beta_2^n m_2)} a_n\big),$$

$$F_{kl}(t) := \sum_{m_1=0}^{M_1} \sum_{m_2=0}^{M_2} 2^{i(\alpha_1 m_1+\alpha_2 m_2)t} (y_k y_l H_{kl}\, \chi_Q)\big(2^{(\beta_1^1 m_1+\beta_2^1 m_2)} a_1, \dots, 2^{(\beta_1^n m_1+\beta_2^n m_2)} a_n\big).$$

It will therefore suffice to establish estimates of the form (2.20) for each of these functions F_0, F_k and F_{kl}. We begin with F_0.

We may choose $r \in \mathbb{N}^\times$ so that every β_i^k can be written as $\beta_i^k = p_i^k/r$, with $p_i^k \in \mathbb{Z}$. Let us assume that there is a least one $\beta_2^k \neq 0$ (otherwise, we find some $\beta_1^k \neq 0$ and may proceed with the roles of the indices $i = 1$ and $i = 2$ interchanged). We then let $p_2 := |\prod_{k:p_2^k\neq 0} p_2^k|$ and $q_k := (p_1^k p_2)/p_2^k$ whenever $p_2^k \neq 0$, so that $q_k \in \mathbb{Z}$. Observe next that we may write every $m_1 \in \mathbb{N}$ uniquely in the form $m_1 = \alpha + j_1 p_2$, with $\alpha \in \{0, \dots, p_2 - 1\}$ and $j_1 \in \mathbb{Z}$. This allows us to decompose $F_0(t) = \sum_{\alpha=0}^{p_2-1} F_0^\alpha(t)$, where $F_0^\alpha(t)$ is defined like $F_0(t)$, but the summation in m_1 is restricted to those m_1 that are congruent to α modulo p_2.

Next, an easy computation shows that if $\beta_2^k \neq 0$, then

$$\beta_1^k(\alpha + j_1 p_2) + \beta_2^k m_2 = \beta_1^k \alpha + \beta_2^k(m_2 + q_k j_1).$$

Therefore, if we write $R_k/a_k = (\operatorname{sgn} a_k) 2^{b_k}$, then the restriction imposed by χ_Q on the coordinate of k leads to the condition

$$\beta_1^k \alpha + \beta_2^k(m_2 + q_k j_1) \leq b_k.$$

This means that m_2 lies in an "interval" of the form $\{0, \dots, d_k - q_k j_1\}$ or $\{e_k - q_k j_1, \dots, M_2\}$ (depending on the sign of β_2^k) for every k such that $\beta_2^k \neq 0$ (by an interval we mean here the set of integer points within a real interval). We may, therefore, decompose the set of j_1s over which we are summing into a finite number of (at most $(n!)^2$) pairwise disjoint intervals J_s such that for each given s there are indices k_s, k'_s such that for $j_1 \in I_s$, m_2 will run through an interval of the form $\{e'_s - u_s j_1, \dots, d'_s - v_s j_1\}$, $\{0, \dots, d'_s - v_s j_1\}$ or $\{e'_s - u_s j_1, \dots, M_2\}$, where $e'_s := e_{k_s}$, $u_s := q_{k_s}$ and $d'_s := d_{k'_s}$, $v_s := q_{k'_s}$. We may thus reduce ourselves to considering, for each fixed s, the corresponding part F^α_s of F^α given by summation over the interval I_s, that is,

$$F^\alpha_s(t) := H(0) \sum_{\{j_1 \in I_s : 0 \le \alpha + j_1 p_2 \le M_1\}} \sum_{m_2 \in I_s} 2^{i(\alpha\alpha_1 + p_2\alpha_1 j_1 + \alpha_2 m_2)t}$$
$$\times \prod_{\{k : \beta_2^k = 0\}} \chi_{[-R_k, R_k]}(2^{\beta_1^k(\alpha + j_1 p_2)} a_k).$$

We may split the sum over $m_2 \in I_s$ into the difference of at most two sums, in which we either sum over $m_2 \in \{0, \dots, M_2\}$, or over $m_2 \in \{0, \dots, f'_s - w_s j_1\}$, with $f'_s = d'_s$ and $w_s = v_s$ or $f'_s = e'_s - 1$ and $w_s = u_s$. Let us look at the latter case, assuming that $f'_s = e'_s - 1$ and $w_s = u_s$.

Evaluating the corresponding geometric sum in m_2, we see that the corresponding contribution is given by

$$F^\alpha_{s,1}(t) = H(0) \sum_{\{j_1 \in I_s : 0 \le \alpha + j_1 p_2 \le M_1\}} 2^{i\alpha\alpha_1 t} \frac{2^{i(\alpha_2 e'_s + j_1(\alpha_1 p_2 - \alpha_2 u_s))t} - 2^{ip_2\alpha_1 j_1 t}}{2^{i\alpha_2 t} - 1}$$
$$\times \prod_{\{k : \beta_2^k = 0\}} \chi_{[-R_k, R_k]}(2^{\beta_1^k(\alpha + j_1 p_2)} a_k).$$

An analogous term arises when $f'_s = d'_s$ and $w_s = v_s$. But, by assumption (2.18), $\alpha_1 p_2 - \alpha_2 u_s = (\alpha_1 \beta_2^{k_s} - \alpha_2 \beta_1^{k_s}) n_{k_s} \neq 0$, where $n_{k_s} := p_2/\beta_2^{k_s} \in \mathbb{Z}^\times$, and the characteristic functions of the intervals $[-R_k, R_k]$ again localize the summation over the j_1s to the summation over some interval, which shows that we may estimate

$$|F^\alpha_{s,1}(t)| \le \frac{C}{|2^{i\alpha_2 t} - 1||2^{i\alpha_1 p_2 t} - 1||2^{i(\alpha_1\beta_2^{k_s} - \alpha_2\beta_1^{k_s})n_{k_s} t} - 1|} \le \frac{C}{|\rho(t)|}.$$

The case where we sum over $m_2 \in \{0, \dots, M_2\}$ is even easier to treat, and we can again estimate the corresponding contribution $F^\alpha_{s,2}(t)$ to $F^\alpha_s(t)$ by the right-hand side of the preceding inequality. This establishes the desired estimate for $F_0(t)$.

We next turn to $F_k(t)$. Given k, let us assume again without loss of generality that $\beta_2^k \neq 0$. Then we may write $m_1, m_2 \in \mathbb{Z}$ in a unique way as

$$m_1 = \alpha + j_1 p_2^k,\ m_2 = j_2 - j_1 p_1^k, \quad \text{with} \quad \alpha \in \{0, \dots, |p_2^k|\}, \tag{2.22}$$

with integers $j_1, j_2 \in \mathbb{Z}$. Observe that then

$$\beta_1^l m_1 + \beta_2^l m_2 = \beta_1^l \alpha + j_2\beta_2^l + j_1(\beta_1^l\beta_2^k - \beta_2^l\beta_1^k)r,$$
$$\alpha_1 m_1 + \alpha_2 m_2 = \alpha_1\alpha + j_2\alpha_2 + j_1(\alpha_1\beta_2^k - \alpha_2\beta_1^k)r.$$

In particular, $\beta_1^k m_1 + \beta_2^k m_2 = \beta_1^k\alpha + j_2\beta_2^k$ does not depend on j_1. Moreover, for given α and j_2, the localizations given by the conditions $|2^{(\beta_1^l m_1 + \beta_2^l m_2)} a_l| \le R_l$, $l \neq k$, reduce the summation over j_1 to the summation over an interval $I(\alpha, j_2)$, and summing a geometric sum with respect to j_1, we thus see that

$$|F_k(t)| \le \frac{C'}{|2^{i(\alpha_1\beta_2^k-\alpha_2\beta_1^k)rt}-1|}\sum_{\alpha=0}^{|p_2^k|}\sum_{\{j_2:|2^{\beta_1^k\alpha+j_2\beta_2^k}a_k|\le R_k\}}|2^{\beta_1^k\alpha+j_2\beta_2^k}a_k| \le \frac{CR_k}{|\rho(t)|}.$$

Consider finally $F_{kl}(t)$, for $k \neq l$. We may simply estimate

$$|F_{kl}(t)| \le C\sum_{(m_1,m_2)\in J_{k,l}}|2^{(\beta_1^k m_1+\beta_2^k m_2)}a_k|\,|2^{(\beta_1^l m_1+\beta_2^l m_2)}a_l|,$$

where J_{kl} is the set of all $(m_1, m_2) \in \mathbb{N}^2$ satisfying $|2^{(\beta_1^k m_1+\beta_2^k m_2)}a_k| \le R_k$ and $|2^{(\beta_1^l m_1+\beta_2^l m_2)}a_l| \le R_l$. By comparing with an integral and changing variables in the integral (recall that by our assumption (2.21) the matrix $\begin{pmatrix}\beta_1^k & \beta_2^k\\ \beta_1^l & \beta_2^l\end{pmatrix}$ is non-degenerate) this leads to the estimate

$$\begin{aligned}|F_{kl}(t)| &\le C''\iint_{I_{k,l}}|2^{(\beta_1^k s_1+\beta_2^k s_2)}a_k|\,|2^{(\beta_1^l s_1+\beta_2^l s_2)}a_l|\,ds_1\,ds_2\\ &\le C'\int_{-\infty}^{\log_2(R_l/|a_l|)}\int_{-\infty}^{\log_2(R_k/|a_k|)}|2^{x_1}a_k||2^{x_2}a_l|\,dx_1\,dx_2 \le CR_kR_l,\end{aligned}$$

where I_{kl} denotes the set of all $(s_1, s_2) \in \mathbb{R}_+^2$ satisfying $|2^{(\beta_1^k s_1+\beta_2^k s_2)}a_k| \le R_k$ and $|2^{(\beta_1^l s_1+\beta_2^l s_2)}a_l| \le R_l$.

This concludes the proof of the lemma under our additional hypotheses (2.21).

Let us finally indicate how to remove assumptions (2.21). To this end, let us write $\beta^j := (\beta_1^j, \beta_2^j)$. In the general case, we may decompose the index set $\{1, \dots, n\}$ into pairwise disjoint subsets $I_1, \dots, I_h$ such that the following hold true.

There are nontrivial vectors $\gamma^k = (\gamma_1^k, \gamma_2^k)$, $k = 1, \dots, h$, in $\mathbb{Q}^2$ and rational numbers $r_j \neq 0$, $j = 1, \dots, n$, such that

(a) if $j \in I_k$, then $\beta^j = r_j\gamma^k$;
(b) for $k \neq l$, the vectors γ^k and γ^l are linearly independent.

Let us accordingly define the vectors $Y_k := (y_j)_{j\in I_k} \in \mathbb{R}^{I_k}$, $k = 1, \dots, h$. We may assume (possibly after a permutation of coordinates) that $y = (Y_1, \dots, Y_h)$. Following the first step of the previous proof, we then decompose $H(y) = H(0) + \sum_{k=1}^h {}^tY_k \cdot H_k(y)$, where now H_k maps into $\mathbb{R}^{I_k}$. Next, we set

$$h_k(Y_k) := H_k(0, \dots, 0, Y_k, 0, \dots, 0) \in \mathbb{R}^{I_k}$$

and apply Taylor's formula in order to write

$$H(y) = H(0) + \sum_{k=1}^h {}^tY_k \cdot h_k(Y_k) + \sum_{k\neq l} {}^tY_k \cdot H_{kl}(y)\cdot Y_l,$$

where here H_{kl} is a matrix-valued function. Correspondingly, we define the functions $F_0(t)$, $F_k(t)$ and $F_{kl}(t)$ as before, only with $y_kh_k(y_k)$ replaced by ${}^tY_k \cdot h_k(Y_k)$ and $y_ky_lH_{kl}(y)$ by ${}^tY_k \cdot H_{kl}(y) \cdot Y_l$, respectively. Notice that terms ${}^tY_k \cdot h_k(Y_k)$ arise only when $h \ge 2$.

The discussion of $F_0(t)$ remains unchanged, and the same applies essentially also to the discussion of $F_{kl}(t)$, because of property (b). Finally, for the estimation of $F_k(t)$, notice that for a given, fixed k, if $j \in I_k$, then by (a) we see that the arguments at which ${}^t Y_k \cdot h_k$ is evaluated are all of the form $2^{r_j(\gamma_1^k m_1 + \gamma_2^k m_2)} a_j$. Therefore, in the coordinates given by α, j_1, j_2 from (2.22), they all will not depend on j_1. We may therefore proceed in the estimation of $F_k(t)$ essentially as before, which concludes the proof of Lemma 2.9, also in the general case. Q.E.D.

2.6 NORMAL FORMS OF ϕ UNDER LINEAR COORDINATE CHANGES WHEN $h_{\text{lin}}(\phi) < 2$

Our discussion of the case where $h_{\text{lin}}(\phi) < 2$ will be based on normal forms of ϕ under linear coordinate changes. In the analytic setting, such normal forms are due to Siersma [Si74], but we shall need them also for smooth, finite-type ϕ. The designation of the type of singularity that we list next corresponds to Arnol'd's classification of singularities in the case of analytic functions (cf. [AGV88] and [Dui74]), that is, in the analytic case, nonlinear analytic changes of coordinates would allow to further reduce ϕ to Arnol'd's normal forms.

Proposition 2.11. *Assume that $h_{\text{lin}}(\phi) < 2$, where ϕ satisfies Assumption (NLA).*
Then, after applying a suitable linear change of coordinates, ϕ can be written in the following form on a sufficiently small neighborhood of the origin:

$$\phi(x_1, x_2) = b(x_1, x_2)(x_2 - \psi(x_1))^2 + b_0(x_1), \tag{2.23}$$

where b, b_0 and ψ are smooth functions and where $\psi(x_1) = cx_1^m + O(x_1^{m+1})$, with $c \neq 0$ and $m \geq 2$. Moreover, we can distinguish two cases:

Case a. $b(0,0) \neq 0$. *Then either*

(i) *b_0 is flat* *(singularity of type A_∞),*
or
(ii) *$b_0(x_1) = x_1^n \beta(x_1)$, where $\beta(0) \neq 0$ and $n \geq 2m+1$* *(singularity of type A_{n-1}).*

In these cases we say that ϕ is of type A.

Case b. $b(0,0) = 0$. *Then we may assume that*

$$b(x_1, x_2) = x_1 b_1(x_1, x_2) + x_2^2 b_2(x_2),$$

where b_1 and b_2 are smooth functions, with $b_1(0,0) \neq 0$.
Moreover, either

(i) *b_0 is flat* *(singularity of type D_∞),*
or
(ii) *$b_0(x_1) = x_1^n \beta(x_1)$, where $\beta(0) \neq 0$ and $n \geq 2m+2$.* *(singularity of type D_{n+1}).*

In these cases we say that ϕ is of type D.

Remark 2.12. (a) *It is easy to see that the principal weight κ and the Newton distance $d = d(\phi)$ for these normal forms are given by*

$$\kappa = \left(\frac{1}{2m}, \frac{1}{2}\right) \quad \text{and} \quad d = \frac{2m}{m+1}, \quad \text{if } \phi \text{ is of type A},$$

$$\kappa = \left(\frac{1}{2m+1}, \frac{m}{2m+1}\right) \quad \text{and} \quad d = \frac{2m+1}{m+1}, \quad \text{if } \phi \text{ is of type D},$$

and by Proposition 1.7 that $h_{\text{lin}}(\phi) = d$, that is, that the coordinates x are linearly adapted.

(b) *Similarly, the coordinates $y_1 := x_1$, $y_2 := x_2 - \psi(x_1)$ are adapted to ϕ, and we can choose ψ as the principal root jet. Comparing this with (1.9), we see that here the leading coefficient b_1 of the principal root jet ψ is given by the constant c.*

(c) *When ϕ has a singularity of type A_∞ or D_∞ and satisfies Condition (R), then necessarily $b_0 \equiv 0$.*

Proof. If $D^2\phi(0,0)$ had full rank 2, then the principal part ϕ_{pr} of ϕ would be a non-degenerate quadratic form, and by Proposition 1.2 one would easily see that the coordinates x would already be adapted to ϕ. This would contradict our assumptions. Therefore, rank $D^2\phi(0,0) \leq 1$. Let us denote by P_n the homogeneous part of degree n of the Taylor polynomial of ϕ, that is, $P_n(x_1, x_2) = \sum_{j+k=n} c_{jk} x_1^j x_2^k$.

Case 1. rank$D^2\phi(0,0) = 1$. In this case, by passing to a suitable linear coordinate system, we may assume that $P_2(x_1, x_2) = a x_2^2$, where $a \neq 0$. Consider the equation

$$\partial_2\phi(x_1, x_2) = 0.$$

By the implicit function theorem, locally it has a unique, smooth solution $x_2 = \psi(x_1)$, that is, $\partial_2\phi(x_1, \psi(x_1)) = 0$. A Taylor series expansion of the function $\phi(x_1, x_2)$ with respect to the variable x_2 around $\psi(x_1)$ then shows that

$$\phi(x_1, x_2) = b(x_1, x_2)(x_2 - \psi(x_1))^2 + b_0(x_1),$$

where b and b_0 are smooth functions and $b(0,0) = \frac{1}{2}\partial_2^2\phi(0,0) = a \neq 0$, whereas $b_0(x_1) = O(x_1^2)$, since $\phi(0,0) = 0$, $\nabla\phi(0,0) = 0$ (this is a special instance of what would follow from a classical division theorem; see, for example, [H90]).

Now, either b_0 is flat, which leads to type A_∞, or otherwise we may write $b_0(x_1) = x_1^n \beta(x_1)$, where $\beta(0) \neq 0$ and $n \geq 2$, which leads to type A_{n-1}.

Observe also that the function ψ cannot be flat, for otherwise the Newton polyhedron of ϕ would be the set $(0,2) + \mathbb{R}_+^2$, in case that b_0 is flat, or its principal edge would be the compact line segment with vertices $(0,2)$ and $(n,0)$. In the latter case, the principal part of ϕ is given by $\phi_{\text{pr}}(x_1, x_2) = a x_2^2 + \beta(0) x_1^n$, so that the maximal multiplicity $n(\phi_{\text{pr}})$ of any real root of ϕ_{pr} along the unit circle is at most 1, whereas the Newton distance is given by $d = 1/(\frac{1}{2} + \frac{1}{n}) \geq 1$. Therefore, in both cases, the coordinates x would already be adapted to ϕ, according to Proposition 1.2. Notice also that the same argument shows that the coordinates y

introduced in (1.10) are adapted to ϕ, so that, in particular, $h = 2$ (in case that b_0 is flat) or $h = 1/(\frac{1}{2} + \frac{1}{n}) < 2$ (if $b_0(x_1) = x_1^n \beta(x_1)$).

In particular, since $\psi(0) = 0$, we can write $\psi(x_1) = cx_1^m + O(x_1^{m+1})$ for some $m \in \mathbb{N}^\times$, where $c \neq 0$. Note that indeed $m \geq 2$, since $P_2(x_1, x_2) = ax_2^2$.

Finally, when $b_0(x_1) = x_1^n \beta(x_1)$, a similar reasoning as before shows that the coordinates x are already adapted if $2m \geq n$, so that under Assumption 1.6 we must have $n \geq 2m + 1$.

Case 2. $D^2\phi(0,0) = 0$. Then $P_2 = 0$, and $P_3 \neq 0$, for otherwise we had $h_{\text{lin}} \geq d \geq 1/(\frac{1}{4} + \frac{1}{4}) = 2$, which would contradict our assumption that $h_{\text{lin}} < 2$. Notice also that $P_3 \neq 0$ is homogeneous of odd degree 3, so that necessarily the multiplicity of roots (cf. (1.7)) satisfies $n(P_3) \geq 1$.

Assume first that $n(P_3) = 1$. Then, passing to a suitable linear coordinate system, we may assume that $P_3(x_1, x_2) = x_1(x_2 - \alpha x_1)(x_2 - \beta x_1)$, where either $\alpha \neq \beta$ are both real or $\alpha = \overline{\beta}$ are nonreal. Then one checks easily that the Newton diagram of P_3 is a compact edge intersecting the bisectrix in its interior and contained in the line given by $\frac{1}{3}t_1 + \frac{1}{3}t_2 = 1$. Consequently, it agrees with the principal face $\pi(\phi)$, so that $P_3 = \phi_{\text{pr}}$. We thus find that the Newton distance d in this linear coordinate system satisfies $d = 3/2 > n(\phi_{\text{pr}})$, so that these coordinates would already be adapted, contradicting our assumptions.

Assume next that $n(P_3) = 3$. Then, in a suitable linear coordinate system, $P_3(x_1, x_2) = x_2^3$. These coordinates are then adapted to P_3, so that $h(P_3) = d(P_3) = 3 > 2$. However, as has been shown in [IKM10], page 217, under Assumption 1.6 this implies that the Taylor support of ϕ is contained in the region where $\frac{1}{6}t_1 + \frac{1}{3}t_2 \geq 1$. This, in turn, implies that $h_{\text{lin}} \geq d \geq 1/(\frac{1}{6} + \frac{1}{3}) = 2$, in contrast to what we assumed.

We have thus seen that, necessarily, $n(P_3) = 2$. Then, after applying a suitable linear change of coordinates, we may assume that $P_3(x_1, x_2) = x_1 x_2^2$, that is,

$$\phi(x_1, x_2) = x_1 x_2^2 + R(x_1, x_2),$$

where R is a smooth function such that $\partial^\alpha R(0,0) = 0$ for $|\alpha| \leq 3$. Consider here the equation

$$\partial_2 \phi(x_1, x_2) = 0$$

with respect to x_2. We claim that it has a smooth solution $x_2 = \psi(x_1)$, with $\psi(0) = \psi'(0) = 0$, near the origin. Indeed, we have $\partial_2 \phi(x_1, x_2) = 2x_1 x_2 + R_1(x_1, x_2)$, where R_1 is a smooth function such that $\partial^\alpha R_1(0,0) = 0$ for $|\alpha| \leq 2$, so that we have to solve the equation

$$2x_1 x_2 + R_1(x_1, x_2) = 0. \tag{2.24}$$

To this end, let us write, for $x_1 \neq 0$, $x_2 = x_1 z$. Then (2.24) is equivalent to

$$2x_1^2 z + R_1(x_1, x_1 z) = 0. \tag{2.25}$$

Clearly, by the properties of R_1, we may factor $R_1(x_1, x_1 z) = x_1^3 g(x_1, z)$, with a smooth function $g(x_1, z)$ defined near the origin, and thus for $x_1 \neq 0$, equation (2.25) is equivalent to

$$2z + x_1 g(x_1, z) = 0.$$

Regard this as an equation near the origin in (x_1, z). We can now apply the implicit function theorem to conclude that locally near the origin this equation has a unique, smooth solution $z = \psi_1(x_1)$. In particular, we find that

$$2x_1^2\psi_1(x_1) + R_1(x_1, x_1\psi_1(x_1)) = 0$$

near the origin in x_1. Setting $\psi(x_1) := x_1\psi_1(x_1)$, we then find that indeed $\psi(0) = \psi'(0) = 0$ and

$$\partial_2\phi(x_1, \psi(x_1)) \equiv 0. \tag{2.26}$$

By means of a Taylor expansion of the function $\phi(x_1, x_2)$ with respect to the variable x_2 around $x_2 = \psi(x_1)$, this implies that

$$\phi(x_1, x_2) = b(x_1, x_2)(x_2 - \psi(x_1))^2 + b_2(x_1)x_2 + b_0(x_1),$$

where b, b_0, and b_2 are smooth functions. Again, we have that $\psi(x_1) = cx_1^m + O(x_1^{m+1})$, with $m \geq 2$. Observe that (2.26) implies that $b_2 = 0$; hence,

$$\phi(x_1, x_2) = b(x_1, x_2)(x_2 - \psi(x_1))^2 + b_0(x_1).$$

Moreover, since $\partial_2^2\phi(0, 0) = 0$, $\partial_1\partial_2^2\phi(0, 0) \neq 0$, $\partial_2^3\phi(0, 0) = 0$, we have that

$$b(0, 0) = 0, \quad \partial_1 b(0, 0) \neq 0 \quad \text{and } \partial_2 b(0, 0) = 0.$$

By Taylor's formula, this implies that

$$b(x_1, x_2) = x_1 b_1(x_1, x_2) + x_2^2 b_2(x_2),$$

where b_1 and b_2 are smooth functions, with $b_1(0, 0) \neq 0$.

In a similar way as in Case 1, one can see that the coordinates from (1.10) are adapted to ϕ. Moreover, if b_0 is flat, which leads to case D_∞, then $h = 2$, and if $b_0(x_1) = x_1^n\beta(x_1)$, which leads to case D_{n+1}, then $h = \frac{2n}{n+1} < 2$. Finally, one also checks easily that the coordinates x in (1.10) are already adapted to ϕ, if $2m + 1 \geq n$, so that under our assumption we must have $n \geq 2m + 2$.

This concludes the proof of Proposition 2.11. Q.E.D.

Corollary 2.13. *Assume that ϕ satisfies Assumption (NLA). By passing to a suitable linear coordinate system, let us also assume that the coordinates x are linearly adapted to ϕ. Then, if $d = d(\phi) < 2$, the critical exponent in Theorem 1.14 is given by $p_c' = 2d + 2$.*

Proof. Proposition 2.11 shows that the principal face $\pi(\phi)$ of the Newton polyhedron of ϕ is a compact edge whose "upper" vertex v is one the following points $(0, 2)$ or $(1, 2)$, which both lie below the line $H := \{(t_1, t_2) : t_2 = 3\}$ within the positive quadrant. On the other hand, $m + 1 \geq 3$. It is then clear from the geometry of the lines H, the line L that contains $\pi(\phi)$, and the line $\Delta^{(m)}$ that $\Delta^{(m)}$ will intersect L above the vertex v. Since, by Varchenko's algorithm, the point v will also be a vertex of the Newton polyhedron of ϕ^a, this easily implies that $h^r(\phi) = d$ (compare Figure 1.4). This proves the claim. Q.E.D.

Chapter Three

Reduction to Restriction Estimates near the Principal Root Jet

We now turn to the proof of Theorem 1.14 (which includes Theorem 1.10). As a first step, we shall reduce considerations to a small, "horn-shaped" neighborhood of the principal root jet ψ (cf. Subsection 1.1.1). This reduction will be achieved by means of a dyadic decomposition of the complementary region, making use of the dilations associated to the principal face $\pi(\phi)$, in combination with Greenleaf's restriction Theorem 1.1 and Littlewood-Paley theory, which will allow us to sum the estimates for the dyadic pieces that we obtain.

It turns out that for the contribution by this complementary region, the restriction estimate (1.1) will hold true even for the possibly wider range of ps given by $p' \geq 2d + 2$. For the case where $d = h_{\text{lin}}(\phi) > 2$, this will be an easy consequence of the fact that every root of ϕ_{pr} that does not agree with the principal root $x_2 = b_1 x_1^{m_1}$ does have multiplicity strictly less than d. The discussion of the case where $d = h_{\text{lin}}(\phi) \leq 2$ will require more refined estimates for oscillatory integrals, based on the normal forms provided by Proposition 2.6.

Recall that our coordinates x are assumed to be linearly adapted but not adapted to ϕ. By decomposing the plane $\mathbb{R}^2$ into two half-planes, we may and shall in the sequel always restrict ourselves to the closed right half plane where $x_1 \geq 0$, that is, we shall assume that the surface carried measure $d\mu = \rho d\sigma$ is of the form

$$\langle \mu, f \rangle = \int_{x_1 \geq 0} f(x, \phi(x))\, \eta(x)\, dx, \qquad f \in C_0(\mathbb{R}^3),$$

where $\eta(x) := \rho(x, \phi(x))\sqrt{1 + |\nabla\phi(x)|^2}$ is smooth and supported in a neighborhood Ω of the origin, which we may assume to be sufficiently small. The contribution by the other half plane where $x_1 \leq 0$ can be treated in an analogous way.

If F is any integrable function defined on Ω, we put

$$\mu^F := (F \otimes 1)\mu, \quad \text{i.e., } \langle \mu^F, f \rangle = \int_{x_1 \geq 0} f(x, \phi(x))\, \eta(x) F(x)\, dx.$$

Recall from (1.9) that $\psi(x_1) = b_1 x_1^m + O(x_1^{m+1})$. We choose a nonnegative bump function $\chi_0 \in C_0^\infty(\mathbb{R})$ that is supported in $[-1, 1]$ and identically 1 on $\left[-\frac{1}{2}, \frac{1}{2}\right]$, and put

$$\rho_1(x_1, x_2) := \chi_0\Big(\frac{x_2 - b_1 x_1^m}{\varepsilon x_1^m}\Big),$$

where $\varepsilon > 0$ is a small parameter to be determined later. Notice that ρ_1 is supported in the κ-homogeneous subset of $\Omega \cap \overline{H^+}$ where

$$|x_2 - b_1 x_1^m| \leq \varepsilon x_1^m, \tag{3.1}$$

which contains the curve $x_2 = \psi(x_1)$ when Ω is sufficiently small. In this chapter, we shall prove the following result, which will allow us to reduce our considerations to a domain of the form (3.1).

PROPOSITION 3.1. *Let $\varepsilon > 0$. If we choose the support of μ sufficiently small, then*

$$\left(\int_S |\widehat{f}|^2 \, d\mu^{1-\rho_1}\right)^{1/2} \leq C_{p,\varepsilon} \|f\|_{L^p(\mathbb{R}^3)}, \qquad f \in \mathcal{S}(\mathbb{R}^3),$$

whenever $p' \geq 2d + 2$. In particular, this estimate is valid whenever $p' \geq p_c'$.

The strategy of the proof will, by and large, follow the one of the proof of Theorem 1.7 in [IM11b]. By $\{\delta_r\}_{r>0}$ we shall again denote the dilations associated to the principal weight κ. We fix a suitable smooth cutoff function $\chi \geq 0$ on $\mathbb{R}^2$ supported in a closed annulus $\mathcal{A} \subset \mathbb{R}^2$ on which $|x| \sim 1$ in such a way that the functions $\chi_k := \chi \circ \delta_{2^k}$ form a partition of unity, and decompose the measure $\mu^{1-\rho_1}$ dyadically as

$$\mu^{1-\rho_1} = \sum_{k \geq k_0} \mu_k, \tag{3.2}$$

where $\mu_k := \mu^{\chi_k(1-\rho_1)}$. Let us extend the dilations δ_r to $\mathbb{R}^3$ by putting

$$\delta_r^e(x_1, x_2, x_3) := (r^{\kappa_1} x_1, r^{\kappa_2} x_2, r x_3).$$

We rescale the measure μ_k by defining $\nu_k := 2^{-k} \mu_k \circ \delta_{2^{-k}}^e$, that is,

$$\langle \nu_k, f \rangle = 2^{|\kappa| k} \langle \mu_k, f \circ \delta_{2^k}^e \rangle = \int_{x_1 \geq 0} f(x, \phi^k(x)) \, \eta(\delta_{2^{-k}} x) \chi(x)(1 - \rho_1(x)) \, dx,$$

with

$$\phi^k(x) := 2^k \phi(\delta_{2^{-k}} x) = \phi_\kappa(x) + \text{error terms of order } O(2^{-\delta k}), \tag{3.3}$$

where $\delta > 0$. Recall here that the principal part ϕ_{pr} of ϕ agrees with ϕ_κ. Let us denote by S^k the smooth hypersurface which is defined as the graph of ϕ^k. Then (3.3) shows that the measures ν_k are supported on S^k, that their total variations are uniformly bounded, i.e., $\sup_k \|\nu_k\|_1 < \infty$, and that they are approaching the surface carried measure ν_∞ on S defined by

$$\langle \nu_\infty, f \rangle := \int_{x_1 \geq 0} f(x, \phi_\kappa(x)) \, \eta(0) \chi(x)(1 - \rho_1(x)) \, dx$$

as $k \to \infty$. Following [IM11b], the key point in proving Proposition 3.1 will be to establish the following uniform estimates for the Fourier transforms of the measures ν_k.

LEMMA 3.2. *If $k_0 \in \mathbb{N}$ is sufficiently large, then there exists a constant $C > 0$ such that*

$$|\widehat{\nu_k}(\xi)| \leq C(1 + |\xi|)^{-1/d} \quad \textit{for every } \xi \in \mathbb{R}^3, \, k \geq k_0.$$

Proof. Assume first that $h_{\text{lin}} = h_{\text{lin}}(\phi) \geq 2$. Then $h(\phi) > 2$ by Assumption (NLA). Thus, in this case, the proof of Lemma 2.3 in [IM11b] shows that indeed the estimate in Lemma 3.2 holds true. The main argument used here is that every root of ϕ_{pr} that does not agree with the principal root $x_2 = b_1 x_1^{m_1}$ must have multiplicity strictly less than $d = d(\phi)$, as can be seen from Corollary 2.3 in [IM11a].

We may, therefore, assume from now on that $h_{\text{lin}} < 2$. Then we may even assume that ϕ is given by one of the normal forms appearing in Proposition 2.11 and that $h_{\text{lin}} = d$ is the Newton distance. Moreover the leading coefficient b_1 of the principal root jet ψ is given by the constant c of Proposition 2.11. We shall work from now on with c in place of b_1 in order to avoid possible confusion with the coefficient function $b_1(x_1, x_2)$. Let us rewrite

$$\widehat{\nu_k}(\xi) = \int_{x_1 \geq 0} e^{-i(\xi_1 x_1 + \xi_2 x_2 + \xi_3 \phi^k(x_1, x_2))} \eta(\delta_{2^{-k}} x) \chi(x)(1 - \rho_1(x))\, dx$$

and observe that, by a partition of unity argument, it will suffice to prove the following.

Given any point $v = (v_1, v_2)$ in the compact set $\mathcal{A} \cap \overline{H^+}$ such that

$$v_2 - c v_1^m \neq 0, \tag{3.4}$$

there is neighborhood V of v such that for every bump function $\chi_v \in C_0^\infty(\mathbb{R}^2)$ supported in V we have

$$|J^{\chi_v}(\xi)| \leq C(1 + |\xi|)^{-1/d} \quad \text{for every } \xi \in \mathbb{R}^3, k \geq k_0, \tag{3.5}$$

where

$$J^{\chi_v}(\xi) := \int_{x_1 \geq 0} e^{-i(\xi_1 x_1 + \xi_2 x_2 + \xi_3 \phi^k(x_1, x_2))} \eta(\delta_{2^{-k}} x) \chi_v(x)\, dx.$$

To prove this, we shall distinguish Cases a and b from Proposition 2.11.

Case a (ϕ of type A). In this case, we see that $\kappa = (\frac{1}{2m}, \frac{1}{2})$ and

$$\phi_\kappa(x_1, x_2) = \phi_{\text{pr}}(x_1, x_2) = b(0, 0)(x_2 - c x_1^m)^2,$$

so that $1/d = 1/2 + 1/2m$. After applying a suitable linear change of coordinates (and possibly complex conjugation to the integral $J^{\chi_v}(\xi)$), we may assume that $b(0, 0) = 1$. Then the Hessian determinant of ϕ_κ is given by

$$\text{Hess}(\phi_\kappa)(x_1, x_2) = -4m(m-1)c x_1^{m-2}(x_2 - c x_1^m).$$

Therefore, by (3.4), if $m = 2$, or $v_1 \neq 0$, then $\text{Hess}(\phi_\kappa)(v) \neq 0$. In this case, in view of (3.3) we can apply the method of stationary phase for phase functions depending on small parameters and easily obtain

$$|J^{\chi_v}(\xi)| \leq C(1 + |\xi|)^{-1} \quad \text{for every } \xi \in \mathbb{R}^3, k \geq k_0,$$

provided V is sufficiently small and k_0 sufficiently large. Since $d \geq 1$, this yields (3.5).

We are left with the case where $m > 2$ and $v_1 = 0$. Since $v = (v_1, v_2) \in \mathcal{A}$, this implies that $v_2 \neq 0$. Putting $\tilde{\phi}^k(y_1, y_2) := \phi^k(y_1, v_2 + y_2)$, we may rewrite $J^{\chi_v}(\xi)$ as

$$J^{\chi_v}(\xi) = e^{-iv_2\xi_2} \int_{y_1 \geq 0} e^{-i(\xi_1 y_1 + \xi_2 y_2 + \xi_3 \tilde{\phi}^k(y_1, y_2))} \eta\big(\delta_{2^{-k}}(y_1, v_2 + y_2)\big) \tilde{\chi}_0(y)\, dy,$$

where $\tilde{\chi}_0$ is now supported in a sufficiently small neighborhood of the origin. But,

$$\begin{aligned}\tilde{\phi}^k(y_1, y_2) &= (v_2 + y_2 - c y_1^m)^2 + O(2^{-\delta k})\\ &= v_2^2 + 2v_2 y_2 + \big(y_2^2 - 2c v_2 y_1^m + c^2 y_1^{2m} - 2c y_2 y_1^m\big) + O(2^{-\delta k}).\end{aligned}$$

The main nonlinear term here is $(y_2^2 - 2cv_2 y_1^m)$, which shows that the phase has a singularity of type A_{m-1} in the sense of Arnol'd.

By means of a linear change of variables in ξ-space, which replaces $\xi_2 + 2v_2\xi_3$ by ξ_2, we may thus reduce this to assuming that the complete phase in the oscillatory integral $J^{\chi_v}(\xi)$ is given by

$$\xi_1 y_1 + \xi_2 y_2 + \xi_3 \left(y_2^2 - 2cv_2 y_1^m + c^2 y_1^{2m} - 2c y_2 y_1^m + O(2^{-\delta k})\right).$$

We claim that

$$|J^{\chi_v}(\xi)| \leq C(1 + |\xi|)^{-(1/2+1/m)} \quad \text{for every } \xi \in \mathbb{R}^3, k \geq k_0,$$

which is even stronger than (3.5).

Indeed, if $|\xi_3| \ll \max\{|\xi_1|, |\xi_2|\}$, then this follows easily by integration by parts, so let us assume that

$$|\xi_3| \geq M \max\{|\xi_1|, |\xi_2|\}$$

for some constant $M > 0$. Then $|\xi_3| \sim |\xi|$. Consequently, by first applying the method of stationary phase to the integration in y_2 and then van der Corput's estimate to the y_1 integration, we obtain the preceding estimate. Observe here that these types of estimates are stable under small, smooth perturbations.

Case b (ϕ of type D). In this case, we see that $\kappa = \left(\frac{1}{2m+1}, \frac{m}{2m+1}\right)$ and

$$\phi_\kappa(x_1, x_2) = \phi_{\text{pr}}(x_1, x_2) = b_1(0,0) x_1 (x_2 - c x_1^m)^2,$$

so that $1/d = (m+1)/(2m+1)$. Again, we may assume without loss of generality that the coefficient function $b_1(x_1, x_2)$ satisfies $b_1(0,0) = 1$, so that

$$\phi_\kappa(x_1, x_2) = x_1 x_2^2 - 2c x_1^{m+1} x_2 + c^2 x_1^{2m+1}.$$

Straightforward computations show that

$$\begin{aligned}\partial_1^2 \phi_\kappa(x) &= -2cm(m+1) x_1^{m-1} x_2 + c^2 2m(2m+1) x_1^{2m-1},\\ \partial_1 \partial_2 \phi_\kappa(x) &= 2x_2 - 2c(m+1) x_1^m, \quad \partial_2^2 \phi_\kappa(x) = 2x_1;\end{aligned}$$

hence,

$$\operatorname{Hess}(\phi_\kappa)(v) := -4(x_2 - c x_1^m)\left(x_2 + c(m^2 - m - 1) x_1^m\right).$$

In view of (3.4), we see that $\operatorname{Hess}(\phi_\kappa)(v) \neq 0$, if $v_2 + c(m^2 - m - 1) v_1^m \neq 0$, so that we can again estimate $J^{\chi_v}(\xi)$ by means of the method of stationary phase.

Let us therefore assume that $\mathrm{Hess}(\phi_\kappa)(v) = 0$, that is,

$$v_2 = -c(m^2 - m - 1)v_1^m. \tag{3.6}$$

Observe that then $v_1 \neq 0$, $v_2 \neq 0$. Denote by

$$P_j(y) := \sum_{|\alpha|=j} \frac{1}{\alpha!}\partial^\alpha \phi_\kappa(v) y^\alpha$$

the homogeneous Taylor polynomial of ϕ_κ of degree j, centered at v. Then clearly

$$P_2(y) = v_1 \left(y_2 + \frac{(v_2 - c(m+1)v_1^m)y_1}{v_1} \right)^2 = v_1 \left(y_2 - cm^2 v_1^{m-1} y_1\right)^2.$$

Moreover, by (3.6)

$$\begin{aligned} P_3(y) &= -y_1 \left(\tfrac{1}{3}c^2 m^2(m^3 - m^2 + 2m + 1)v_1^{2m-2} y_1^2 - cm(m+1)v_1^{m-1} y_1 y_2 + y_2^2\right) \\ &= -y_1\, Q(y). \end{aligned}$$

Passing to the linear coordinates $z_1 := y_1,\ z_2 := y_2 - cm^2 v_1^{m-1} y_1$, one finds that

$$P_2 = v_1 z_2^2, \quad P_3 = -z_1 \tilde{Q}(z),$$

where $\tilde{Q} = z_2^2 + 2\beta_1 z_1 z_2 + \beta_2 z_1^2$ is again a quadratic form. Moreover, straightforward computations show that

$$\beta_2 = \frac{c^2}{3} m^2 (m-1)(m^2-1) v_1^{2m-2} \neq 0.$$

Applying Taylor's formula, we thus find that, in the coordinates z,

$$\begin{aligned} \tilde{\phi}(z) :&= \phi_\kappa(v_1 + y_1, v_2 + y_2) = c_0 + c_1 z_1 + c_2 z_2 + (v_1 z_2^2 - \beta_2 z_1^3) \\ &\quad -(z_1 z_2^2 + 2\beta_1 z_1^2 z_2) + O(|z|^4). \end{aligned}$$

Let us put $\phi^v(z) := \phi(z) - (c_0 + c_1 z_1 + c_2 z_2)$, so that $\phi^v(0,0) = 0,\ \nabla\phi^v(0,0) = 0$. Then one finds that the principal part of ϕ^v is given by

$$\phi^v_{\mathrm{pr}}(z) = v_1 z_2^2 - \beta_2 z_1^3, \qquad \text{where } \beta_2 \neq 0.$$

We can now argue in a very similar way as in the previous case. Indeed, by passing from the variables (x_1, x_2) in the integral defining $J^{\chi_v}(\xi)$ to the variables (z_1, z_2), and then applying first the method of stationary phase to the integration in z_2, and subsequently van der Corput's estimate to the z_1 integration (in the case where $|\xi_3| \geq M \max\{|\xi_1|, |\xi_2|\}$), we obtain the estimate

$$|J^{\chi_v}(\xi)| \leq C(1 + |\xi|)^{-(1/2+1/3)} \quad \text{for every } \xi \in \mathbb{R}^3, k \geq k_0.$$

Again, this is a stronger estimate than (3.5), since here

$$\frac{1}{d} = \frac{1}{2} + \frac{1}{4m+2} \leq \frac{1}{2} + \frac{1}{3}.$$

Q.E.D.

We can now conclude the proof of Proposition 3.1. According to Greenleaf's Theorem 1.1, the estimates in Lemma 3.2 imply the restriction estimates

$$\left(\int |\hat{f}(x)|^2 \, dv_k(x) \right)^{1/2} \leq C\|f\|_p, \qquad f \in \mathcal{S}(\mathbb{R}^3), \tag{3.7}$$

if $p' \geq 2d+2$, and the proof of Theorem 1 in [Gl81] reveals that the constant C can be chosen independently of k.

Let us rescale these estimates by putting

$$f_{(r)}(x) := r^{|\kappa|/2} f(\delta^e_r x), \quad r > 0,$$

for any function f on $\mathbb{R}^3$. Then $\widehat{f_{(r)}} = r^{-|\kappa|/2-1}\, \widehat{f \circ \delta^e_{r^{-1}}}$, and (3.7) implies

$$\int |\hat{f}(x)|^2 \, d\mu_k(x) = \int |\widehat{f_{(2^{-k})}}(x)|^2 \, d\nu_k(x) \leq C^2 2^{(|\kappa|/2+1)k} \| f \circ \delta^e_{2^k} \|_p^2;$$

hence,

$$\int |\hat{f}(x)|^2 \, d\mu_k(x) \leq C^2 \| f \|_p^2, \tag{3.8}$$

with a constant C that does not depend on k.

Fix a cutoff function $\tilde{\chi} \in C_0^\infty(\mathbb{R}^2)$ supported in an annulus centered at the origin such that $\tilde{\chi} = 1$ on the support of χ, and define dyadic decomposition operators Δ'_k by

$$\widehat{\Delta'_k f}(x) := \tilde{\chi}(\delta_{2^k} x')\, \hat{f}(x', x_3).$$

Then $\int |\hat{f}(x)|^2 d\mu_k(x) = \int |\widehat{\Delta'_k f}(x)|^2 d\mu_k(x)$, so that (3.8) yields

$$\int |\hat{f}(x)|^2 d\mu_k(x) \leq C^2 \, \| \widehat{\Delta'_k f} \|_p^2,$$

for any $k \geq k_0$. In combination with Minkowski's inequality, this implies

$$\left(\int |\hat{f}(x)|^2 d\mu^{1-\rho_1}(x) \right)^{1/2} = \left(\sum_{k \geq k_0} \int |\hat{f}(x)|^2 d\mu_k(x) \right)^{1/2} \leq C \left(\sum_{k \geq k_0} \| \Delta'_k f \|_p^2 \right)^{1/2}$$

$$= C \left(\left(\sum_{k \geq k_0} \left(\int |\Delta'_k f(x)|^p dx \right)^{2/p} \right)^{p/2} \right)^{1/p} \leq C \left\| \left(\sum_{k \geq k_0} |\Delta'_k f(x)|^2 \right)^{1/2} \right\|_{L^p(\mathbb{R}^3)},$$

since $p < 2$. Thus, by Littlewood-Paley theory [S93], we obtain the estimate in Proposition 3.1.

Proposition 3.1 shows that we are left with proving a Fourier restriction estimate for the measure μ^{ρ_1}, which is supported in a small neighborhood of the form (3.1) of the principal root jet, that is, with the verification of the following.

Proposition 3.3. *Assume that ϕ satisfies the assumptions of Theorem 1.14. If $\varepsilon > 0$ is sufficiently small, then we have*

$$\left(\int_S |\widehat{f}|^2 \, d\mu^{\rho_1} \right)^{1/2} \leq C_{p,\varepsilon} \| f \|_{L^p(\mathbb{R}^3)}, \qquad f \in \mathcal{S}(\mathbb{R}^3),$$

whenever $p' \geq p'_c$.

Notice that by interpolation with the trivial L^1-L^2-restriction estimate, it will suffice to prove this estimate for the endpoint $p = p_c$.

We shall distinguish between the cases where $h_{\text{lin}} < 2$ and $h_{\text{lin}} \geq 2$ since their treatments will require somewhat different approaches.

For $h_{\text{lin}} \geq 2$, our ultimate approach will require many different steps and rather intricate interpolation arguments. Luckily enough, a substantially simpler approach is available for $h_{\text{lin}} \geq 5$, which is based on restriction estimates for curves with nonvanishing torsion originating from the seminal work by S. W. Drury [Dru85]. Since the discussion of this approach will allow us to explain some of the basic ideas that will also be needed when $h_{\text{lin}} < 5$ but in a less complicated setting, we shall first explain this simpler approach. The proof for the general case $h_{\text{lin}} \geq 2$ will finally occupy the last three chapters.

Chapter Four

Restriction for Surfaces with Linear Height below 2

In this chapter and the following one we shall study the case where $h_{\rm lin}(\phi) < 2$. We may—and shall—then assume that ϕ is given by one of the normal forms listed in Proposition 2.11. Recall also from Corollary 2.13 that for $h_{\rm lin}(\phi) < 2$, we have $p'_c = 2d + 2$. Moreover, since we are assuming Condition (R) holds, the term b_0 in (2.23) vanishes identically if ϕ is of type A_∞ or D_∞ (cf. Remark 2.12(c)).

In order to prove Proposition 3.3 in this case, in a first step we shall follow the approach of the previous chapter and perform a dyadic decomposition of the domain (3.1) and the corresponding measure $d\mu^{\rho_1}$ by means of the dilations associated to the principal weight κ. Littlewood-Paley theory then again shows that it suffices to prove uniform Fourier restriction estimates for the corresponding dyadic constituents $d\mu_k$ of the measure $d\mu^{\rho_1}$, which, in return, are equivalent to similar estimates for a family of rescaled and normalized measures $d\nu_k$ (compare (4.1)). The latter measures will be supported where $x_1 \sim 1$ and $|x_2| \sim 1$.

Next, given a measure $d\nu_k$, which will actually depend on small parameters $\delta_1, \delta_2, \delta_3$ that are fractional powers of 2^{-k} (which is why we shall from then on denote these measure by $d\nu_\delta$, in place of $d\nu_k$), we shall perform yet another Littlewood-Paley decomposition, this time with respect to the variable x_3. This will allow us to restrict to subdomains on which $|\phi| \sim 2^{-2j}$, $j \geq j_0$. Then, depending on the size of $2^{2j}\delta_3$, we shall have to distinguish three different situations, and the size condition on $|\phi|$ will help us to obtain important additional information on the support of these further localized measures $d\nu_{\delta,j}$.

Since the decay of the Fourier transforms of these measures is strongly nonisotropic, in a last step we shall perform a dyadic decomposition in each of the frequency variables ξ_1, ξ_2, and ξ_3 dual to x_1, x_2, and x_3, which will lead to complex measures $d\nu_j^\lambda$ into which the measure $d\nu_{\delta,j}$ will be decomposed, with dyadic numbers $\lambda_i = 2^{k_i}$, $i = 1, 2, 3$, forming the vector $\lambda = (\lambda_1, \lambda_2, \lambda_3)$. Following Tomas' T^*T-argument, we shall then estimate the norms of the operators T_j^λ of convolution with $\widehat{\nu_j^\lambda}$, as operators from L^p to $L^{p'}$ and, finally, control the sum of the norms of these operators over all dyadic λ. This program will necessitate the distinction of various subcases. In the end, a few cases in which we cannot sum these operator norms will remain open, and we shall collect all these remaining cases in Proposition 4.2. In all these cases we will have that ϕ is of type A and $m = 2$.

The treatment of these remaining open cases in the course of the proof of Proposition 4.2 will require even more refined methods, in particular, certain interpolation arguments, in order to capture the endpoint $p = p_c$ and in some cases a deeper analysis based on further frequency localizations in terms of the distance to certain "Airy cones." All this will be carried out in Chapter 5.

4.1 PRELIMINARY REDUCTIONS BY MEANS OF LITTLEWOOD-PALEY DECOMPOSITIONS

In a first step, we shall follow the arguments from the preceding chapter and decompose the measure μ^{ρ_1} dyadically by means of the dilations associated to the principal weight κ. Applying subsequent rescalings, we may then reduce ourselves by means of Littlewood-Paley theory to proving a uniform estimate analogous to estimate (3.7) for the renormalized measures ν_k, $k \in \mathbb{N}$, which are now given by

$$\langle \nu_k, f\rangle := 2^{|\kappa|k}\langle \mu_k, f\circ \delta^e_{2^k}\rangle = \int_{x_1\geq 0} f(x,\phi^k(x))\,\eta(\delta_{2^{-k}}x)\chi(x)\rho_1(x_1,x_2)\,dx.$$

The functions ϕ^k are again defined by (3.3). Observe that

$$x_1 \sim 1 \sim |x_2|$$

in the support of the integrand. Recall also from (1.11) that

$$\phi(x_1,x_2) = \phi^a(x_1, x_2 - \psi(x_1)),$$

where according to (1.9) we may write

$$\psi(x_1) = x_1^m\omega(x_1), \quad (m\geq 2),$$

with a smooth function ω satisfying $\omega(0)\neq 0$.

What we then need to prove is that for $\varepsilon > 0$ sufficiently small, there are constants $C_\varepsilon > 0$ and $k_0 \in \mathbb{N}$ such that for every $k \geq k_0$

$$\Big(\int |\widehat{f}|^2\,d\nu_k\Big)^{1/2} \leq C_\varepsilon \|f\|_{L^{p_c}(\mathbb{R}^3)}, \qquad f\in\mathcal{S}(\mathbb{R}^3), \tag{4.1}$$

with a constant C_ε not depending on k.

In order to prove this estimate, observe that by (3.3) and (2.23), our rescaled phase function ϕ^k can be written in the form $\phi^k(x) = \phi(x,\delta)$, with

$$\phi(x,\delta) := \tilde{b}(x_1,x_2,\delta_1,\delta_2)(x_2 - x_1^m\omega(\delta_1 x_1))^2 + \delta_3 x_1^n\beta(\delta_1 x_1), \tag{4.2}$$

where we have put

$$\delta = (\delta_1,\, \delta_2\,, \delta_3) := (2^{-\kappa_1 k}, 2^{-\kappa_2 k}, 2^{-(n\kappa_1-1)k}).$$

Notice that δ consists of small parameters that tend to 0 as k tends to infinity. Moreover, $\tilde{b}$ is a smooth function in all its variables, given by

$$\tilde{b}(x_1,x_2,\delta_1,\delta_2) := \begin{cases} b(\delta_1x_1,\delta_2x_2), & \text{for } \phi \text{ of type } A,\\ x_1b_1(\delta_1x_1,\delta_2x_2) + \delta_1^{2m-1}x_2^2b_2(\delta_2x_2), & \text{for } \phi \text{ of type } D.\end{cases} \tag{4.3}$$

Here we have used that $\delta_2 = \delta_1^m$ (cf. Remark 2.12). We shall indeed ignore the particular dependence of δ on k and regard $\phi(x,\delta)$ as a smooth function of x and δ

for abitrary (sufficiently small) $\delta_j \geq 0$. Note that $\delta_3 := 0$ when ϕ is of type A_∞ or D_∞. Notice also that

$$|\tilde{b}(x_1, x_2, 0, 0)| \sim 1.$$

It is thus easily seen by means of a partition of unity argument that it will suffice to prove the following proposition in order to verify (4.1).

Proposition 4.1. *Let $\phi(x, \delta)$ be as in (4.2). Then, for every point $v = (v_1, v_2)$ such that $v_1 \sim 1$ and $v_2 = v_1^m \omega(0)$, there exists a neighborhood V of v in H^+ such that for every cutoff function $\eta \in \mathcal{D}(V)$, the measure ν_δ given by*

$$\langle \nu_\delta, f \rangle := \int f(x, \phi(x, \delta))\, \eta(x_1, x_2)\, dx$$

satisfies a restriction estimate

$$\Big(\int |\widehat{f}|^2\, d\nu_\delta \Big)^{1/2} \leq C_\eta \| f \|_{L^{p_c}(\mathbb{R}^3)}, \qquad f \in \mathcal{S}(\mathbb{R}^3), \tag{4.4}$$

provided δ is sufficiently small, with a constant C_η that depends only on the C^l-norm of η, for some $l \in \mathbb{N}$.

In order to prove this proposition, we shall perform yet another dyadic decomposition, this time with respect to the x_3-variable, in order to restrict ourselves to level ranges of the function $\phi(x, \delta)$.

A straightforward modification of the Littlewood-Paley argument at the end of Chapter 3 then allows us to reduce the proof to establishing uniform restriction estimates for the following family of measures:

$$\langle \nu_{\delta,j}, f \rangle := \int f(x, \phi(x, \delta))\, \chi_1(2^{2j} \phi(x, \delta)) \eta(x_1, x_2)\, dx. \tag{4.5}$$

Here, $\chi_1 \in \mathcal{D}(\mathbb{R})$ is again a suitable fixed, nonnegative smooth bump function supported in, say, the set $(-2, -\frac{1}{2}) \cup (\frac{1}{2}, 2)$ such that $\chi_1 \equiv 1$ in a neighborhood of the points -1 and 1. Notice that $\nu_{\delta,j}$ is supported where $|\phi(x, \delta)| \sim 2^{-2j}$. That is, in place of (4.4), it will be sufficient to prove an analogous uniform estimate

$$\Big(\int |\widehat{f}|^2\, d\nu_{\delta,j} \Big)^{1/2} \leq C_\eta \| f \|_{L^{p_c}(\mathbb{R}^3)}, \qquad f \in \mathcal{S}(\mathbb{R}^3), \tag{4.6}$$

for all $j \in \mathbb{N}$ sufficiently large, say, $j \geq j_0$, where the constant C_η depends neither on δ nor on j. Notice that by (4.2),

$$2^{2j} \phi(x, \delta) = 2^{2j} \tilde{b}(x_1, x_2, \delta_1, \delta_2) \Big(x_2 - x_1^m \omega(\delta_1 x_1) \Big)^2 + 2^{2j} \delta_3 x_1^n \beta(\delta_1 x_1),$$

In order to verify (4.6), we shall, therefore, distinguish three cases, depending on the size of $2^{2j} \delta_3$.

4.1.1 The situation where $2^{2j} \delta_3 \gg 1$

Observe first that if j is sufficiently large, then by (4.2) and since $x_1 \sim 1$, $\nu_{\delta,j} = 0$ unless $\tilde{b}(v, \delta_1, \delta_2)$ and $\beta(0)$ have opposite signs. So, let us, for instance, assume that $\tilde{b}(x_1, x_2, \delta_1, \delta_2) > 0$ and $\beta(\delta_1 x_1) < 0$ on the support of η. Then $\tilde{\beta} := -\beta > 0$, and we may rewrite

$$2^{2j}\phi(x,\delta) = 2^{2j}\tilde{b}(x_1, x_2, \delta_1, \delta_2)(x_2 - x_1^m\omega(\delta_1 x_1))^2 - 2^{2j}\delta_3 x_1^n\tilde{\beta}(\delta_1 x_1).$$

We introduce new coordinates y by putting $y_1 := x_1$ and $y_2 := 2^{2j}\phi(x_1, x_2, \delta)$. Solving for x_2, one easily finds that

$$x_2 = \tilde{b}_1\Big(y_1, \sqrt{2^{-2j}y_2 + \delta_3 y_1^n\tilde{\beta}(\delta_1 y_1)}, \delta_1, \delta_2\Big) \times \sqrt{2^{-2j}y_2 + \delta_3 y_1^n\tilde{\beta}(\delta_1 y_1)} + y_1^m\omega(\delta_1 y_1),$$

where $\tilde{b}_1$ is smooth and has properties similar to $\tilde{b}$. Moreover, by the support properties of the amplitude $\chi_1(2^{2j}\phi(x,\delta))\eta(x_1, x_2)$, we see that for the new coordinates we also have $y_1 \sim 1 \sim y_2$, and changing to the coordinates (y_1, y_2), we may rewrite

$$\langle \nu_{\delta,j}, f\rangle = \frac{2^{-2j}}{\sqrt{\delta_3}}\int f(y_1, \phi(y,\delta,j), 2^{-2j}y_2)\, a(y,\delta,j)\,\chi_1(y_1)\chi_1(y_2)\,dy,$$

with a cutoff function χ_1 as before, as well as where $a(y,\delta,j)$ is smooth in y and δ, with C^l-norms uniformly bounded in δ and j, and where the new phase function $\phi(x,\delta,j)$ is given by

$$\phi(x,\delta,j) := \tilde{b}_1\Big(x_1, \sqrt{2^{-2j}x_2 + \delta_3 x_1^n\tilde{\beta}(\delta_1 x_1)}, \delta_1, \delta_2\Big) \times \sqrt{2^{-2j}x_2 + \delta_3 x_1^n\tilde{\beta}(\delta_1 x_1)} + x_1^m\omega(\delta_1 x_1).$$

We have renamed the variable y to become x here, since if we define the normalized measure $\tilde{\nu}_{\delta,j}$ by

$$\langle \tilde{\nu}_{\delta,j}, f\rangle := \int f(x_1, \phi(x,\delta,j), x_2)\, a(x,\delta,j)\,\chi_1(x_1)\chi_1(x_2)\,dx, \tag{4.7}$$

then the restriction estimate (4.6) for the measure $\nu_{\delta,j}$ is equivalent to the following restriction estimate for the measure $\tilde{\nu}_{\delta,j}$:

$$\int |\widehat{f}|^2\, d\tilde{\nu}_{\delta,j} \le C_\eta\sqrt{\delta_3}\, 2^{2j(1-2/p_c')}\, \|f\|^2_{L^{p_c}(\mathbb{R}^3)}, \qquad f \in \mathcal{S}(\mathbb{R}^3), \tag{4.8}$$

for all $j \in \mathbb{N}$ sufficiently large, say $j \ge j_0$, where the constant C_η depends neither on δ nor on j.

Formula (4.7) shows that the Fourier transform of the measure $\tilde{\nu}_{\delta,j}$ can be expressed as an oscillatory integral

$$\widehat{\tilde{\nu}_{\delta,j}}(\xi) = \int e^{-i\Phi(x,\delta,j,\xi)}a(x,\delta,j)\,\chi_1(x_1)\chi_1(x_2)\,dx, \tag{4.9}$$

where the complete phase function Φ is given by

$$\Phi(x,\delta,j,\xi) := \xi_2\phi(x,\delta,j) + \xi_3 x_2 + \xi_1 x_1.$$

In order to establish the restriction estimates (4.8), we shall finally perform dyadic frequency decompositions of the measure $\tilde{\nu}_{\delta,j}$ in each of the three coordinates. To this end, we again fix a suitable smooth cutoff function $\chi_1 \geq 0$ on $\mathbb{R}$ supported in $(-2, -\frac{1}{2}) \cup (\frac{1}{2}, 2)$ such that the functions $\chi_k(t) := \chi_1(2^{1-k}t)$, $k \in \mathbb{N} \setminus \{0\}$, in combination with a suitable smooth function χ_0 supported in $(-1, 1)$, form a partition of unity, that is,

$$\sum_{k=0}^{\infty} \chi_k(t) = 1 \qquad \text{for all } t \in \mathbb{R}.$$

For every multi-index $k = (k_1, k_2, k_3) \in \mathbb{N}^3$, we let

$$\chi_k(\xi) := \chi_{k_1}(\xi_1)\chi_{k_2}(\xi_2)\chi_{k_3}(\xi_3)$$

and, finally, define the smooth functions $\nu_{k,j}$ by

$$\widehat{\nu_{k,j}}(\xi) := \chi_k(\xi)\widehat{\tilde{\nu}_{\delta,j}}(\xi).$$

In order to simplify the notation, we have suppressed here the dependency of this smooth function on the small parameters given by δ. We then find that

$$\tilde{\nu}_{\delta,j} = \sum_{k \in \mathbb{N}^3} \nu_{k,j}, \tag{4.10}$$

in the sense of distributions. To simplify the subsequent discussion, we shall concentrate on those measures $\nu_{k,j}$ for which none of its k_i components are zero, since the remaining cases where for instance k_i is zero can be dealt with in the same way as the corresponding cases where $k_i \geq 1$ is small.

Now, if $1 \leq \lambda_i = 2^{k_i - 1}$, $i = 1, 2, 3$, are dyadic numbers, we shall accordingly write ν_j^λ in place of $\nu_{k,j}$, that is,

$$\widehat{\nu_j^\lambda}(\xi) = \chi_1\Big(\frac{\xi_1}{\lambda_1}\Big)\chi_1\Big(\frac{\xi_2}{\lambda_2}\Big)\chi_1\Big(\frac{\xi_3}{\lambda_3}\Big)\widehat{\tilde{\nu}_{\delta,j}}(\xi). \tag{4.11}$$

Note that

$$|\xi_i| \sim \lambda_i, \qquad \text{on supp}\, \widehat{\nu_j^\lambda}. \tag{4.12}$$

Moreover, by (4.7),

$$\begin{aligned} \nu_j^\lambda(x) = \lambda_1\lambda_2\lambda_3 \int & \check{\chi}_1\,(\lambda_1(x_1 - y_1))\ \check{\chi}_1\,(\lambda_2(x_2 - \phi(y, \delta, j))) \\ & \times \check{\chi}_1\,(\lambda_3(x_2 - y_2))\ a(y, \delta, j)\,\chi_1(y_1)\chi_1(y_2)\,dy, \end{aligned} \tag{4.13}$$

where $\check{f}$ denotes the inverse Fourier transform of f.

We begin by estimating the Fourier transform of ν_j^λ. To this end, we first integrate in x_1 in (4.7) and then in x_2, assuming that (4.12) holds true. We shall concentrate on those ν_j^λ for which

$$\lambda_1 \sim \lambda_2 \sim \sqrt{\delta_3}2^{2j}\lambda_3. \tag{4.14}$$

In all other cases, the phase Φ in (4.9) has no critical point on the support of the amplitude, and we obtain much faster Fourier decay estimates by repeated integrations by parts in x_1, respectively, x_2, and the corresponding terms can be considered as error terms. Observe also that

$$\left|\frac{\partial^2}{\partial x_2^2}\Phi(x,\delta,j,\xi)\right| \sim \lambda_2\delta_3^{-3/2}2^{-4j}$$

on the support of the amplitude. We therefore distinguish two subcases.

Case 1. $1 \le \lambda_1 \lesssim \delta_3^{3/2}2^{4j}$. In this case we cannot gain from the integration in x_2, but, by applying van der Corput's lemma of order $M = 2$ (or the method of stationary phase) in x_1, we obtain

$$\|\widehat{\nu_j^\lambda}\|_\infty \lesssim \lambda_1^{-1/2}. \tag{4.15}$$

Case 2. $\lambda_1 \gg \delta_3^{3/2}2^{4j}$. Then, by first applying the method of stationary phase to the integration in x_1 and subsequently applying the classical van der Corput Lemma 2.1 of order $M = 2$ to the integration in x_2, we obtain

$$\|\widehat{\nu_j^\lambda}\|_\infty \lesssim \lambda_1^{-1/2}(\lambda_2\delta_3^{-3/2}2^{-4j})^{-1/2} \lesssim \delta_3^{3/4}2^{2j}\lambda_1^{-1}. \tag{4.16}$$

Next, from (4.13), by making use of the first and the third factor of the integrand, we trivially obtain the following estimate for the L^∞-norm of ν_j^λ :

$$\|\nu_j^\lambda\|_\infty \lesssim \lambda_2 \sim \lambda_1 \tag{4.17}$$

in Case 1 as well as in Case 2. All these estimates are uniform in δ for δ sufficiently small.

For each of the measures ν_j^λ, we can now obtain suitable restriction estimates by applying the usual approach. Let us denote by $T_{\delta,j}$ the convolution operator

$$T_{\delta,j} : \varphi \mapsto \varphi * \widehat{\tilde{\nu}_{\delta,j}},$$

and, similarly, by T_j^λ the convolution operator

$$T_j^\lambda : \varphi \mapsto \varphi * \widehat{\nu_j^\lambda}.$$

Formally, by (4.10), $T_{\delta,j}$ decomposes as

$$T_{\delta,j} = \sum_{k\in\mathbb{N}^3} T_j^{2^k} \tag{4.18}$$

if 2^k represents the vector $2^k := (2^{k_1}, 2^{k_2}, 2^{k_3})$ (with a suitably modified definition of $T_j^{2^k}$ when one of the components k_i is zero). If we denote by $\|T\|_{p\to q}$ the norm of T as an operator from L^p to L^q, then clearly $\|T_j^\lambda\|_{1\to\infty} = \|\widehat{\nu_j^\lambda}\|_\infty$ and $\|T_j^\lambda\|_{2\to 2} = \|\nu_j^\lambda\|_\infty$.

The estimates (4.15)–(4.17) thus yield the following bounds:

$$\|T_j^\lambda\|_{1\to\infty} \lesssim \begin{cases} \lambda_1^{-1/2}, & \text{if } 1 \le \lambda_1 \lesssim \delta_3^{3/2}2^{4j}, \\ \delta_3^{3/4}2^{2j}\lambda_1^{-1}, & \text{if } \lambda_1 \gg \delta_3^{3/2}2^{4j}, \end{cases}$$

and $\|T_j^\lambda\|_{2\to 2} \lesssim \lambda_1$. Interpolating the estimates (4.16) and (4.17), and defining the critical interpolation parameter $\theta = \theta_c$ by $1/p_c' = (1-\theta)/\infty + \theta/2 = \theta/2$, that is,

$$\theta := \frac{2}{p_c'},$$

we find that

$$\|T_j^\lambda\|_{p_c \to p_c'} \lesssim \begin{cases} \lambda_1^{(3\theta-1)/2}, & \text{if } 1 \le \lambda_1 \lesssim \delta_3^{3/2} 2^{4j}, \\ \delta_3^{3(1-\theta)/4} 2^{2(1-\theta)j} \lambda_1^{2\theta-1}, & \text{if } \lambda_1 \gg \delta_3^{3/2} 2^{4j}, \end{cases} \tag{4.19}$$

where according to Remark 2.12,

$$\theta = \begin{cases} \dfrac{m+1}{3m+1}, & \text{if } \phi \text{ is of type } A, \\ \dfrac{m+1}{3m+2}, & \text{if } \phi \text{ is of type } D, \end{cases} \tag{4.20}$$

since $p_c' = 2d + 2$. Observe that in particular,

$$\tfrac{1}{3} < \theta \le \tfrac{3}{7}, \tag{4.21}$$

and $\theta = \frac{3}{7}$ if and only if $m = 2$ and ϕ is of type A. The latter case will turn out to be the most difficult one.

Observe next that the main contributions to the series (4.18) come from those dyadic $\lambda = 2^k$ for which $\lambda_1 \sim \lambda_2 \sim \sqrt{\delta_3} 2^{2j} \lambda_3$. Under these relations, for λ_1 given, λ_2 and λ_3 may only vary in a finite set whose cardinality is bounded by a fixed number. This shows that, up to an easily bounded error term,

$$\|T_{\delta,j}\|_{p_c \to p_c'} \lesssim \sum_{\lambda_1 = 2}^{\delta_3^{3/2} 2^{4j}} \lambda_1^{(3\theta-1)/2} + \sum_{\lambda_1 > \delta_3^{3/2} 2^{4j}} \delta_3^{3(1-\theta)/4} 2^{2(1-\theta)j} \lambda_1^{(2\theta-1)}.$$

Here, and in the sequel, summation over λ_1, λ_2, and so on, will always mean that we are summing over dyadic numbers λ_1, λ_2, and so on, only.

Now, by (4.21), $2\theta - 1 < 0$ and $0 < 3\theta - 1 \le 1$, which yields

$$\|T_{\delta,j}\|_{p_c \to p_c'} \lesssim \delta_3^{3(3\theta-1)/4} 2^{(3\theta-1)2j}.$$

Recall that, by the standard T^*T-argument of Tomas, applied to the restriction operator $Rf := \hat{f} d\tilde{\nu}_{\delta,j}$, we have

$$\int |\widehat{f}|^2 \, d\tilde{\nu}_{\delta,j} \le (2\pi)^{-3} \|T_{\delta,j}\|_{p_c \to p_c'} \|f\|_{p_c}^2,$$

and thus we need to prove that

$$\delta_3^{3(3\theta-1)/4} 2^{(3\theta-1)2j} \le C\sqrt{\delta_3}\, 2^{2j(1-2/p_c')}$$

in order to verify that the restriction estimate (4.8) holds true for $p = p_c = 2d + 2$. However, since $2/p'_c = \theta$, the previous estimate is equivalent to

$$2^{2j(4\theta-2)} \le C\delta_3^{(5-9\theta)/4}.$$

But, since $2^{2j}\delta_3 \gg 1$ and $2\theta - 1 < 0$, we see that $2^{2j(4\theta-2)} \le C\,\delta_3^{2-4\theta}$, and therefore we have to verify only that $2 - 4\theta \ge (5 - 9\theta)/4$, that is, $7\theta \le 3$, which is true according to (4.21).

We thus have verified the restriction estimate (4.6) in this subcase.

4.2 RESTRICTION ESTIMATES FOR NORMALIZED RESCALED MEASURES WHEN $2^{2j}\delta_3 \lesssim 1$

There remains the case $2^{2j}\delta_3 \le C$, where C is a fixed, possibly large constant.

In this situation, we perform the change of variables $(x_1, x_2) \mapsto (x_1, x_2 + x_1^m\omega(\delta_1 x_1))$ and subsequently scale in x_2 by the factor 2^{-j}. This allows us to rewrite the measure $\nu_{\delta,j}$ given by (4.5) as

$$\langle \nu_{\delta,j}, f\rangle = 2^{-j}\int f\Big(x_1, 2^{-j}x_2 + x_1^m\omega(\delta_1 x_1), 2^{-2j}\phi^a(x,\delta,j)\Big)\, a(x,\delta,j)\,dx,$$

where

$$\phi^a(x,\delta,j) := \tilde b(x_1, 2^{-j}x_2 + x_1^m\omega(\delta_1x_1), \delta_1, \delta_2)x_2^2 + 2^{2j}\delta_3 x_1^n\beta(\delta_1 x_1) \tag{4.22}$$

and

$$a(x,\delta,j) := \chi_1(\phi^a(x,\delta,j))\,\eta(x_1, 2^{-j}x_2 + x_1^m\omega(\delta_1 x_1)). \tag{4.23}$$

Let us introduce here the normalized measures $\tilde\nu_{\delta,j}$ given by

$$\langle \tilde\nu_{\delta,j}, f\rangle := \int f(x_1, 2^{-j}x_2 + x_1^m\omega(\delta_1x_1), \phi^a(x,\delta,j))\, a(x,\delta,j)\,dx. \tag{4.24}$$

Then, it is easy to see by means of a scaling of the variable x_3 by the factor 2^{-2j} that the restriction estimate (4.6) for the measure $\nu_{\delta,j}$ is equivalent to the following restriction estimate for the measure $\tilde\nu_{\delta,j}$:

$$\int_S |\widehat f|^2\, d\tilde\nu_{\delta,j} \le C_\eta\, 2^{(1-4/p'_c)j}\,\|f\|^2_{L^{p_c}(\mathbb{R}^3)} = C_\eta\, 2^{(1-2\theta)j}\,\|f\|^2_{L^{p_c}(\mathbb{R}^3)}, \qquad f \in \mathcal{S}(\mathbb{R}^3), \tag{4.25}$$

for all $j \in \mathbb{N}$ sufficiently large, say $j \ge j_0$, where the constant C_η depends neither on δ nor on j.

In order to prove (4.25), we again distinguish two subcases.

4.2.1 The situation where $2^{2j}\delta_3 \ll 1$

Notice that here the phase $\phi^a(x,\delta,j)$ is a small perturbation of $\tilde b(v_1, 0, 0, 0)x_2^2$, where $\tilde b(v_1, 0, 0, 0) \sim 1$. This shows also that in the new coordinates appearing in

(4.24), we have $x_1 \sim 1 \sim |x_2|$ on the support of the amplitude a, which in turn implies

$$\frac{\partial}{\partial x_2}\phi^a(x,\delta,j) \sim 1. \tag{4.26}$$

We can thus write

$$\widehat{\tilde{\nu}_{\delta,j}}(\xi) = \int e^{-i\Phi(x,\delta,j,\xi)} a(x,\delta,j)\chi_1(x_1)\chi_1(x_2)\,dx, \tag{4.27}$$

where the complete phase function Φ is now given by

$$\Phi(x,\delta,j,\xi) := \xi_3\phi^a(x,\delta,j) + 2^{-j}\xi_2 x_2 + \xi_2 x_1^m \omega(\delta_1 x_1) + \xi_1 x_1, \tag{4.28}$$

with ϕ^a given by (4.22), and where χ_1 has similar properties as before.

As in the previous subcase, we perform dyadic frequency decompositions of the measure $\tilde{\nu}_{\delta,j}$ by means of the functions in (4.10) and define the measure ν_j^λ as in (4.11). Then in the present situation we have

$$\begin{aligned}\nu_j^\lambda(x) = \lambda_1\lambda_2\lambda_3 \int & \check{\chi}_1(\lambda_1(x_1-y_1))\ \check{\chi}_1(\lambda_2(x_2-2^{-j}y_2-y_1^m\omega(\delta_1 y_1)))\\ &\times\check{\chi}_1(\lambda_3(x_3-\phi^a(y,\delta,j)))\,a(y,\delta,j)\,\chi_1(y_1)\chi_1(y_2)\,dy.\end{aligned} \tag{4.29}$$

We begin by estimating the Fourier transform of ν_j^λ. To this end, we first integrate in x_2 in (4.27) and then in x_1. We may assume that (4.12) holds true. Then the phase function Φ has no critical point in x_2 unless $\lambda_3 \sim 2^{-j}\lambda_2$; similarly, if we assume that $\lambda_3 \sim 2^{-j}\lambda_2$, then there is no critical point with respect to x_1 unless $\lambda_2 \sim \lambda_1$. We shall, therefore, concentrate on those ν_j^λ for which

$$\lambda_1 \sim \lambda_2 \quad \text{and} \quad 2^{-j}\lambda_2 \sim \lambda_3. \tag{4.30}$$

In all other cases, we obtain much faster Fourier decay estimates by repeated integrations by parts, so that the corresponding terms can be considered as error terms.

Case 1. $1 \le \lambda_1 \le 2^j$. In this case the phase function has essentially no oscillation in the x_2 variable. But, by applying van der Corput's lemma of order $M = 2$ (or the method of stationary phase) in x_1, we obtain in view of (4.30) that

$$\|\widehat{\nu_j^\lambda}\|_\infty \lesssim \lambda_1^{-1/2}. \tag{4.31}$$

Case 2. $\lambda_1 > 2^j$. Observe that in this case, our assumptions imply that $\delta_3 2^{2j}\lambda_3 \ll \lambda_3 \ll \lambda_2$ if $j \ge j_0 \gg 1$. Moreover, depending on the signs of the ξ_i, we may have no critical point or exactly one nondegenerate critical point with respect to each of the variables x_2 and x_1. So, integrating by parts, respectively applying the method of stationary phase in the presence of a critical point, first in x_2 and then in x_1, we obtain

$$\|\widehat{\nu_j^\lambda}\|_\infty \lesssim \lambda_3^{-1/2}\lambda_1^{-1/2} \sim 2^{j/2}\lambda_1^{-1}. \tag{4.32}$$

Next, we estimate the L^∞-norm of ν_j^λ. To this end, notice that (4.26) shows that we may change coordinates in (4.29) by putting $(z_1, z_2) := (y_1, \phi^a(y_1, y_2, \delta, j))$.

Since the Jacobian of this coordinate change is of order 1, we thus obtain that

$$|\nu_j^\lambda(x)| \lesssim \lambda_1\lambda_2\lambda_3 \iint \left|\check{\chi}_1(\lambda_1(x_1 - z_1))\, \check{\chi}_1\,(\lambda_3(x_3 - z_2))\; \tilde{a}(z, \delta, j)\right| dz_1\, dz_2;$$

hence

$$\|\nu_j^\lambda\|_\infty \lesssim \lambda_2 \sim \lambda_1 \tag{4.33}$$

in Case 1 as well as in Case 2.

For the operators T_j^λ that appear in this subcase, the estimates (4.31)–(4.33) thus yield the following bounds:

$$\|T_j^\lambda\|_{1\to\infty} \lesssim \begin{cases} \lambda_1^{-1/2}, & \text{if } 1 \le \lambda_1 \le 2^j, \\ 2^{j/2}\lambda_1^{-1}, & \text{if } \lambda_1 > 2^j, \end{cases}$$

and $\|T_j^\lambda\|_{2\to 2} \lesssim \lambda_1$. Interpolating these estimates, we find that

$$\|T_j^\lambda\|_{p_c\to p_c'} \lesssim \begin{cases} \lambda_1^{(3\theta-1)/2}, & \text{if } 1 \le \lambda_1 \le 2^j, \\ 2^{[(1-\theta)/2]j}\lambda_1^{2\theta-1}, & \text{if } \lambda_1 > 2^j, \end{cases} \tag{4.34}$$

where θ is again given by (4.20).

Now, in view of (4.30), the main contribution to the series (4.18) comes here from those dyadic $\lambda = 2^k$ for which $\lambda_1 \sim \lambda_2$ and $2^{-j}\lambda_2 \sim \lambda_3$. Thus, up to an easily bounded error term, the operator $T_{\delta,j}$ arising in this subcase can be estimated by

$$\|T_{\delta,j}\|_{p_c\to p_c'} \lesssim \sum_{\lambda_1=2}^{2^j} \lambda_1^{(3\theta-1)/2} + \sum_{\lambda_1=2^{j+1}}^{\infty} 2^{[(1-\theta)/2]j}\lambda_1^{2\theta-1} \lesssim 2^{[(3\theta-1)/2]j} \le 2^{(1-2\theta)j},$$

since, by (4.21) we have $2\theta - 1 < 0$ and $(3\theta - 1)/2 \le 1 - 2\theta$.

This verifies the restriction estimate (4.25) and thus concludes the proof of Proposition 4.1 in this subcase as well.

4.2.2 The situation where $2^{2j}\delta_3 \sim 1$

Notice that in this situation we can no longer conclude that $x_2 \sim 1$ on the support of the amplitude $a(x, \delta, j)$, only that $|x_2| \lesssim 1$, whereas it is still the case that $x_1 \sim 1$. Also observe that here the cases A_∞ and D_∞ are excluded, since in these cases $\delta_3 = 0$.

Putting

$$\sigma := 2^{2j}\delta_3 \quad \text{and} \quad b^\sharp(x, \delta, j) := \tilde{b}(x_1, 2^{-j}x_2 + x_1^m\omega(\delta_1 x_1), \delta_1, \delta_2),$$

we may rewrite the complete phase in (4.28) as

$$\begin{aligned}\Phi(x, \delta, j, \xi) = {} & \xi_1 x_1 + \xi_2 x_1^m \omega(\delta_1 x_1) + \xi_3 \sigma x_1^n \beta(\delta_1 x_1) \\ & + 2^{-j}\xi_2 x_2 + \xi_3 b^\sharp(x, \delta, j)\, x_2^2,\end{aligned} \tag{4.35}$$

where $\sigma \sim 1$ and $|b^{\pm}(x,\delta,j)| \sim 1$, and in place of (4.29) we obtain

$$\begin{aligned} \nu_j^\lambda(x) = \lambda_1\lambda_2\lambda_3 \int & \check{\chi}_1(\lambda_1(x_1-y_1))\,\check{\chi}_1(\lambda_2(x_2-2^{-j}y_2-y_1^m\omega(\delta_1 y_1))) \\ & \times \check{\chi}_1(\lambda_3(x_3-b^{\pm}(y,\delta,j)\,y_2^2-\sigma y_1^n\beta(\delta_1 y_1))) \\ & \times a(y,\delta,j)\chi_1(y_1)\chi_0(y_2)\,dy. \end{aligned} \tag{4.36}$$

Here, we have suppressed the dependence of ν_j^λ on the parameter σ in order to simplify the notation. Also observe that we then may drop the parameter δ_3 from the definition of δ, that is, we may assume that $\delta = (\delta_1, \delta_2)$, since only σ depends on δ_3. Recall from (4.23) that $a(y,\delta,j)$ is supported where $y_1 \sim 1$ and $|y_2| \lesssim 1$.

Observe that by Lemma 2.1(b), we have $\left|\int \check{\chi}_1\left(\lambda_3(c-t^2)\right)dt\right| \le C\lambda_3^{-1/2}$, with a constant C that is independent of c. Thus, making use of the localizations given for the integration in y_2 from the third factor, respectively, second factor, in the integrand of (4.36) and then for the integration in y_1 by the first factor, it is easy to see that

$$\|\nu_j^\lambda\|_\infty \lesssim \min\{\lambda_2\lambda_3^{1/2}, 2^j\lambda_3\} = \lambda_3^{1/2}\min\{\lambda_2, 2^j\lambda_3^{1/2}\}. \tag{4.37}$$

In order to estimate $\widehat{\nu_j^\lambda}(\xi)$, we may again assume that (4.12) holds true. In the oscillatory integral defining $\widehat{\nu_j^\lambda}(\xi)$, we shall first perform the integration in x_1. If one of the quantities λ_1, λ_2, or λ_3 is much bigger than the other two, we see that we have no critical point in x_1 on the support of the amplitude, so that the corresponding terms can again be viewed as error terms. Let us therefore assume that all three λ_is are of comparable size or that two of them are of comparable size and the third one is much smaller. We shall begin with the latter situation and distinguish various possibilities.

Case 1. $\lambda_1 \sim \lambda_3$ and $\lambda_2 \ll \lambda_1$. In this case, we apply the method of stationary phase to the integration in x_1 and, subsequently, van der Corput's estimate to the x_2-integration and obtain

$$\|\widehat{\nu_j^\lambda}\|_\infty \lesssim \lambda_1^{-1/2}\lambda_3^{-1/2} \sim \lambda_1^{-1}.$$

1.1. The subcase where $\lambda_2 \le 2^j\lambda_1^{1/2}$. Then, by (4.37), $\|\nu_j^\lambda\|_\infty \lesssim \lambda_2\lambda_1^{1/2}$, and we obtain by interpolation, in a similar way as before, that

$$\|T_j^\lambda\|_{p_c\to p_c'} \lesssim \lambda_1^{(3\theta-2)/2}\lambda_2^\theta.$$

Here, $\frac{3\theta-2}{2} < 0$ because of (4.21). Notice next that if $2^j\lambda_1^{1/2} \le \lambda_1$, that is, if $\lambda_1 \ge 2^{2j}$, then by our assumptions $\lambda_2 \le 2^j\lambda_1^{1/2}$, and if $\lambda_1 < 2^{2j}$, then we may use that $\lambda_2 \le \lambda_1$. We thus find that the contribution $T_{\delta,j}^I$ of the operators T_j^λ, with λ satisfying the assumptions of this subcase, to $T_{\delta,j}$ can be estimated by

$$\|T^I_{\delta,j}\|_{p_c\to p_c'} \lesssim \sum_{\lambda_1=2}^{2^{2j}} \sum_{\lambda_2=2}^{\lambda_1} \lambda_1^{(3\theta-2)/2}\lambda_2^{\theta} + \sum_{\lambda_1=2^{2j+1}}^{\infty} \sum_{\lambda_2=2}^{2^j\lambda_1^{1/2}} \lambda_1^{(3\theta-2)/2}\lambda_2^{\theta}$$

$$\lesssim \sum_{\lambda_1=2}^{2^{2j}} \lambda_1^{(5\theta-2)/2} + \sum_{\lambda_1=2^{2j+1}}^{\infty} 2^{\theta j}\lambda_1^{2\theta-1}.$$

But, we have seen that $2\theta - 1 < 0$, so that

$$\|T^I_{\delta,j}\|_{p_c\to p_c'} \lesssim \max\{j, 2^{(5\theta-2)j}\}.$$

Now, if $5\theta - 2 > 0$, then, again because of (4.21), $\|T^I_{\delta,j}\|_{p_c\to p_c'} \lesssim 2^{(5\theta-2)j} \leq 2^{(1-2\theta)j}$. And, if $5\theta-2 \leq 0$, then $\|T^I_{\delta,j}\|_{p_c\to p_c'} \lesssim j \lesssim 2^{(1-2\theta)j}$, that is,

$$\|T^I_{\delta,j}\|_{p_c\to p_c'} \lesssim 2^{(1-2\theta)j}. \tag{4.38}$$

1.2. The subcase where $\lambda_2 > 2^j\lambda_1^{1/2}$. Then, by (4.37), $\|v^\lambda_j\|_\infty \lesssim 2^j\lambda_1$, and we obtain by interpolation, in a similar way as before, that

$$\|T^\lambda_j\|_{p_c\to p_c'} \lesssim 2^{\theta j}\lambda_1^{2\theta-1}.$$

Observing that we have $2^j\lambda_1^{1/2} < \lambda_2 \leq \lambda_1$ and then also $\lambda_1 > 2^{2j}$, we see that the contribution $T^{II}_{\delta,j}$ of the operators T^λ_j, with λ satisfying the assumptions in this subcase, to $T_{\delta,j}$ can be estimated by

$$\|T^{II}_{\delta,j}\|_{p_c\to p_c'} \lesssim 2^{\theta j} \sum_{\lambda_1=2^{2j}}^{\infty} \sum_{2^j\lambda_1^{1/2}<\lambda_2\leq\lambda_1} \lambda_1^{2\theta-1} \lesssim 2^{\theta j} \sum_{\lambda_1=2^{2j}}^{\infty} (\log_2\lambda_1 - 2j)\,\lambda_1^{2\theta-1}$$

$$\lesssim 2^{\theta j} \sum_{k=2j}^{\infty} (k-2j)\,2^{(2\theta-1)k} \lesssim 2^{(5\theta-2)j} \sum_{k=0}^{\infty} k\,2^{(2\theta-1)k} \lesssim 2^{(5\theta-2)j},$$

so that also

$$\|T^{II}_{\delta,j}\|_{p_c\to p_c'} \lesssim 2^{(1-2\theta)j}. \tag{4.39}$$

Case 2. $\lambda_2 \sim \lambda_3$ and $\lambda_1 \ll \lambda_2$. Here, we can estimate $\widehat{v^\lambda_j}$ in the same way as in the previous case and obtain $\|\widehat{v^\lambda_j}\|_\infty \lesssim \lambda_2^{-1/2}\lambda_3^{-1/2} \sim \lambda_2^{-1}$. Moreover, by (4.37), we have $\|v^\lambda_j\|_\infty \lesssim \lambda_2 \min\{\lambda_2^{1/2}, 2^j\}$. Both these estimates are independent of λ_1. We therefore consider the sum over all v^λ_j such that $\lambda_1 \ll \lambda_2$ by putting

$$\sigma_j^{\lambda_2,\lambda_3} := \sum_{\lambda_1\ll\lambda_2} v^\lambda_j.$$

This means that

$$\widehat{\sigma_j^{\lambda_2,\lambda_3}}(\xi) = \chi_0\Big(\frac{\xi_1}{\lambda_2}\Big)\chi_1\Big(\frac{\xi_2}{\lambda_2}\Big)\chi_1\Big(\frac{\xi_3}{\lambda_3}\Big)\,\widehat{\nu_{\delta,j}}(\xi),$$

where now χ_0 is smooth and compactly supported in an interval $[-\varepsilon, \varepsilon]$, where $\varepsilon > 0$ is sufficiently small. In particular, $\sigma_j^{\lambda_2,\lambda_3}(x)$ is again given by the expression (4.36), only with the first factor $\check{\chi}_1(\lambda_1(x_1 - y_1))$ in the integrand replaced by $\check{\chi}_0(\lambda_2(x_1 - y_1))$ and λ_1 replaced by λ_2. Thus we obtain the same type of estimates,

$$\|\widehat{\sigma_j^{\lambda_2,\lambda_3}}\|_\infty \lesssim \lambda_2^{-1}, \quad \|\sigma_j^{\lambda_2,\lambda_3}\|_\infty \lesssim \lambda_2 \min\{\lambda_2^{1/2}, 2^j\}. \tag{4.40}$$

Denote by $T_j^{\lambda_2,\lambda_3}$ the operator of convolution with $\widehat{\sigma_j^{\lambda_2,\lambda_3}}$.

2.1. The subcase where $\lambda_2 \le 2^{2j}$. Then we have $\|\sigma_j^{\lambda_2,\lambda_3}\|_\infty \lesssim \lambda_2^{3/2}$, and interpolating this with the first estimate in (4.40), we obtain

$$\|T_j^{\lambda_2,\lambda_3}\|_{p_c\to p_c'} \lesssim \lambda_2^{\theta-1}\lambda_2^{3\theta/2} = \lambda_2^{(5\theta-2)/2}.$$

We thus find that the contribution $T_{\delta,j}^{III}$ of the operators T_j^{λ}, with λ satisfying the assumptions of this subcase to $T_{\delta,j}$, can be estimated by

$$\|T_{\delta,j}^{III}\|_{p_c\to p_c'} \lesssim \sum_{\lambda_2=2}^{2^{2j}} \lambda_2^{(5\theta-2)/2}.$$

Arguing in a similar way as in Subcase 1.1, this implies that

$$\|T_{\delta,j}^{III}\|_{p_c\to p_c'} \lesssim 2^{(1-2\theta)j}. \tag{4.41}$$

2.2. The subcase where $\lambda_2 > 2^{2j}$. Then, by (4.40), $\|\sigma_j^{\lambda_2,\lambda_3}\|_\infty \lesssim 2^j\lambda_2$, and we obtain by interpolation, in a similar way as before, that

$$\|T_j^{\lambda_2,\lambda_3}\|_{p_c\to p_c'} \lesssim \lambda_2^{\theta-1}2^{\theta j}\lambda_2^{\theta} = 2^{\theta j}\lambda_2^{2\theta-1},$$

where, according to (4.21), $2\theta - 1 < 0$. We thus find that the contribution $T_{\delta,j}^{IV}$ of the operators T_j^{λ}, with λ satisfying the assumptions of this subcase, to $T_{\delta,j}$ can be estimated by

$$\|T_{\delta,j}^{IV}\|_{p_c\to p_c'} \lesssim 2^{\theta j}\sum_{\lambda_2=2^{2j}}^{\infty} \lambda_2^{2\theta-1} \lesssim 2^{(5\theta-2)j}.$$

As before, this implies that

$$\|T_{\delta,j}^{IV}\|_{p_c\to p_c'} \lesssim 2^{(1-2\theta)j}. \tag{4.42}$$

Case 3. $\lambda_1 \sim \lambda_2$ and $\lambda_3 \ll \lambda_1$. Notice that the phase Φ has no critical point with respect to x_2 when $2^{-j}\lambda_2 \gg \lambda_3$, so we shall concentrate on the case where $\lambda_2 \lesssim 2^j\lambda_3$. Then we can estimate $\widehat{\nu_j^\lambda}$ in the same way as in the previous cases and obtain

$$\|\widehat{\nu_j^\lambda}\|_\infty \lesssim \lambda_1^{-1/2}\lambda_3^{-1/2}. \tag{4.43}$$

3.1. The subcase where $\lambda_3^{1/2} \gtrsim \lambda_1 2^{-j}$. Then $(2^{-j}\lambda_1)^2 \lesssim \lambda_3 \ll \lambda_1$, and hence we may assume that $\lambda_1 \le 2^{2j}$; from (4.37) and the previous estimate for $\widehat{\nu_j^\lambda}$, we obtain by interpolation

$$\|T_j^\lambda\|_{p_c \to p_c'} \lesssim \lambda_1^{(3\theta-1)/2} \lambda_3^{(2\theta-1)/2}.$$

We thus find that the contribution $T_{\delta,j}^V$ of the operators T_j^λ, with λ satisfying the assumptions of this subcase, to the operator $T_{\delta,j}$ can be estimated by

$$\|T_{\delta,j}^V\|_{p_c \to p_c'} \lesssim \sum_{\lambda_1=2}^{2^{2j}} \sum_{(2^{-j}\lambda_1)^2 \lesssim \lambda_3 \ll \lambda_1} \lambda_1^{(3\theta-1)/2} \lambda_3^{(2\theta-1)/2} \lesssim 2^{(1-2\theta)j} \sum_{\lambda_1=2}^{2^{2j}} \lambda_1^{(7\theta-3)/2}$$

(recall that $2\theta - 1 < 0$, according to (4.21)). If $\theta < \frac{3}{7}$, this implies the desired estimate

$$\|T_{\delta,j}^V\|_{p_c \to p_c'} \lesssim 2^{(1-2\theta)j} \qquad (\text{ if } \theta < \tfrac{3}{7}). \tag{4.44}$$

However, if $\theta = \frac{3}{7}$, that is, if ϕ is of finite type A and $m = 2$, we get only the estimate

$$\|T_{\delta,j}^V\|_{p_c \to p_c'} \lesssim j 2^{(1-2\theta)j}. \tag{4.45}$$

In order to improve on this estimate, we shall have to apply a complex interpolation argument. There will be a few more cases that require such an interpolation argument, and we shall collect them all in Proposition 4.2 (see also Section 5.3). We also remark that

$$\sum_{2 \le \lambda_1 \lesssim 2^j} \sum_{(2^{-j}\lambda_1)^2 \lesssim \lambda_3 \ll \lambda_1} \lambda_1^{(3\theta-1)/2} \lambda_3^{(2\theta-1)/2} \lesssim 2^{[(3\theta-1)/2]j} \lesssim 2^{(1-2\theta)j},$$

so that we need only to control the terms with $\lambda_1 \gg 2^j$.

3.2. The subcase where $\lambda_3^{1/2} \ll \lambda_1 2^{-j}$. Then we have

$$\lambda_3 \ll \min\{\lambda_1, (2^{-j}\lambda_1)^2\},$$

which implies that necessarily $\lambda_1 \gg 2^j$, and interpolation yields in this case that

$$\|T_{\delta,j}^\lambda\|_{p_c \to p_c'} \lesssim 2^{\theta j} \lambda_1^{-(1-\theta)/2} \lambda_3^{(3\theta-1)/2}.$$

First, assume that $\lambda_1 > 2^{2j}$. Then $\lambda_1 = \min\{\lambda_1, (2^{-j}\lambda_1)^2\}$, so that we shall use $\lambda_3 \ll \lambda_1$. Denoting by $T_{\delta,j}^{VI,1}$ the sum of the operators T_j^λ, with λ satisfying the assumptions of this subcase and $\lambda_1 > 2^{2j}$, and recalling that $3\theta - 1 > 0$ and $2\theta - 1 < 0$, we see that

$$\|T_{\delta,j}^{VI,1}\|_{p_c \to p_c'} \lesssim 2^{\theta j} \sum_{\lambda_1=2^{2j}}^{\infty} \sum_{\lambda_3=2}^{\lambda_1} \lambda_1^{-(1-\theta)/2} \lambda_3^{(3\theta-1)/2} \lesssim 2^{(5\theta-2)j} \le 2^{(1-2\theta)j}. \tag{4.46}$$

There remains the case where $2^j \ll \lambda_1 \le 2^{2j}$. Then $\lambda_3 \ll (2^{-j}\lambda_1)^2$. Denoting by $T^{VI,2}_{\delta,j}$ the sum of the operators T^λ_j, with λ satisfying the assumptions of this subcase and $2^j \ll \lambda_1 \le 2^{2j}$, and recalling that $3\theta - 1 > 0$ and $2\theta - 1 < 0$, we see that

$$\|T^{VI,2}_{\delta,j}\|_{p_c \to p_c'} \lesssim 2^{\theta j} \sum_{\lambda_1=2^j}^{2^{2j}} \sum_{\lambda_3=2}^{(2^{-j}\lambda_1)^2} \lambda_1^{-(1-\theta)/2} \lambda_3^{(3\theta-1)/2} \lesssim 2^{(1-2\theta)j}$$

$$\times \sum_{\lambda_1=2}^{2^{2j}} \lambda_1^{(7\theta-3)/2} \lesssim 2^{(1-2\theta)j}, \tag{4.47}$$

provided $\theta < \frac{3}{7}$. If $\theta = \frac{3}{7}$, we pick up an additional factor j as in (4.45):

$$\|T^{VI,2}_{\delta,j}\|_{p_c \to p_c'} \lesssim j 2^{j/7} = j 2^{(1-2\theta)j}. \tag{4.48}$$

In order to improve on this estimate, we shall have to again apply a complex interpolation argument (cf. Section 5.3).

What is left is Case 4.

Case 4. $\lambda_1 \sim \lambda_2 \sim \lambda_3$. Here, we can first apply the method of stationary phase to the integration in x_2. This produces a phase function in x_1, which is of the form $\phi_1(x_1) = \xi_1 x_1 + \xi_2(\omega(0)x_1^m + \text{error}) + \xi_3(\sigma\beta(0)x_1^n + \text{error})$, with small error terms of order $O(|\delta| + 2^{-j})$. We assume again that (4.12) holds true. Then ϕ_1 has a singularity of Airy type, which implies that the oscillatory integral with phase ϕ_1 that we have arrived at decays with order $O(|\lambda|^{-1/3})$. Indeed, we have $n \ge 2m + 1$ and $m \ge 2$, and since $x_1 \sim 1$, it is easy to see by studying the linear system of equations $y_j = \phi_1^{(j)}(x_1)$, $j = 1, 2, 3$, that there exist constants $0 < c_1 \le c_2$ that do not depend on ξ and $x_1 \sim 1$ such that

$$c_1|\xi| \le \sum_{j=1}^{3} |\phi_1^{(j)}(x_1)| \le c_2|\xi|.$$

Hence, our claim follows from Lemma 2.1. We thus find that

$$\|\widehat{\nu^\lambda_j}\|_\infty \lesssim \lambda_1^{-1/2}\lambda_1^{-1/3} = \lambda_1^{-5/6}.$$

4.1. The subcase where $\lambda_1 > 2^{2j}$. Then, by (4.37) $\|\nu^\lambda_j\|_\infty \lesssim 2^j\lambda_1$, and we obtain

$$\|T^\lambda_j\|_{p_c \to p_c'} \lesssim 2^{\theta j} \lambda_1^{(11\theta-5)/6}.$$

The estimates in (4.21) show that $11\theta - 5 < 0$, which implies that the contribution $T^{VII}_{\delta,j}$ of the operators T^λ_j, with λ satisfying the assumptions of this subcase, to $T_{\delta,j}$ can again be estimated by

$$\|T_{\delta,j}^{VII}\|_{p_c\to p_c'} \lesssim \sum_{\lambda_1=2^{2j}}^{\infty} 2^{\theta j}\lambda_1^{(11\theta-5)/6} \lesssim 2^{(14\theta-5)j/3} \lesssim 2^{(1-2\theta)j}, \qquad (4.49)$$

provided that $\theta \le \frac{2}{5}$. According to (4.20), this is true, with the only exception being the case where ϕ is of type A and $m = 2$.

Observe also that if $m = 2$, then $\theta = \frac{3}{7}$ and $p_c' = \frac{14}{3}$, so that $\|T_j^{\lambda}\|_{p_c\to p_c'} \lesssim 2^{3j/7}\lambda_1^{-1/21}$, and

$$\sum_{\lambda_1>2^{6j}} 2^{3j/7}\lambda_1^{-1/21} \lesssim 2^{j/7} = 2^{(1-2\theta)j}.$$

This leaves open the sum over the terms with $\lambda_1 \le 2^{6j}$, in the case where ϕ is of type A and $m = 2$.

4.2. The subcase where $\lambda_1 \le 2^{2j}$. Then, by (4.37) $\|\nu_j^{\lambda}\|_\infty \lesssim \lambda_1^{3/2}$, and we obtain

$$\|T_j^{\lambda}\|_{p_c\to p_c'} \lesssim \lambda_1^{(14\theta-5)/6}.$$

We thus find that the contribution $T_{\delta,j}^{VIII}$ of the operators T_j^{λ}, with λ satisfying the assumptions of this subcase, to $T_{\delta,j}$ can be estimated by

$$\|T_{\delta,j}^{VIII}\|_{p_c\to p_c'} \lesssim \sum_{\lambda_1=2}^{2^{2j}} \lambda_1^{(14\theta-5)/6}.$$

If $14\theta - 5 \le 0$, then we immediately obtain the desired estimate $\|T_{\delta,j}^{VIII}\|_{p_c\to p_c'} \lesssim j \lesssim 2^{(1-2\theta)j}$, so assume that $14\theta - 5 > 0$. Then $\|T_{\delta,j}^{VIII}\|_{p_c\to p_c'} \lesssim 2^{[(14\theta-5)/3]j}$, and arguing as before (compare (4.49)), we see that

$$\|T_{\delta,j}^{VIII}\|_{p_c\to p_c'} \lesssim 2^{(1-2\theta)j}, \qquad (4.50)$$

unless ϕ is of type A and $m = 2$. But, recall that the case A_∞ was excluded here, so that ϕ is of type A_{n-1}, with finite $n \ge 5$ (compare Proposition 2.11).

The estimates (4.38)–(4.44), (4.46)–(4.47), (4.49), and (4.50) show that estimate (4.25) also holds true in the situation of this section, which completes the proof of Proposition 4.1, with the exception of the case where ϕ is of type A_{n-1}, with finite $n \ge 5$ and $m = 2$, in which we still need to improve on the estimates (4.45) and (4.48) in Subcases 3.1 and 3.2; moreover, we need to find stronger estimates for the cases where $\lambda_1 \sim \lambda_2 \sim \lambda_3$ when $\lambda_1 \le 2^{6j}$. Observe also that in Case 3, we have that $\lambda_1 \sim \lambda_2$, and thus we may assume that $\lambda_2 = 2^K\lambda_1$, where K is from a finite set of integers. This allows us to assume that $\lambda_2 = 2^K\lambda_1$ for a given, fixed integer K, and for the sake of simplicity, we shall even assume that $K = 0$, so that $\lambda_1 = \lambda_2$ (the other cases can be treated in exactly the same way). In a similar way, we may and shall assume that $\lambda_1 = \lambda_2 = \lambda_3$ in Case 4.

Thus, in order to complete the proof of Proposition 4.1, and hence that of Theorem 1.14 when $h_{\text{lin}}(\phi) < 2$, what remains to be proven is the following.

PROPOSITION 4.2. *Assume that ϕ is of type A_{n-1}, with $m = 2$ and finite $n \geq 5$, so that $p'_c = \frac{14}{3}$ and $\theta := 2/p'_c = \frac{3}{7}$. Then the following hold true, provided $j, M \in \mathbb{N}$ are sufficiently large and δ is sufficiently small:*

(a) Let

$$\nu^{V}_{\delta,j} := \sum_{\lambda_1=2^{M+j}}^{2^{2j}} \sum_{\lambda_3=(2^{-M-j}\lambda_1)^2}^{2^{-M}\lambda_1} \nu_j^{(\lambda_1,\lambda_1,\lambda_3)},$$

and denote by $T^{V}_{\delta,j}$ the convolution operator $\varphi \mapsto \varphi * \widehat{\nu^{V}_{\delta,j}}$. Then

$$\|T^{V}_{\delta,j}\|_{14/11\to 14/3} \leq C\, 2^{j/7}. \tag{4.51}$$

(b) Let

$$\nu^{VI}_{\delta,j} := \sum_{\lambda_1=2^{M+j}}^{2^{2j}} \sum_{\lambda_3=2}^{(2^{-M-j}\lambda_1)^2} \nu_j^{(\lambda_1,\lambda_1,\lambda_3)},$$

and denote by $T^{VI}_{\delta,j}$ the convolution operator $\varphi \mapsto \varphi * \widehat{\nu^{VI}_{\delta,j}}$. Then

$$\|T^{VI}_{\delta,j}\|_{14/11\to 14/3} \leq C\, 2^{j/7}. \tag{4.52}$$

(c) Let

$$\nu^{VII}_{\delta,j} := \sum_{\lambda_1=2}^{2^{6j}} \nu_j^{(\lambda_1,\lambda_1,\lambda_1)}$$

and denote by $T^{VII}_{\delta,j}$ the convolution operator $\varphi \mapsto \varphi * \widehat{\nu^{VII}_{\delta,j}}$. Then

$$\|T^{VII}_{\delta,j}\|_{14/11\to 14/3} \leq C\, 2^{j/7}. \tag{4.53}$$

Here, the constant C depends neither on δ nor on j.

The proof of Proposition 4.2 will be given in the next chapter.

REMARK 4.3. *Recall that in the situation of Subsection 4.2.2, we had replaced the small parameter δ_3 by the parameter $\sigma = 2^{2j}\delta_3 \sim 1$. If ϕ is now of type A_{n-1} and $m = 2$, it will often be convenient in the sequel to augment our former vector $\delta = (\delta_1, \delta_2)$ by the parameter*

$$\delta_0 := 2^{-j} \ll 1,$$

that is, we redefine δ to become $\delta := (\delta_0, \delta_1, \delta_2)$. In this way, we may replace the parameter $j \in \mathbb{N}$ by the small parameter δ_0. Observe that according to (4.3) *and* (4.35), *we may then rewrite, in* (4.36),

$$b^{\sharp}(y, \delta_1, \delta_2, j) = b_0(y, \delta) := \begin{cases} b^a(\delta_1 y_1, \delta_0\delta_2 y_2), & \text{if } \phi \text{ is of type A},\\ \delta_1^{-1} b^a(\delta_1 y_1, \delta_0\delta_2 y_2), & \text{if } \phi \text{ is of type D}, \end{cases}$$

where $b^a(y_1, y_2) := b(y_1, y_2 + y_1^m\omega(y_1))$ expresses b in adapted coordinates.

Then, by (4.36), *we may write*

$$\nu_j^\lambda(x) =: \nu_\delta^\lambda(x) = \lambda_1\lambda_2\lambda_3 \int \check{\chi}_1\Big(\lambda_1(x_1 - y_1)\Big)\, \check{\chi}_1\Big(\lambda_2(x_2 - \delta_0 y_2 - y_1^m \omega(\delta_1 y_1))\Big)$$
$$\times \check{\chi}_1\Big(\lambda_3(x_3 - b_0(y,\delta)\, y_2^2 - \sigma y_1^n \beta(\delta_1 y_1))\Big)\, \eta(y,\delta)\, dy, \quad (4.54)$$

where $\eta \in C^\infty(\mathbb{R}^2 \times \mathbb{R}^3)$ *is supported if* $y_1 \sim 1$ *and* $|y_2| \lesssim 1$ *(and, say,* $|\delta| \leq 1$*) and where* χ_1 *is a smooth cutoff function supported near* 1. *Notice that the measure* ν_δ^λ *indeed also depends on* $\sigma \sim 1$, *but we shall suppress this dependency in order to simplify the notation.*

Chapter Five

Improved Estimates by Means of Airy-Type Analysis

We next turn to the proof of Proposition 4.2. Recall from the discussion in the previous chapter that for the operators $T^V_{\delta,j}$ and $T^{VI}_{\delta,j}$ appearing in parts (a) and (b) of that proposition, we had already established the conjectured $L^p \to L^{p'}$ estimates, with only the exception of the endpoint $p = p_c$. It may, therefore, not come as a surprise that this endpoint result can be established by means of Stein's interpolation theorem for analytic families of operators (see [SW71]), as will be shown in Sections 5.2 and 5.3. Indeed, we shall construct analytic families of complex measure μ_ζ, for ζ in the complex strip Σ given by $0 \le \mathrm{Re}\,\zeta \le 1$, by introducing complex coefficients in the sums defining the measures $\nu^V_{\delta,j}$ and $\nu^{VI}_{\delta,j}$, respectively (our approach is somewhat different from more classical ones, for instance, in the work by Stein and Tomas, but it is better adapted to the dyadic frequency decompositions that we have performed and appears to be even more elementary). These coefficients will be chosen as exponentials of suitable affine-linear expression in ζ in such a way that, in particular, $\mu_{\theta_c} = \nu^V_{\delta,j}$, respectively, $\mu_{\theta_c} = \nu^{VI}_{\delta,j}$. As it will turn out, the main problem will then consist in establishing suitable uniform bounds for the measure μ_ζ when ζ lies on the right boundary line of Σ (compare estimates (5.49) and (5.57)).

The arguments required to prove these uniform estimates for the functions $\mu_{1+it}, t \in \mathbb{R}$, will turn out to be rather involved. In particular, we shall have to invoke our uniform estimates for oscillatory sums from Lemmas 2.7 and 2.9, in combination with suitable estimates for particular classes of integrals of sublevel type. We shall see several further instances of such kind of arguments in the last two chapters, in which we shall deal with phase functions for which the function $\phi(x_1, x_2) = (x_2 - x_1^2)^3 + x_1^n$ can be viewed as a prototype. These phase functions will turn out to generate the highest degree of complexity among all, and thus the corresponding arguments in Chapters 7 and 8 will become even more involved.

As for part (c) in Proposition 4.2, recall from our discussion in Chapter 4 that we not only missed the endpoint estimate for the operator $T^{VI}_{\delta,j}$, but we even have to close a large gap between the desired estimate and the actual estimate given by (4.49), in the case where ϕ is of type A and $m = 2$. The reason for this gap lies in the fact that the Fourier transform of the measure ν^λ_δ can be expressed as a one-dimensional integral of Airy type (after an easy application of the method of stationary phase in the second variable), whose decay rate will be smallest along a certain "Airy cone."

We shall, therefore, perform yet another dyadic frequency decomposition into a region close to the given Airy cone and dyadic regions in which the distance to the cone is of a fixed order of size. The corresponding measures will be denoted by $\nu^\lambda_{\delta,Ai}$ and $\nu^\lambda_{\delta,l}$ (compare (5.12)). Here, Lemma 2.2 will become important. Such a type of decomposition is familiar from other contexts, for instance, from work by Greenleaf and Seeger [GS99] on oscillatory integrals with folding canonical relations.

A major problem will then consist in deriving enough information, in particular on the L^∞-norms, of the functions $\nu^\lambda_{\delta,Ai}$ and $\nu^\lambda_{\delta,l}$ from knowledge of their Fourier transforms, in particular, from the support properties of $\widehat{\nu^\lambda_{\delta,Ai}}$ and $\widehat{\nu^\lambda_{\delta,l}}$ (compare estimates (5.15) and (5.24)). This will necessitate further decompositions of the measures $\mu^\lambda_{\delta,l}$ in Subsection 5.2.2. The estimates that we can establish in this way will then allow us to verify the conjectured $L^p \to L^{p'}$ estimates for the operator $T^{VII}_{\delta,j}$, except for the endpoint $p = p_c$. In order to capture also the endpoint, we shall again have to apply a complex interpolation argument of a similar kind as described before. This will be done in Subsection 5.2.

We shall begin with the discussion of part (c) in Proposition 4.2, since this case causes even more challenges than the discussion of parts (a) and (b) (which will be treated in the last Section 5.2).

5.1 AIRY-TYPE DECOMPOSITIONS REQUIRED FOR PROPOSITION 4.2(C)

In order to prove estimate (4.52) in Proposition 4.2, we recall that $\sigma \sim 1$ and that we are assuming that

$$2 \le \lambda_1 = \lambda_2 = \lambda_3 \le 2^{6j}.$$

In order to simplify the notation, we shall in the sequel denote by λ the common value of $\lambda_1 = \lambda_2 = \lambda_3$ and put

$$s_1 := \frac{\xi_1}{\xi_3}, \quad s_2 := \frac{\xi_2}{\xi_3}, \quad s_3 := \frac{\xi_3}{\lambda}, \tag{5.1}$$

so that $|s_1| \sim |s_2| \sim |s_3| \sim 1$ and

$$\xi = \lambda s_3 (s_1, s_2, 1).$$

In view of the special role that s_3 will play, we shall write

$$s := (s_1, s_2, s_3), \qquad s' := (s_1, s_2).$$

Correspondingly, we shall rewrite the complete phase in (4.35) as

$$\Phi(x, \delta_1, \delta_2, j, \xi) = \lambda s_3 \tilde{\Phi}(x, \delta, \sigma, s_1, s_2),$$

where

$$\begin{aligned}\tilde{\Phi}(x, \delta, \sigma, s_1, s_2) := {} & s_1 x_1 + s_2 x_1^2 \omega(\delta_1 x_1) + \sigma x_1^n \beta(\delta_1 x_1) \\ & + \delta_0 s_2 x_2 + x_2^2 b_0(x, \delta),\end{aligned} \tag{5.2}$$

with $b_0(x, \delta)$ defined as in Remark 4.3. Recall also that $\omega(0) \neq 0$, $\beta(0) \neq 0$, and $b_0(x, 0) = b(0, 0) \neq 0$.

According to (4.54), we then have

$$\widehat{\nu_j^\lambda}(\xi) = \widehat{\nu_\delta^\lambda}(\xi) = \chi_1(s_1s_3)\chi_1(s_2s_3)\chi_1(s_3) \int e^{-i\lambda s_3\tilde{\Phi}(x,\delta,\sigma,s_1,s_2)}\,\tilde{a}(x,\delta)\,dx,$$

where the amplitude $\tilde{a}(x,\delta) := a(x,\delta)\chi_1(x_1)\chi_0(x_2)$ is a smooth function of x supported where $x_1 \sim 1$ and $|x_2| \lesssim 1$ and whose derivatives are uniformly bounded with respect to the parameters δ. Moreover, if T_δ^λ denotes the convolution operator

$$T_\delta^\lambda \varphi := \varphi * \widehat{\nu_\delta^\lambda},$$

then we see that the estimate (4.52) can be rewritten as

$$\Big\| \sum_{2\le\lambda\le\delta_0^{-6}} T_\delta^\lambda \Big\|_{14/11\to 14/3} \le C\,\delta_0^{-1/7}. \tag{5.3}$$

We shall need to understand the precise behavior of $\widehat{\nu_\delta^\lambda}(\xi)$. To this end, consider the integration with respect to x_2 in the corresponding integral. Notice that there always is a critical point x_2^c with respect to x_2. Writing $x_2 = \delta_0 s_2 t$ and applying the implicit function theorem to t, we find that

$$x_2^c = \delta_0 s_2 Y_2(\delta_1 x_1, \delta_2, \delta_0 s_2), \tag{5.4}$$

where Y_2 is smooth and of size $|Y_2| \sim 1$. Notice also that $Y_2(0,0,0) = -1/(2b(0,0))$ when $\delta = 0$. We let

$$\Psi(x_1,\delta,\sigma,s_1,s_2) := \tilde{\Phi}(x_1,x_2^c,\sigma,s_1,s_2) = \tilde{\Phi}(x_1,\delta_0 s_2 Y_2(\delta_1 x_1,\delta_2,\delta_0 s_2),\sigma,s_1,s_2). \tag{5.5}$$

Applying the method of stationary phase with parameters to the x_2-integration (see, e.g., [S93]) and ignoring the region away from the critical point x_2^c, which leads to better estimates by means of integrations by parts, we find that we may assume that

$$\widehat{\nu_\delta^\lambda}(\xi) = \lambda^{-1/2}\chi_1(s_1s_3)\chi_1(s_2s_3)\chi_1(s_3) \int e^{-i\lambda s_3\Psi(y_1,\delta,\sigma,s')}\,a_0(y_1,s,\delta;\,\lambda)\,\chi_1(y_1)\,dy_1\,, \tag{5.6}$$

where χ_1 is a smooth cutoff function supported, say, in the interval $[\frac{1}{2}, 2]$.

Moreover, $a_0(y_1,s,\delta;\,\lambda)$ is smooth and *uniformly a classical symbol* of order 0 with respect to λ. By this we mean that it is a classical symbol of order zero for every given parameter (here these are y_1, s_1, s_2, s_3 and δ), and the constants in the symbol estimates are uniformly controlled for these parameters. It will be important to observe that this implies that $\frac{\partial}{\partial\lambda}a_0(y_1,s,\delta;\,\lambda)$ is even a symbol of order -2 with respect to λ, uniformly in y_1, s, δ (the latter property will become relevant later!).

We shall need more precise information on the phase Ψ. Indeed, in the subsequent lemmata, we shall establish two different presentations of Ψ, both of which will become relevant.

Lemma 5.1. *For $|x_1| \lesssim 1$, we may write*

$$\Psi(x_1,\delta,\sigma,s_1,s_2) = s_1x_1 + s_2x_1^2\omega(\delta_1x_1) + \sigma x_1^n\beta(\delta_1x_1) + (\delta_0s_2)^2Y_3(\delta_1x_1,\delta_2,\delta_0s_2),$$

where Y_3 is smooth and $Y_3(\delta_1x_1,\delta_2,\delta_0s_2) = c_0 + O(|\delta|)$, with $c_0 := -1/4b$ $(0,0) \neq 0$.

Proof. We have

$$\Psi(x_1,\delta,\sigma,s_1,s_2)=s_1x_1+s_2x_1^2\omega(\delta_1x_1)+\sigma x_1^n\beta(\delta_1x_1)+\delta_0s_2x_2^c+(x_2^c)^2b_0(x_1,x_2^c,\delta),$$

so that, by definition,

$$Y_3(\delta_1x_1,\delta_2,\delta_0s_2):=Y_2(\delta_1x_1,\delta_2,\delta_0s_2)+Y_2(\delta_1x_1,\delta_2,\delta_0s_2)^2b_0(x_1,x_2^c,\delta),$$

where for $\delta=0$ we have

$$Y_3(0,0,0)=Y_2(0,0,0)+Y_2(0,0,0)^2b_0(0,0,0)=-\frac{1}{4b(0,0)}\neq 0,$$

because $Y_2(0,0,0)=-1/(2b(0,0))$. Q.E.D.

Next, we shall verify that Ψ has indeed a singularity of Airy type with respect to the variable x_1. To this end, let us first consider the case where $\delta=0$. Then

$$\Psi(x_1,0,\sigma,s_1,s_2):=s_1x_1+s_2x_1^2\omega(0)+\sigma x_1^n\beta(0),$$

and depending again on the signs of $s_2\omega(0)$ and $\beta(0)$, the first derivative (with respect to x_1)

$$\Psi'(x_1,0,\sigma,s_1,s_2)=s_1+2s_2\omega(0)x_1+n\sigma\beta(0)x_1^{n-1}$$

may have a critical point, or not. If not, Ψ will have at worst nondegenerate critical points, and this case can be treated again by the method of stationary phase, respectively, integrations by parts. We shall therefore concentrate on the case where Ψ' does have a critical point x_1^c, which will then be given explicitly by

$$x_1^c=x_1^c(0,\sigma,s_2):=\left(-\frac{2\omega(0)}{n(n-1)\sigma\beta(0)}s_2\right)^{1/(n-2)}.$$

Let us assume that $s_2>0$ (the case where it is negative can be treated similarly). By scaling in x_1, we may and shall assume for simplicity that

$$-\frac{2\omega(0)}{n(n-1)\sigma\beta(0)}=1 \qquad (\text{and } s_2\sim 1). \tag{5.7}$$

Then $x_1^c(0,\sigma,s_2)=s_2^{1/(n-2)}$, and $|\Psi'''(x_1^c,0,\sigma,s_1,s_2)|\sim 1$. Thus, the implicit function theorem shows that for δ sufficiently small, there is a unique critical point $x_1^c=x_1^c(\delta,\sigma,s_2)$ of Ψ' depending smoothly on δ, σ and s_2, that is,

$$\Psi''(x_1^c(\delta,\sigma,s_2),\delta,\sigma,s_1,s_2)=0.$$

Lemma 5.2. *The phase Ψ given by (5.5) can be developed locally around the critical point x_1^c of Ψ' in the form*

$$\Psi(x_1^c(\delta,\sigma,s_2)+y_1,\delta,\sigma,s_1,s_2)=B_0(s',\delta,\sigma)-B_1(s',\delta,\sigma)y_1+B_3(s_2,\delta,\sigma,y_1)y_1^3,$$

where B_0, B_1 and B_3 are smooth functions and where $|B_3(s_2,\delta,\sigma,y_1)|\sim 1$, and indeed

$$B_3(s_2,\delta,\sigma,0)=s_2^{(n-3)/(n-2)}G_4(s_2,\delta,\sigma),$$

where G_4 *is smooth and satisfies*

$$G_4(s_2, 0, \sigma) = \frac{n(n-1)(n-2)}{6}\sigma\beta(0).$$

Moreover, we may write

$$\begin{cases} x_1^c(\delta, \sigma, s_2) &= s_2^{1/(n-2)} G_1(s_2, \delta, \sigma), \\ B_0(s', \delta, \sigma) &= s_1 s_2^{1/(n-2)} G_1(s_2, \delta, \sigma) - s_2^{n/(n-2)} G_2(s_2, \delta, \sigma), \\ B_1(s', \delta, \sigma) &= -s_1 + s_2^{(n-1)/(n-2)} G_3(s_2, \delta, \sigma), \end{cases} \tag{5.8}$$

with smooth functions G_1, G_2, *and* G_3 *satisfying*

$$\begin{cases} G_1(s_2, 0, \sigma) &= 1, \\ G_2(s_2, 0, \sigma) &= \dfrac{n^2-n-2}{2}\sigma\beta(0), \\ G_3(s_2, 0, \sigma) &= n(n-2)\sigma\beta(0). \end{cases} \tag{5.9}$$

Notice that all the numbers in (5.9) *are nonzero, since we assume* $n \geq 5$. *Moreover, if we put* $G_5 := G_1 G_3 - G_2$, *then we have*

$$G_3(s_2, 0, \sigma) \neq 0 \quad \textit{and} \quad G_5(s_2, 0, \sigma) = \frac{n^2-3n+2}{2}\sigma\beta(0) \neq 0. \tag{5.10}$$

Proof. The first statements in (5.8) and (5.9) are obvious. Next, by (5.5) and (5.4) we have

$$\begin{aligned} B_0(s', \delta, \sigma) = \Psi(x_1^c(\delta, \sigma, s_2), \delta, \sigma, s_1, s_2) &= s_1 s_2^{1/(n-2)} G_1(s_2, \delta, \sigma) \\ &+ s_2^{n/(n-2)}(G_1(s_2, \delta, \sigma)^2 \omega(\delta_1 x_1^c) + \sigma G_1(s_2, \delta, \sigma)^n \beta(\delta_1 x_1^c) \\ &+ \delta_0^2 s_2^{(n-4)/(n-2)} Y_3(\delta_1 x_1^c, \delta_2, \delta_0 s_2)), \end{aligned}$$

where x_1^c is given by the first identity in (5.8). In combination with (5.7), we thus obtain the second identity in (5.8) and the third in (5.9) because $s_2 \sim 1$.

Similarly,

$$\begin{aligned} -B_1(s', \delta, \sigma) &= \Psi'(x_1^c(\delta, \sigma, s_2), \delta, \sigma, s_1, s_2) \\ &= s_1 + 2s_2 x_1^c \omega(\delta_1 x_1^c) + n\sigma(x_1^c)^{n-1}\beta(\delta_1 x_1^c) + O(|\delta|), \end{aligned}$$

which in view of (5.7) easily implies the last identities in (5.8) and (5.9). Finally, when $y_1 = 0$, then

$$\begin{aligned} 6B_3(s_2, \delta, \sigma, 0) &= \Psi'''(x_1^c(\delta, \sigma, s_2), \delta, \sigma, s_1, s_2) \\ &= n(n-1)(n-2)\sigma\beta(0)(x_1^c)^{n-3} + O(|\delta|), \end{aligned}$$

which shows that $|B_3(s_2, \delta, \sigma, y_1)| \sim 1$ for $|y_1|$ sufficiently small. The proof of (5.10) is straightforward. Q.E.D.

Translating the coordinate y_1 in (5.6) by x_1^c, Lemma 5.2 then allows us to rewrite (5.6) in the following form:

$$\widehat{\nu_\delta^\lambda}(\xi) = \lambda^{-1/2}\chi_1(s_1 s_3)\chi_1(s_2 s_3)\chi_1(s_3)\, e^{-i\lambda s_3 B_0(s',\delta,\sigma)}$$
$$\times \int e^{-i\lambda s_3(B_3(s_2,\delta,\sigma,y_1)y_1^3 - B_1(s',\delta,\sigma)y_1)}\, a_0(y_1, s, \delta;\, \lambda)\, \chi_0(y_1)\, dy_1. \quad (5.11)$$

Here, χ_0 is a smooth cutoff function supported in sufficiently small neighborhood of the origin, and $a_0(y_1, s, \delta;\, \lambda)$ is again a smooth function (possible different from the one in (5.6)), which is uniformly a classical symbol of order 0 with respect to λ.

We shall now make use of Lemma 2.2, with $B = 3$. Let us apply this lemma and the subsequent remark to the oscillatory integral in (5.11). Putting $u := B_1(s, \delta, \sigma)$, in view of this lemma we shall decompose the frequency support of ν_δ^λ furthermore into the domain where $\lambda^{2/3}|B_1(s, \delta, \sigma)| \lesssim 1$ (this is essentially a conic region in ξ-space (cf. (5.1)), which will be called the "region near the Airy cone," defined by $B_1 = 0$), and the remaining domain into the conic regions where $(2^{-l}\lambda)^{2/3}$ $|B_1(s, \delta, \sigma)| \sim 1$, for $M_0 \le 2^l \le \frac{\lambda}{M_1}$, where $M_0, M_1 \in \mathbb{N}$ are sufficiently large. Notice also that the *Airy cone* is given by the equation $B_1 = 0$, that is, by

$$s_1 = s_2^{(n-1)/(n-2)} G_3(s_2, \delta, \sigma)$$

(recall here from (5.1) that the s_j are functions $s_j(\xi)$ of the frequency variables ξ, which are homogeneous of degree 0).

More precisely, we choose smooth cutoff functions χ_0 and χ_1 such that $\chi_0 = 1$ on a sufficiently large neighborhood of the origin and $\chi_1(t)$ is supported where $|t| \sim 1$ and $\sum_{l\in\mathbb{Z}} \chi_1(2^{-2l/3}t) = 1$ on $\mathbb{R}\setminus\{0\}$ and define the functions $\nu_{\delta,Ai}^\lambda$ and $\nu_{\delta,l}^\lambda$ by

$$\widehat{\nu_{\delta,Ai}^\lambda}(\xi) := \chi_0(\lambda^{2/3} B_1(s', \delta, \sigma))\, \widehat{\nu_\delta^\lambda}(\xi),$$
$$\widehat{\nu_{\delta,l}^\lambda}(\xi) := \chi_1((2^{-l}\lambda)^{2/3} B_1(s', \delta, \sigma))\, \widehat{\nu_\delta^\lambda}(\xi), \qquad M_0 \le 2^l \le \frac{\lambda}{M_1},$$

so that

$$\nu_\delta^\lambda = \nu_{\delta,Ai}^\lambda + \sum_{M_0 \le 2^l \le \lambda/M_1} \nu_{\delta,l}^\lambda. \quad (5.12)$$

Denote by $T_{\delta,Ai}^\lambda$ and $T_{\delta,l}^\lambda$ the convolution operators

$$T_{\delta,Ai}^\lambda \varphi := \varphi * \widehat{\nu_{\delta,Ai}^\lambda}, \qquad T_{\delta,l}^\lambda \varphi := \varphi * \widehat{\nu_{\delta,l}^\lambda}.$$

Since $\delta_0 = 2^{-j}$, we note that in order to prove (5.3) and thus Proposition 4.2, it will suffice to prove the following estimate: if $p_c := \frac{14}{13}$, then

$$\sum_{2\le\lambda\le\delta_0^{-6}} \|T_{\delta,Ai}^\lambda\|_{p_c\to p_c'} + \Bigg\| \sum_{M_0\le 2^l\le\lambda/M_1} \sum_{2\le\lambda\le\delta_0^{-6}} T_{\delta,l}^\lambda \Bigg\|_{p_c\to p_c'} \le C\,\delta_0^{-1/7}, \quad (5.13)$$

provided δ is sufficiently small and $M_0, M_1 \in \mathbb{N}$ are sufficiently large.

5.1.1 Estimation of $T^{\lambda}_{\delta,Ai}$

We first consider the region near the Airy cone and prove the following.

Lemma 5.3. *There are constants C_1, C_2 such that*

$$\|\widehat{\nu^{\lambda}_{\delta,Ai}}\|_\infty \leq C_1 \lambda^{-5/6}, \tag{5.14}$$

$$\|\nu^{\lambda}_{\delta,Ai}\|_\infty \leq C_2 \lambda^{7/6}, \tag{5.15}$$

uniformly in σ and δ, provided λ is sufficiently large and δ is sufficiently small.

Notice that by interpolation (again with $\theta = \frac{3}{7}$), these estimates imply that

$$\|T^{\lambda}_{\delta,Ai}\|_{p_c \to p_c'} \lesssim (\lambda^{-5/6})^{4/7}(\lambda^{7/6})^{3/7} = \lambda^{1/42},$$

so that

$$\sum_{2 \leq \lambda \leq \delta_0^{-6}} \|T^{\lambda}_{\delta,Ai}\|_{p_c \to p_c'} \lesssim \delta_0^{-1/7}, \tag{5.16}$$

which is exactly the estimate that we need (cf. (5.13)).

Let us turn to the proof of Lemma 5.3. The first estimate (5.14) is immediate from (5.11) and Lemma 2.2 (with $M = 3$).

In order to prove the second estimate, observe first that by Lemma 2.2(a) and the subsequent remark, we may write

$$\chi_0(\lambda^{2/3} B_1(s,\delta,\sigma)) \int e^{-i\lambda s_3 (B_3(s_2,\delta,\sigma,y_1) y_1^3 - B_1(s',\delta,\sigma) y_1)} a_0(y_1, s, \delta; \lambda)\, \chi_0(y_1)\, dy_1$$
$$= \lambda^{-1/3} \chi_0(\lambda^{2/3} B_1(s',\delta,\sigma))\, g(\lambda^{2/3} B_1(s',\delta,\sigma), \lambda, \delta, \sigma, s),$$

where g is a smooth function whose derivates of any order are uniformly bounded on its natural domain.

Applying the Fourier inversion formula to $\nu^{\lambda}_{\delta,Ai}$, (5.11) and this identity yield that

$$\nu^{\lambda}_{\delta,Ai}(x) = \iint \lambda^{-1/2} \lambda^{-1/3} \chi_0(\lambda^{2/3} B_1(s',\delta,\sigma))\, \chi_1(s_1 s_3)\chi_1(s_2 s_3)\chi_1(s_3)\, e^{i\xi\cdot x}$$
$$\times e^{-i\lambda s_3 B_0(s',\delta,\sigma)}\, g(\lambda^{2/3} B_1(s',\delta,\sigma), \xi_3, \delta, \sigma, s)\, d\xi.$$

We again change coordinates from $\xi = (\xi_1, \xi_2, \xi_3)$ to (s_1, s_2, s_3) according to (5.1).

We then find that

$$\nu^{\lambda}_{\delta,Ai}(x) = \lambda^{13/6} \int e^{-i\lambda s_3 (B_0(s',\delta,\sigma) - s_1 x_1 - s_2 x_2 - x_3)} \chi_0(\lambda^{2/3} B_1(s',\delta,\sigma))$$
$$\times\, g\left(\lambda^{2/3} B_1(s',\delta,\sigma), \lambda, \delta, \sigma, s\right) \tilde{\chi}_1(s)\, ds_1 ds_2 ds_3, \tag{5.17}$$

with a smooth function g, and where

$$\tilde{\chi}_1(s) := \chi_1(s_1 s_3)\chi_1(s_2 s_3)\chi_1(s_3)\, s_3^2$$

localizes to a region where $s_j \sim 1$, $j = 1, 2, 3$.

Observe first that when $|x| \gg 1$, then we easily obtain by means of integration by parts that

$$|\nu^{\lambda}_{\delta,Ai}(x)| \le C_N \lambda^{-N}, \quad N \in \mathbb{N}, \text{ if } |x| \gg 1.$$

Indeed, when $|x_1| \gg 1$, then we integrate by parts repeatedly in s_1 to see this, and a similar argument applies when $|x_2| \gg 1$, where we use the s_2-integration. Observe that in each step, we gain a factor λ^{-1} and lose at most $\lambda^{2/3}$. Finally, when $|x_1| + |x_2| \lesssim 1$ and $|x_3| \gg 1$, then we can integrate by parts in s_3 in order to establish this estimate.

We may therefore assume in the sequel that $|x| \lesssim 1$. Then we perform yet another change of coordinates, passing from $s' = (s_1, s_2)$ to (z, s_2), where

$$z := \lambda^{2/3} B_1(s', \delta, \sigma).$$

Applying (5.8), we find that

$$z = \lambda^{\frac{2}{3}}(-s_1 + s_2^{(n-1)/(n-2)} G_3(s_2, \delta, \sigma))$$

so that

$$s_1 = s_2^{(n-1)(n-2)} G_3(s_2, \delta, \sigma) - \lambda^{-2/3} z. \tag{5.18}$$

In combination with (5.8), we obtain that

$$B_0(s, \delta, \sigma) = -\lambda^{-2/3} z\, s_2^{1/(n-2)} G_1(s_2, \delta, \sigma) + s_2^{n/n-2} G_5(s_2, \delta, \sigma). \tag{5.19}$$

We may thus rewrite

$$\begin{aligned}\nu^{\lambda}_{\delta,Ai}(x) = \lambda^{3/2} \int & e^{-i\lambda s_3 \Phi(z, s_2, x_1, \delta, \sigma)} \\ & \times g\left(z, \lambda, \delta, \sigma, s_2^{(n-1)/(n-2)} G_3(s_2, \delta, \sigma) - \lambda^{-2/3} z, s_2, s_3\right) \\ & \times \tilde{\chi}_1\left(s_2^{(n-1)/(n-2)} G_3(s_2, \delta, \sigma) - \lambda^{-2/3} z, s_2\right) \chi_0(z)\, dz\, ds_2\, ds_3,\end{aligned} \tag{5.20}$$

where

$$\begin{aligned}\Phi(z, s_2, x_1, \delta, \sigma) := & s_2^{n/(n-2)} G_5(s_2, \delta, \sigma) - s_2^{(n-1)(n-2)} G_3(s_2, \delta, \sigma) x_1 - s_2 x_2 - x_3 \\ & + \lambda^{-2/3} z\,(x_1 - s_2^{1/(n-2)} G_1(s_2, \delta, \sigma)).\end{aligned} \tag{5.21}$$

Recall that $n \ge 5$ and, from (5.10), that $G_3(s_2, \delta, \sigma) \ne 0$ and $G_5(s_2, \delta, \sigma) \ne 0$ when $\delta = 0$. Moreover, the exponents $n/(n-2)$, $(n-1)/(n-2)$, and 1 of s_2, which appear in Φ (regarding the last term in (5.21) as an error term), are all different. Recall also that we assume $|x| \lesssim 1$. It is then easily seen that this implies that, when $\delta = 0$,

$$\sum_{j=1}^{3} |\partial^j_{s_2} \Phi(z, s_2, x_1, \delta, \sigma)| \sim 1 \quad \text{for every } s_2 \sim 1,$$

uniformly in z and σ. The same type of estimate then remains valid for δ sufficiently small. We may thus apply the van der Corput type Lemma 2.1 to the s_2-integration in (5.20), which in combination with Fubini's theorem, yields

$$\|\nu^{\lambda}_{\delta,Ai}\|_\infty \le C\lambda^{3/2}\lambda^{-1/3}$$

and, hence, (5.15). This concludes the proof of Lemma 5.3.

5.1.2 Estimation of $T^{\lambda}_{\delta,l}$

We next regard the region away from the Airy cone. The study of this region will require substantially more refined techniques. Let us first note that by (5.6) and Fourier inversion we have

$$\nu^{\lambda}_{\delta,l}(x) = \lambda^3 \lambda^{-1/2} \iint \chi_1(s_1 s_3)\chi_1(s_2 s_3)\chi_1(s_3)\, \chi_1((2^{-l}\lambda)^{2/3} B_1(s', \delta, \sigma))$$
$$\times\, e^{-i\lambda s_3(\Psi(y_1,\delta,\sigma,s_1,s_2)-s_1x_1-s_2x_2-x_3)}\, a(y_1, \delta, \sigma, s;\, \lambda)\, \chi_1(y_1)\, dy_1\, ds, \quad (5.22)$$

where the amplitude a has properties similar to those of a_0.

In order to indicate the problems that we have to face here, let us state (yet without proof) an analogue to Lemma 5.3, which we believe gives essentially optimal estimates (the proof will be established later in the course of the proof of the even more refined estimates of the next section).

Lemma 5.4. *There is a constant C so that*

$$\|\widehat{\nu^{\lambda}_{\delta,l}}\|_\infty \le C 2^{-l/6} \lambda^{-5/6}, \quad (5.23)$$

$$\|\nu^{\lambda}_{\delta,l}\|_\infty \le C \min\left\{\lambda^{7/6} 2^{l/3}, \frac{\lambda}{\delta_0}\right\}, \quad (5.24)$$

uniformly in σ and δ, provided δ is sufficiently small.

In order to apply this lemma, let us write $\lambda = 2^r$, $r \in \mathbb{N}$. Then, according to (5.23), we have

$$\|\widehat{\nu^{\lambda}_{\delta,l}}\|_\infty \lesssim 2^{-(5r+l)/6}, \quad (5.25)$$

For $k \in \mathbb{N}$, we therefore define

$$\nu_{\delta,k} := \sum_{I_k} \nu^{2^r}_{\delta,l},$$

where $I_k := \{(r,l) \in \mathbb{N}^2 : 5r + l = k,\, 2^r \le \delta_0^{-6}\}$ (if $I_k = \emptyset$, then by definition $\nu_{\delta,k} := 0$). Then

$$\sum_{M_0 \le 2^l \le \lambda/M_1}\ \sum_{2 \le \lambda \le \delta_0^{-6}} \nu^{\lambda}_{\delta,l} = \sum_{k\in\mathbb{N}} \nu_{\delta,k}, \quad (5.26)$$

and we have the following consequence of Lemma 5.4.

Corollary 5.5. *There is a constant C so that*

$$\|\widehat{\nu_{\delta,k}}\|_\infty \le C\, 2^{-k/6}; \quad (5.27)$$

$$\|\nu_{\delta,k}\|_\infty \le C\, 2^{2k/9}\, \delta_0^{-1/3}, \quad (5.28)$$

uniformly in σ and δ, provided δ is sufficiently small.

Proof. The first estimate (5.27) follows immediately from (5.25) because the supports of the functions $\{\widehat{\nu^{2^r}_{\delta,l}}\}_{r,l}$ are essentially disjoint.

Next, we decompose $I_k = I_k^1 \cup I_k^2$, where

$$\begin{aligned} I_k^1 &:= \{(r,l) \in \mathbb{N}^2 : 5r + l = k,\ 2^{r+2l} \le \delta_0^{-6}\}, \\ I_k^2 &:= \{(r,l) \in \mathbb{N}^2 : 5r + l = k,\ \delta_0^{-6} < 2^{r+2l},\ 2^r \le \delta_0^{-6}\}. \end{aligned}$$

Notice that according to (5.24), for $(r,l) \in I_k^1$ we have $\lambda^{7/6} 2^{l/3} \le \frac{\lambda}{\delta_0}$; hence, $\|\nu_\delta^\lambda\|_\infty \lesssim 2^{7r/6} 2^{l/3} = 2^{2k/9} 2^{(r+2l)/18}$, whereas for $(r,l) \in I_k^2$ we have $\|\nu_\delta^\lambda\|_\infty \lesssim 2^r/\delta_0 = (2^{2k/9}/\delta_0) 2^{-(r+2l)/9}$, so that

$$\|\nu_{\delta,k}\|_\infty \le C\, 2^{\frac{2}{9}k} \sum_{(r,l)\in I_k^1} 2^{(r+2l)/18} + \frac{2^{2k/9}}{\delta_0} \sum_{(r,l)\in I_k^2} 2^{-(r+2l)/9}.$$

Comparing the latter sums with one-dimensional geometric series and using that $2^{r+2l} \le \delta_0^{-6}$ in the first sum and $2^{r+2l} > \delta_0^{-6}$ in the second sum, we obtain (5.28).
Q.E.D.

Let us denote by $T_{\delta,k}$ the convolution operator $\varphi \mapsto \varphi * \widehat{\nu_{\delta,k}}$. Interpolating the estimates in the preceding lemma, again with parameter $\theta_c := \frac{3}{7}$, we obtain

$$\|T_{\delta,k}\|_{p_c \to p_c'} \lesssim \delta_0^{-1/7}$$

uniformly in k, whereas for $1 \le p < p_c$ we get $\|T_{\delta,k}\|_{p\to p'} \lesssim 2^{-\varepsilon k} \delta_0^{-1/7}$ for some $\varepsilon > 0$ that depends on p, so that by (5.12), (5.16), and (5.26),

$$\Big\| \sum_{2 \le \lambda \le \delta_0^{-6}} T_\delta^\lambda \Big\|_{p\to p'} \lesssim \delta_0^{-\frac{1}{7}} + \sum_{k\in\mathbb{N}} \|T_{\delta,k}\|_{p\to p'} \lesssim \delta_0^{-\frac{1}{7}}.$$

We thus barely fail to establish the estimate (5.13) at the critical exponent $p = p_c$.

In order to prove the estimate (5.13) also at the endpoint $p = p_c$, we need to apply an interpolation argument. As mentioned in Section 2.4, in the majority of situations we shall make use of Stein's interpolation theorem for analytic families of operators, but in a few cases we can alternatively apply the real interpolation result of Proposition 2.6, which is based on ideas of Bak and Seeger [BS11] and whose assumptions are much easier to verify.

Nevertheless, we shall also encounter several situations in which we don't know how to adapt the methods from [BS11] but that still can by treated by means of complex interpolation. The latter applies also to the proof of the endpoint estimate in Proposition 4.2(c). Indeed, what seems to prevent the application of the real interpolation method is that the (complex) measures $\nu_{\delta,k}$ arise from the positive measure ν_δ by means of spectral localizations to certain frequency regions, that is, $\nu_{\delta,k} = \nu_\delta * \psi_{\delta,k}$, and the obstacle in applying the method from [BS11] is that there is no uniform bound for the L^1-norms of the functions $\psi_{\delta,k}$ as k tends to infinity.

The proofs, based on complex interpolation, are technically involved, and our arguments outlined in the next section can be viewed as prototypical for other proofs of the same kind appearing in later chapters.

5.2 THE ENDPOINT IN PROPOSITION 4.2(C): COMPLEX INTERPOLATION

We keep the notation of the previous section. According to (5.11) and Lemma 5.2, we may write (recalling that $\xi = \lambda s_3(s_1, s_2, 1)$)

$$\widehat{\nu_{\delta,l}^{\lambda}}(\xi) := \lambda^{-\frac{1}{2}} \chi_1((2^{-l}\lambda)^{2/3} B_1(s', \delta, \sigma)) \tilde{\chi}_1(s)\, e^{-i\lambda s_3 B_0(s',\delta,\sigma)}\, J(\lambda, s, \delta, \sigma), \quad (5.29)$$

where we recall that $\tilde{\chi}_1$ localizes to a region where $s_j \sim 1,\ j = 1, 2, 3$, and where

$$J(\lambda, s, \delta, \sigma) := \int e^{-i\lambda s_3 \tilde{\Psi}_0(y_1,\delta,\sigma,s_1,s_2)}\, a_0(y_1, s, \delta;\ \lambda)\, \chi_0(y_1)\, dy_1$$

with

$$\tilde{\Psi}_0(y_1, \delta, \sigma, s_1, s_2) := B_3(s_2, \delta, \sigma, y_1) y_1^3 - B_1(s', \delta, \sigma) y_1.$$

Since B_1 is of size $(2^l/\lambda)^{2/3}$, we scale by the factor $(2^l/\lambda)^{1/3}$ in the integral defining $J(\lambda, s, \delta, \sigma)$ by putting $y_1 = (2^l/\lambda)^{1/3} u_1$ and obtain

$$J(\lambda, s, \delta, \sigma) = (2^l\lambda^{-1})^{1/3} \int e^{-is_3 2^l \Psi_0(u_1,s',\delta,\lambda,l)}\, a_0((2^l\lambda^{-1})^{1/3}u_1,\ s,\ \delta,\ \lambda) \times \chi_0((2^l\lambda^{-1})^{1/3}u_1)\, du_1$$

with

$$\Psi_0(u_1, s', \delta, \lambda, l) := B_3(s_2,\ \delta, \sigma, (2^l\lambda^{-1})^{1/3}u_1)\, u_1^3 - (2^l\lambda^{-1})^{-2/3} B_1(s',\ \delta, \sigma)\, u_1.$$

Observe that the coefficients of u_1 and of u_1^3 in Ψ_0 are both of size 1, so that Ψ_0 will have no critical point with respect to u_1 unless $|u_1| \sim 1$.

We may therefore choose a smooth cutoff function $\chi_1 \in C_0^\infty(\mathbb{R})$ supported away from 0 so that Ψ_0 has no critical point outside the support of χ_1, and decompose

$$J := J(\lambda, s, \delta, \sigma) = J_1 + J_\infty,$$

where $J_1 = J_1(\lambda, s, \delta, \sigma)$ is given by

$$J_1 := (2^l\lambda^{-1})^{1/3} \int e^{-is_3 2^l \Psi_0(u_1,s',\delta,\lambda,l)}\, a_0((2^l\lambda^{-1})^{1/3}u_1,\ s,\ \delta,\ \lambda) \times \chi_0((2^l\lambda^{-1})^{1/3}u_1)\, \chi_1(u_1)\, du_1.$$

Accordingly, in view of (5.29) we may decompose

$$\nu_{\delta,l}^{\lambda} = \nu_{l,1}^{\lambda} + \nu_{l,\infty}^{\lambda},$$

where the summands are defined by

$$\widehat{\nu_{l,1}^{\lambda}}(\xi) := \lambda^{-1/2} \chi_1((2^{-l}\lambda)^{2/3} B_1(s',\ \delta, \sigma)) \tilde{\chi}_1(s)\, e^{-i\lambda s_3 B_0(s',\delta,\sigma)}\, J_1(\lambda, s, \delta, \sigma),$$

$$\widehat{\nu_{l,\infty}^{\lambda}}(\xi) := \lambda^{-1/2} \chi_1((2^{-l}\lambda)^{2/3} B_1(s',\ \delta, \sigma)) \tilde{\chi}_1(s)\, e^{-i\lambda s_3 B_0(s',\delta,\sigma)}\, J_\infty(\lambda, s, \delta, \sigma)$$

(we have dropped the dependence on δ in order to simplify the notation).

Let us first consider the contribution given by the $\nu^\lambda_{l,\infty}$: By means of integration by parts, we easily obtain that for every $N \in \mathbb{N}$, we have $|J_\infty| \lesssim (2^l\lambda^{-1})^{1/3}2^{-lN}$; hence,

$$\|\widehat{\nu^\lambda_{l,\infty}}\|_\infty \lesssim \lambda^{-1/2}(2^l\lambda^{-1})^{1/3}2^{-lN} \qquad \forall N \in \mathbb{N}. \tag{5.30}$$

Next, we may assume that we have chosen $\tilde{\chi}_1$ so that the Fourier inversion formula for $\nu^\lambda_{l,\infty}(x)$ reads

$$\nu^\lambda_{l,\infty}(x) = \lambda^3 \int_{\mathbb{R}^3} e^{i\lambda s_3(s_1x_1+s_2x_2+x_3)}\,\widehat{\nu^\lambda_{l,\infty}}(\xi)\,ds$$

(with $\xi = \lambda s_3(s_1, s_2, 1)$). We then use the change of variables from $s' = (s_1, s_2)$ to (z, s_2), where now

$$z := (2^{-l}\lambda)^{2/3}B_1(s', \delta, \sigma),$$

and find that (compare (5.18))

$$s_1 = s_2^{(n-1)/(n-2)}G_3(s_2, \delta, \sigma) - (2^{-l}\lambda)^{-2/3}z, \tag{5.31}$$

and, in particular (compare (5.19)),

$$B_0(s, \delta, \sigma) = -(2^{-l}\lambda)^{-2/3}z\, s_2^{1/(n-2)}G_1(s_2, \delta, \sigma) + s_2^{n/(n-2)}G_5(s_2, \delta, \sigma). \tag{5.32}$$

Notice that $2^l/\lambda \le 1/M_1 \ll 1$. And, if we plug in the previous formula for $\widehat{\nu^\lambda_{l,\infty}}$ and write $\nu^\lambda_{l,\infty}(x)$ as an oscillatory integral with respect to the variables u_1, z, s_2, s_3, we see that the complete phase is of the form

$$\begin{aligned}-\lambda s_3(s_2^{n/(n-2)}&G_5(s_2, \delta, \sigma) - x_1 s_2^{(n-1)/(n-2)}G_3(s_2, \delta, \sigma)\\ &-s_2x_2 - x_3 + O(2^l\lambda^{-1}(1+|u_1|^3))),\end{aligned}$$

where according to (5.10), $|G_5| \sim 1$. Observe that the localization given by the function χ_0 in the definition of $J(\lambda, s, \delta, \sigma)$ implies that $2^l\lambda^{-1}|u_1|^3 \ll 1$. Again, first applying N integrations by parts with respect to u_1 and then van der Corput's lemma (with $M = 3$) for the integration in s_2, also taking into account the Jacobian of our change of coordinates to z, we see that

$$\|\nu^\lambda_{l,\infty}\|_\infty \lesssim \lambda^3\lambda^{-1/2}(2^l\lambda^{-1})^{1/3}2^{-lN}\,(2^{-l}\lambda)^{-2/3}\,\lambda^{-1/3} = \lambda^{7/6}2^{-l(N-1)}.$$

Interpolating between this estimate and (5.30), with $\theta_c = \frac{3}{7}$, we see that the convolution operator $T^\lambda_{l,\infty}$, which maps φ to $\varphi * \widehat{\nu^\lambda_{l,\infty}}$, can be estimated by

$$\|T^\lambda_{l,\infty}\|_{p_c\to p_c'} \lesssim \lambda^{(-5/6)(4/7)+(7/6)(3/7)}2^{-l} = \lambda^{1/42}2^{-l},$$

if we choose $N = 2$. This implies the desired estimate

$$\sum_{M_0\le 2^l\le\lambda/M_1}\ \sum_{2\le\lambda\le\delta_0^{-6}} \|T^\lambda_{l,\infty}\|_{p_c\to p_c'} \lesssim \delta_0^{-1/7}.$$

5.2.1 The operators $T_{l,1}^{\lambda}$

We now turn to the investigation of the convolution operator $T_{l,1}^{\lambda}$, which maps φ to $\varphi * \widehat{\nu_{l,1}^{\lambda}}$. According to (5.13), what we need to prove is that the operator

$$T_1 := \sum_{M_0 \le 2^l \le \lambda/M_1} \sum_{2 \le \lambda \le \delta_0^{-6}} T_{l,1}^{\lambda}$$

satisfies

$$\|T_1\|_{p_c \to p_c'} \lesssim \delta_0^{-1/7}, \tag{5.33}$$

with a bound that is independent of δ and σ.

Now, if the phase Ψ_0 has no critical point on the support of χ_1, then we can estimate J_1 in the same way as J_∞ before and can handle the operators $T_{l,1}^{\lambda}$ as we did for the $T_{l,\infty}^{\lambda}$. Let us therefore assume in the sequel that Ψ_0 does have a critical point $u_1^c \in \operatorname{supp} \chi_1$, so that $|u_1| \sim 1$.

Applying the method of stationary phase, we then get $|J_1| \lesssim (2^l\lambda^{-1})^{1/3} 2^{-l/2}$; hence,

$$\|\widehat{\nu_{l,1}^{\lambda}}\|_\infty \lesssim \lambda^{-1/2}(2^l\lambda^{-1})^{1/3}2^{-l/2} = \lambda^{-5/6}2^{-l/6} = 2^{-k/6}, \tag{5.34}$$

where we have used the same abbreviations,

$$\lambda := 2^r, \; k = k(r,l) := 5r + l,$$

as in the previous section.

In view of this estimate, we define for ζ in the complex strip $\Sigma := \{\zeta \in \mathbb{C} : 0 \le \operatorname{Re}\zeta \le 1\}$ the following analytic family of measures

$$\mu_\zeta(x) := \gamma(\zeta)\, \delta_0^{\zeta/3} \sum_{M_0 \le 2^l \le 2^r/M_1} \sum_{2 \le 2^r \le \delta_0^{-6}} 2^{[k(r,l)(3-7\zeta)]/18} \nu_{l,1}^{2^r},$$

where

$$\gamma(\zeta) := \frac{2^{7(\zeta-1)/2} - 1}{2^{-2} - 1},$$

and denote by T_ζ the operator of convolution with $\widehat{\mu_\zeta}$. Observe that for $\zeta = \theta_c = \frac{3}{7}$, we have $T_{\theta_c} = \delta_0^{1/7} T_1$, so that by Stein's interpolation theorem [SW71], (5.33) will follow if we can prove the following estimates on the boundaries of the strip Σ : $\|T_{it}\|_{1\to\infty} \le C$ and $\|T_{1+it}\|_{2\to 2} \le C$, where the constant C is independent of $t \in \mathbb{R}$ and the parameters δ, σ (provided δ is sufficiently small). Equivalently, we shall prove that

$$\|\widehat{\mu_{it}}\|_\infty \le C \qquad \forall t \in \mathbb{R}, \tag{5.35}$$

$$\|\mu_{1+it}\|_\infty \le C \qquad \forall t \in \mathbb{R}. \tag{5.36}$$

Since the supports of the functions $\{\widehat{\nu_{l,1}^{2^r}}\}$ are almost disjoint for l, r in the given range, we see that the first estimate (5.35) is an immediate consequence of (5.34).

The main problem will consist in estimating $\|\mu_{1+it}\|_\infty$. To this end, observe that, again by Fourier inversion, we have (with $\xi = \lambda s_3(s_1, s_2, 1)$)

$$\nu^\lambda_{l,1}(x) = \lambda^3 \int_{\mathbb{R}^3} e^{i\lambda s_3(s_1x_1+s_2x_2+x_3)} \widehat{\nu^\lambda_{l,1}}(\xi)\, ds.$$

Using once again the change of variables from s_1 to z, so that $z = (2^{-l}\lambda)^{2/3} B_1(s', \delta, \sigma)$ and $s_1 = s_2^{(n-1)/(n-2)} G_3(s_2, \delta, \sigma) - (2^{-l}\lambda)^{-2/3} z$, we find that (compare (5.31), (5.32))

$$\nu^\lambda_{l,1}(x) = \lambda^{3/2} 2^l \int e^{-is_3\Phi_1(x,u_1,z,s_2,\delta,\lambda,l)}\, a\left((2^l\lambda^{-1})^{1/3}u_1, z, s_2, s_3, \delta;\ \lambda\right)$$
$$\times \chi_1(u_1)\chi_1(z)\chi_1(s_2)\chi_1(s_3)\, du_1\, dz\, ds_2\, ds_3, \tag{5.37}$$

where $\Phi_1 = \Phi_1(x, u_1, z, s_2, \delta, \lambda, l)$ is given by

$$\Phi_1 := 2^l(B_3(s_2, \delta, \sigma, (2^l\lambda^{-1})^{1/3}u_1)\, u_1^3 - z\, u_1)$$
$$+\lambda(s_2^{n/(n-2)} G_5(s_2, \delta, \sigma) - s_2^{(n-1)(n-2)} G_3(s_2, \delta, \sigma)\, x_1 - s_2x_2 - x_3)$$
$$+\lambda(2^l\lambda^{-1})^{2/3} z\, (x_1 - s_2^{1/(n-2)} G_1(s_2, \delta, \sigma)). \tag{5.38}$$

Moreover, $a(v, u_1, z, s_2, s_3, \delta;\ \lambda)$ is a smooth function that is uniformly a classical symbol of order 0 with respect to λ.

Notice that, in order to simplify the notation, here we have suppressed the dependence on σ, which we shall also do in the sequel.

5.2.2 Preliminary reductions

Assume now, first, that $|x| \gg 1$. If $|x_1| \ll |(x_2, x_3)|$, then we easily see by means of integrations by parts in (5.37) with respect to the variables s_2 or s_3 that $|\nu_{l,1}(x)| \lesssim \lambda^{-N}$ for every $N \in \mathbb{N}$, and if $|x_1| \gtrsim |(x_2, x_3)|$, then we easily obtain $|\nu_{l,1}(x)| \lesssim (\lambda(2^l\lambda^{-1})^{2/3})^{-N}$, by means of integrations by parts in z. Since $2^l \le \lambda$, it follows easily that there are constants $A \ge 1$ and C such that $\sup_{|x|\ge A} \sup_{t\in\mathbb{R}} |\mu_{1+it}(x)| \le C$, uniformly in δ and σ.

From now on we shall therefore assume that $|x| \le A$. For such x fixed, we decompose the support of $\chi_1(s_2)$ into the subset L_{II} of all s_2 such that

$$\varepsilon(2^l\lambda^{-1})^{1/3} < |x_1 - s_2^{1/(n-2)} G_1(s_2, \delta, \sigma)| < \frac{1}{\varepsilon}(2^l\lambda^{-1})^{1/3}$$

and the complementary subsets L_I, where $|x_1 - s_2^{1/(n-2)} G_1(s_2, \delta, \sigma)| \ge (2^l\lambda^{-1})^{1/3}/\varepsilon$, and L_{III}, where $|x_1 - s_2^{1/(n-2)} G_1(s_2, \delta, \sigma)| \le \varepsilon(2^l\lambda^{-1})^{1/3}$. Here, $\varepsilon > 0$ will be a sufficiently small fixed number.

If we restrict the set of integration in (5.37) to these subsets with respect to the variable s_2, we obtain corresponding measures $\nu^\lambda_{l,I}$, $\nu^\lambda_{l,II}$ and $\nu^\lambda_{l,III}$ into which $\nu^\lambda_{l,1}$ decomposes, that is,

$$\nu^\lambda_{l,1} = \nu^\lambda_{l,I} + \nu^\lambda_{l,II} + \nu^\lambda_{l,III}.$$

Observe also that $|\lambda(2^l\lambda^{-1})^{2/3}\, (x_1 - s_2^{1/(n-2)} G_1(s_2, \delta, \sigma))| \ge 2^l$ if and only if $|x_1 - s_2^{1/(n-2)} G_1(s_2, \delta, \sigma)| \ge (2^l\lambda^{-1})^{1/3}$.

Thus, if $s_2 \in L_I$, the last term in (5.38) becomes dominant as a function of z, provided we choose ε sufficiently small. Consequently, the phase has no critical point as a function of z, and applying N integrations by parts in z, we may estimate

$$|\nu^\lambda_{l,I}(x)| \lesssim \lambda^{3/2} 2^l \int_{\{s_2:\lambda^{1/3}2^{2l/3}|x_1 - s_2^{1/(n-2)}G_1(s_2,\delta,\sigma)| \geq C2^l,\, |s_2|\sim 1\}} \frac{ds_2}{\left(\lambda^{1/3}2^{2l/3}\,|x_1 - s_2^{1/(n-2)}G_1(s_2,\delta,\sigma)|\right)^N}$$

$$\lesssim \lambda^{3/2} 2^l \int_{\lambda^{1/3}2^{2l/3}|v| \geq C2^l} \frac{dv}{(\lambda^{1/3}2^{2l/3}\,|v|)^N}$$

$$\lesssim \lambda^{3/2}2^l\,(\lambda^{1/3}2^{2l/3})^{-1}2^{(1-N)l} = \lambda^{7/6}2^{(4/3-N)l}.$$

Similarly, if $s_2 \in L_{III}$, the first term in (5.38) becomes dominant as a function of z, and thus N integrations by parts in z and the fact that the s_2-integral is restricted to a set of size $(2^l\lambda^{-1})^{\frac{1}{3}}$ yield the same estimate:

$$|\nu^\lambda_{l,III}(x)| \lesssim \lambda^{3/2}2^l 2^{-Nl}(2^l\lambda^{-1})^{1/3} = \lambda^{7/6}2^{(4/3-N)l}.$$

This implies the desired estimate

$$\Big|\gamma(1+it)\,\delta_0^{(1+it)/3} \sum_{M_0\leq 2^l\leq 2^r/M_1} \sum_{2\leq 2^r\leq\delta_0^{-6}} 2^{k(r,l)(3-7(1+it))/18}(\nu^{2^r}_{l,I} + \nu^{2^r}_{l,III})(x)\Big|$$

$$\lesssim \delta_0^{1/3} \sum_{M_0\leq 2^l\leq\lambda/M_1,\ 1\ll\lambda<\delta_0^{-6}} \lambda^{-10/9}2^{-2l/9}(|\nu^\lambda_{l,I}(x)| + |\nu^\lambda_{l,III}(x)|)$$

$$\lesssim \delta_0^{1/3} \sum_{M_0\leq 2^l,\ \lambda<\delta_0^{-6}} \lambda^{1/18}2^{(10/9-N)l} \lesssim 1,$$

if we choose $N \geq 2$.

5.2.3 The region where $|x_1 - s_2^{1/(n-2)}G_1(s_2,\delta,\sigma)| \sim (2^l\lambda^{-1})^{1/3}$

We are thus left with the measures $\nu^\lambda_{l,II}(x)$ and the corresponding family of measures

$$\mu^{II}_{1+it}(x) := \gamma(1+it)\,\delta_0^{1+it/3} \sum_{M_0\leq 2^l\leq 2^r/M_1} \sum_{2\leq 2^r\leq\delta_0^{-6}} 2^{-k(r,l)(4+7it)/18}\nu^{2^r}_{l,II}.$$

In order to establish the estimate (5.36), we still need prove that there is a constant C such that

$$|\mu^{II}_{1+it}(x)| \leq C, \tag{5.39}$$

where C is independent of t, x, δ, and σ. Note that $\partial_{s_2}(s_2^{1/(n-2)}G_1(s_2, \delta, \sigma)) \sim 1$ because $s_2 \sim 1$ and $G_1(s_2, 0, \sigma) = 1$ (compare (5.9)). Therefore, the relation $|x_1 - s_2^{1/(n-2)}G_1(s_2, \delta, \sigma)| \sim (2^l\lambda^{-1})^{1/3}$ can be rewritten as $|s_2 - \tilde{G}_1(x_1, \delta, \sigma)| \sim (2^l\lambda^{-1})^{1/3}$, where $\tilde{G}_1$ is again a smooth function. If we write

$$s_2 = (2^l\lambda^{-1})^{1/3}v + \tilde{G}_1(x_1, \delta, \sigma),$$

then this means that $|v| \sim 1$. We shall therefore change variables from s_2 to v in the sequel.

In these new variables, the phase function $\Phi_1 = \Phi_1(x, u_1, z, s_2, \delta, \lambda, l)$ is given by

$$\Phi_2(x, u_1, z, v, \delta, \lambda, l) := \Phi_1(x, u_1, z, (2^l\lambda^{-1})^{1/3}v + \tilde{G}_1(x_1, \delta, \sigma), \delta, \lambda, l).$$

This is a function of the form

$$\begin{aligned}\Phi_2 = 2^l(\tilde{B}_3((2^l\lambda^{-1})^{1/3}u_1,\ (2^l\lambda^{-1})^{1/3}v, x_1, \delta, \sigma)u_1^3 \\ -z(u_1 - H(v, x_1, (2^l\lambda^{-1})^{1/3}, \delta, \sigma)) + R(v, x, \delta, \lambda),\end{aligned}$$

where $\tilde{B}_3$, H and R are smooth and where $R(v, x, \delta, \lambda)$ is the sum of all terms not depending on u_1 and z. Moreover, $|\tilde{B}_3| \sim 1$. Note also that $u_1 \sim 1$, $|v| \sim 1$. More precisely, after this change of variables, $\nu_{l,II}^\lambda(x)$ assumes the form

$$\begin{aligned}\nu_{l,II}^\lambda(x) = \lambda^{7/6}2^{4l/3} \int e^{-is_3\Phi_2(x,u_1,z,v,\delta,\lambda,l)}\, a((2^l\lambda^{-1})^{1/3}u_1, z, v, x_1, s_3, \delta;\ \lambda) \\ \times \chi_1(u_1)\chi_1(z)\tilde{\chi}_1(v)\chi_1(s_3)\, du_1 dz dv\, ds_3,, \qquad (5.40)\end{aligned}$$

where a is again smooth and uniformly a classical symbol of order 0 with respect to λ (in order to simplify our notation, we shall here and in the sequel usually denote such symbols by the letter a, even if they may be different from one instance of occurance to another). Moreover, $\tilde{\chi}_1(v)$ is smooth and supported in a region where $|v| \sim 1$.

Observe next that the function Φ_2 has at worst a nondegenerate critical point (u_1^c, z^c) with respect to the variables (u_1, z) and that the Hessian matrix at such a point is of the form $2^l\begin{pmatrix}\alpha & -1\\ -1 & 0\end{pmatrix}$, where $|\alpha| \lesssim 1$, so that in particular the Hessian determinant is of size 2^{2l}. If there is no critical point, we can again integrate by parts and obtain estimates that are stronger than needed, so let us assume that there is a critical point. We may then apply the method of stationary phase to the integration in the variables (u_1, z). This leads to the following new expression for $\nu_{l,II}^\lambda(x)$:

$$\nu_{l,II}^\lambda = \lambda^{7/6}2^{l/3}\mu_l^\lambda, \quad \text{with} \qquad (5.41)$$

$$\mu_l^\lambda(x) := \int e^{-i\lambda s_3\Phi_3(x,v,\delta,\lambda,l)}\, a((2^l\lambda^{-1})^{1/3}, v, x_1, s_3, \delta;\ \lambda, 2^l)\tilde{\chi}_1(v)\chi_1(s_3)\, dv\, ds_3,$$

up to an error term that is of order $\lambda^{7/6}2^{-2l/3}$ and that will, therefore, be ignored (compare the discussion in Subsection 5.2.2). Here, a is again a smooth function that is uniformly a classical symbol of order 0 with respect to each of the last two variables. Moreover, the phase is given by $\Phi_3(x, v, \delta, \lambda, l) := (1/\lambda)\Phi_2(x, u_1^c, z^c, v, \delta, \lambda, l)$.

Notice that (5.41) already implies the first estimate in (5.24).

In order to compute $\Phi_3(x, v, \delta, \lambda, l)$ more explicitly, observe first that the value of a function at a critical point is invariant under changes of coordinates. Since we had switched from the coordinates (y_1, s_1), in which Φ_1 is given by the function

$$\Phi_0(x, y_1, s_1, s_2, \delta, \lambda) := \Psi(y_1, \delta, \sigma, s_1, s_2) - s_1x_1 - s_2x_2 - x_3$$

(compare (5.22)) to the coordinates (u_1, z), this means that we can also write

$$\Phi_3(x, v, \delta, \lambda, l) = \Phi_0(x, y_1^c, s_1^c, s_2, \delta, \lambda),$$

where (y_1^c, s_1^c) denotes the critical point of Φ_0 with respect to the variables (y_1, s_1). This formula turns out to be better suited, since, according to Lemma 5.1, we may write

$$\begin{aligned}\Phi_0(x, y_1, s_1, s_2, \delta, \lambda) = &\, s_1y_1 + s_2y_1^2\omega(\delta_1y_1) + \sigma y_1^n\beta(\delta_1y_1)\\ &+(\delta_0s_2)^2Y_3(\delta_1y_1, \delta_2, \delta_0s_2) - s_1x_1 - s_2x_2 - x_3.\end{aligned}$$

To this phase, we can apply the following lemma (with $\xi := y_1, \eta := s_1$, and $\zeta = x_1$), whose proof is straightforward.

Lemma 5.6. *Let $\phi = \phi(\xi, \eta)$ be a smooth, real function on $\mathbb{R}^2$ of the form*

$$\phi(\xi, \eta) = \xi\eta + f(\xi) - \eta\zeta,$$

with $\zeta \in \mathbb{R}$. Then ϕ has a unique critical point given by $(\xi^c, \eta^c) := (\zeta, -f'(\zeta))$ and $\phi(\xi^c, \eta^c) = f(\zeta)$.

This yields

$$\begin{aligned}\Phi_3(x, v, \delta, \lambda, l) = &\, s_2x_1^2\omega(\delta_1x_1) + \sigma x_1^n\beta(\delta_1x_1)\\ &+(\delta_0s_2)^2Y_3(\delta_1x_1, \delta_2, \delta_0s_2) - s_2x_2 - x_3,\end{aligned}$$

and, passing back to the coordinate v in place of s_2, we obtain

$$\begin{aligned}\Phi_3(x, v, \delta, \lambda, l) = &\, ((2^l\lambda^{-1})^{1/3}v + \tilde{G}_1(x_1, \delta, \sigma))\,x_1^2\omega(\delta_1x_1) + \sigma x_1^n\beta(\delta_1x_1)\\ &+ \delta_0^2((2^l\lambda^{-1})^{1/3}v + \tilde{G}_1(x_1, \delta, \sigma))^2\\ &\times Y_3(\delta_1x_1, \delta_2, \delta_0((2^l\lambda^{-1})^{1/3}v + \tilde{G}_1(x_1, \delta, \sigma)))\\ &- ((2^l\lambda^{-1})^{1/3}v + \tilde{G}_1(x_1, \delta, \sigma))x_2 - x_3.\end{aligned}$$

Expanding this with respect to $(2^l\lambda^{-1})^{1/3}v$, we see that Φ_3 is of the form

$$\begin{aligned}\Phi_3(x, v, \delta, \lambda, l) = &\, \tilde{B}_0(x, \delta, \sigma) + (2^l\lambda^{-1})^{1/3}\tilde{B}_1(x, \delta, \sigma)v\\ &+ \delta_0^2(2^l\lambda^{-1})^{2/3}\tilde{B}_2(x, \delta_0((2^l\lambda^{-1})^{1/3}v, \delta, \sigma)\,v^2, \qquad (5.42)\end{aligned}$$

with smooth functions $\tilde{B}_j(x, \delta, \sigma)$ and where $|\tilde{B}_2(x, \delta_0((2^l\lambda^{-1})^{\frac{1}{3}}v, \delta, \sigma)| \sim 1$. Recall also that $|v| \sim 1$, and notice that when $\delta_0 = 0$, then Φ_3 is a quadratic polynomial in v; thus, for δ_0 sufficiently small, Φ_3 is a small perturbation of this quadratic polynomial.

Observe that if $(\lambda 2^{2l}\delta_0^6)^{1/3} = \lambda\delta_0^2(2^l\lambda^{-1})^{2/3} \gg 1$, then we can apply van der Corput's lemma in v (with $M = 2$), which yields the estimate

$$|\nu_{l,II}^{\lambda}(x)| \lesssim \lambda^{7/6}2^{l/3}(\lambda\delta_0^2(2^l\lambda^{-1})^{2/3})^{1/2} = \frac{\lambda}{\delta_0}.$$

(Notice that this already verifies the second estimate in (5.24)!)

We shall therefore distinguish between the cases where $\lambda 2^{2l} \lesssim \delta_0^{-6}$ and where $\lambda 2^{2l} \gg \delta_0^{-6}$.

Observe also that $2^{7r/6}2^{l/3}2^{-k4/18} = 2^{(r+2l)/18}$. It will therefore be convenient to put $m := r + 2l$, so that $r = m - 2l$. We may then rewrite (compare (5.41))

$$\mu_{1+it}^{II}(x) = \gamma(1+it)\,\delta_0^{(1+it)/3} \sum_{M_2 \le 2^m \le \varepsilon_1\delta_0^{-18}} 2^{m(1-35it)/18}$$

$$\times \sum_{\max\{M_0,\delta_0^3 2^{m/2}\} \le 2^l \le \varepsilon_0 2^{m/3}} 2^{it(7l/2)}\,\mu_l^{2^{m-2l}}(x),$$

where $M_2 := M_1M_0^3$, $\varepsilon_1 := M_1^{-2}$ and $\varepsilon_0 := M_1^{-1/3}$. Notice also that the condition $\lambda 2^{2l} \lesssim \delta_0^{-6}$ then reads as $2^m \lesssim \delta_0^{-6}$. We shall therefore decompose

$$\mu_{1+it}^{II} = \mu_{1+it}^{II,1} + \mu_{1+it}^{II,2},$$

where

$$\mu_{1+it}^{II,1}(x) := \gamma(1+it)\,\delta_0^{(1+it)/3} \sum_{M_2 \le 2^m \le M_0^2\delta_0^{-6}} 2^{m(1-35it)/18} \sum_{M_0 \le 2^l \le \varepsilon_0 2^{m/3}} 2^{it(7l/2)}\, f_{m,x}(2^l),$$

$$\mu_{1+it}^{II,2}(x) := \gamma(1+it)\,\delta_0^{(1+it)/3} \sum_{M_0^2\delta_0^{-6} < 2^m \le \varepsilon_1\delta_0^{-18}} 2^{m(1-35it)/18} \sum_{\delta_0^3 2^{m/2} \le 2^l \le \varepsilon_0 2^{m/3}} 2^{it7l/2}\, f_{m,x}(2^l),$$

with $f_{m,x}(2^l) := \mu_l^{2^{m-2l}}(x)$. Recall from (5.41) and (5.42) that

$$f_{m,x}(2^l) = \int e^{-is_3\tilde{\Phi}_3(x,v,\delta,m,2^l)}\, a(2^{l-m/3}, v, x_1, s_3, \delta;\ 2^{m-2l}, 2^l)\tilde{\chi}_1(v)\chi_1(s_3)\, dv\, ds_3, \tag{5.43}$$

$$\tilde{\Phi}_3 := 2^{m-2l}\tilde{B}_0(x,\delta,\sigma) + 2^{-l}2^{2m/3}\tilde{B}_1(x,\delta,\sigma)v$$
$$+\delta_0^2 2^{m/3}\tilde{B}_2(x,\delta_0 2^{-m/3}2^l v,\delta,\sigma)\,v^2.$$

In several cases the summation in l will require the use of some cancellation properties. Here, Lemma 2.7 will come at hand.

5.2.4 Estimation of $\mu_{1+it}^{II,1}(x)$: The contribution by those m for which $2^m \le M_0^2\delta_0^{-6}$

For such m we have $\delta_0^2 2^{m/3} \lesssim 1$, so that the last term in (5.43) in the phase $\tilde{\Phi}_3$ can be included into the amplitude of $f_{m,x}$, and we may rewrite $f_{m,x}(2^l)$ as an oscillatory integral of the form

$$f_{m,x}(2^l) = \int e^{-is_3\Phi_4(x,v,\delta,m,2^l)}\, a\left(\delta_0^2 2^{m/3}, 2^{l-m/3}, v, x_1, s_3, \delta;\ 2^{m-2l}, 2^l\right) \times \tilde{\chi}_1(v)\chi_1(s_3)\, dv\, ds_3,$$

where

$$\Phi_4(x, v, \delta, m, 2^l) := 2^{m-2l}\tilde{B}_0(x, \delta, \sigma) + 2^{-l}2^{2m/3}\tilde{B}_1(x, \delta, \sigma)v.$$

Observe also that it will here suffice to prove that

$$\left|\gamma(1+it) \sum_{M_0 \le 2^l \le \varepsilon_0 2^{m/3}} 2^{it(7l/2)} f_{m,x}(2^l)\right| \le C, \tag{5.44}$$

with C independent of m, x, t, and so on, because this will immediately imply that $|\mu^{II,1}_{1+it}(x)| \le C'$.

Now, recall first that a is a classical symbol of order 0 with respect to both last variables, so that we may write

$$a\left(\delta_0^2 2^{m/3}, 2^{l-m/3}, v, x_1, s_3, \delta;\ 2^{m-2l}, 2^l\right) = g\left(\delta_0^2 2^{m/3}, 2^{l-m/3}, v, x_1, s_3, \delta\right) + O((2^{m-2l})^{-1} + 2^{-l}),$$

where the first term g is the leading homogeneous term of order 0 of a and, hence, a smooth function of all its variables, and the constant in the error term is independent of the other variables appearing here.

Since we are summing only over those ls for which $m - 2l \ge 0$ and $l \ge 0$, we see that the contributions by the term $O((2^{m-2l})^{-1} + 2^{-l})$ in (5.44) can be estimated in the desired way. With a slight abuse of notation, let us therefore assume from now on that

$$f_{m,x}(2^l) = \int e^{-is_3\Phi_4(x,v,\delta,m,2^l)}\, g\left(\delta_0^2 2^{m/3}, 2^{l-m/3}, v, x_1, s_3, \delta\right) \times \tilde{\chi}_1(v)\chi_1(s_3)\, dv\, ds_3. \tag{5.45}$$

Given x, consider first those l for which $|2^{-l}2^{2m/3}\tilde{B}_1(x, \delta, \sigma)| \ge 1$. Integration by parts in v then implies that

$$|f_{m,x}(2^l)| \le \frac{C}{|2^{-l}2^{2m/3}\tilde{B}_1(x, \delta, \sigma)|}.$$

Summing a geometric series, we thus see that $\sum_l |f_{m,x}(2^l)| \lesssim 1$ for the sum over these ls.

Similarly, if we consider those l for which $|2^{-l}2^{2m/3}\tilde{B}_1(x, \delta, \sigma)| < 1$ and $|2^{m-2l}\tilde{B}_0(x, \delta, \sigma)| \gg 1$, by means of an integration by parts in s_3, we find that

$$|f_{m,x}(2^l)| \le \frac{C}{|2^{m-2l}\tilde{B}_0(x, \delta, \sigma)|},$$

and, again, the resultant sum in l is uniformly bounded.

We may therefore restrict ourselves in the sequel to the set of those l for which $|2^{-l}2^{2m/3}\tilde{B}_1(x,\delta,\sigma)| \lesssim 1$ and $|2^{m-2l}\tilde{B}_0(x,\delta,\sigma)| \lesssim 1$. In this case, (5.45) shows that

$$f_{m,x}(2^l) = H\left(2^{-2l}2^m\tilde{B}_0(x,\delta,\sigma), 2^{-l}2^{2m/3}\tilde{B}_1(x,\delta,\sigma), 2^l 2^{-m/3}\right),$$

where H is a smooth function of its (bounded) variables. Indeed, H will also depend on m, x, δ, and so on, but in such a way that its C^1-norm on compact sets is uniformly bounded. This shows that the contribution of the ls that we are are considering here to the sum in (5.44) leads to a sum of the form (2.15) in Lemma 2.7, with $\alpha := \frac{7}{2}$ and where the cuboid Q is defined by the following set of restrictions for suitable $R_1, R_2 > 0$:

$$|y_1| = |2^{-2l}2^m\tilde{B}_0(x,\delta,\sigma)| \le R_1, \quad |y_2| = |2^{-l}2^{2m/3}\tilde{B}_1(x,\delta,\sigma)| \le R_1,$$
$$|y_3| = |2^l 2^{-m/3}| \le \varepsilon_0.$$

Finally, since $\gamma(1+it) = (2^{i7t/2}-1)/(2^{-2}-1)$, we see that (5.44) is an immediate consequence of Lemma 2.7.

5.2.5 Estimation of $\mu_{1+it}^{II,2}(x)$: The contribution by those m for which $2^m > M_0^2\delta_0^{-6}$

For such m we have $\delta_0^2 2^{m/3} \gg 1$.

We shall have to distinguish three further subcases. Let us first consider the situation where $2^{2m/3}2^{-l}|\tilde{B}_1(x,\delta,\sigma)| \gg \delta_0^2 2^{m/3}$ in (5.43). An integration by parts in v then shows that

$$|f_{m,x}(2^l)| \lesssim (2^{2m/3}2^{-l}|\tilde{B}_1(x,\delta,\sigma)|)^{-1}.$$

The summation over those l for which $2^{2m/3}2^{-l}|\tilde{B}_1(x,\delta,\sigma)| \gg \delta_0^2 2^{m/3}$ can therefore be estimated by a constant times $\left(\delta_0^2 2^{m/3}\right)^{-1}$, so that the contribution of the corresponding $f_{m,x}(2^l)$ to $\mu_{1+it}^{II,2}(x)$ can be estimated by

$$C\delta_0^{1/3}\sum_{2^m > M_0^2\delta_0^{-6}} 2^{m/18}\left(\delta_0^2 2^{m/3}\right)^{-1} \lesssim 1. \tag{5.46}$$

Assume next that $2^{-l}2^{2m/3}|\tilde{B}_1(x,\delta,\sigma)| \lesssim \delta_0^2 2^{m/3}$, but $2^{m-2l}|\tilde{B}_0(x,\delta,\sigma)| \gg \delta_0^2 2^{m/3}$. Then an integration by parts in s_3 shows that

$$|f_{m,x}(2^l)| \lesssim (2^{m-2l}|\tilde{B}_0(x,\delta,\sigma)|)^{-1},$$

so that we can argue in the same way is in the preceding subcase to see that the contribution of the corresponding $f_{m,x}(2^l)$ to $\mu_{1+it}^{II,2}(x)$ is again uniformly bounded with respect to t, x, δ, and σ.

We may thus assume that $2^{2m/3}2^{-l}|\tilde{B}_1(x,\delta,\sigma)| \lesssim \delta_0^2 2^{m/3}$ and 2^{m-2l} $|\tilde{B}_0(x,\delta,\sigma)| \lesssim \delta_0^2 2^{m/3}$. Then we may rewrite $f_{m,x}(2^l)$ in (5.43) as

$$f_{m,x}(2^l) = \int e^{-is_3\delta_0^2 2^{m/3}\Phi_5(x,v,\delta,m,2^l)}\, a(2^{l-m/3}, v, x_1, s_3, \delta;\, 2^{m-2l}, 2^l)$$
$$\times\, \tilde{\chi}_1(v)\chi_1(s_3)\, dv\, ds_3,$$

where

$$\Phi_5 := \tilde{B}_2(x, \delta_0 2^{l-m/3}v, \delta, \sigma)\, v^2 + \delta_0^{-2} 2^{m/3-l} \tilde{B}_1(x, \delta, \sigma)v + 2^{2m/3-2l}\delta_0^{-2}\tilde{B}_0(x, \delta, \sigma).$$

Observe also that here $|\tilde{\Phi}_5(x, v, \delta, m, 2^l)| \lesssim 1$.

Let us first consider those l for which $2^{2m/3}2^{-l}|\tilde{B}_1(x, \delta, \sigma)| \ll \delta_0^2 2^{m/3}$. Then the coefficient of Φ_5 of the linear term in v is small, so that we may change variables in the integral from v to $\Phi_5(x, v, \delta, m, v)$ (as a new variable), which then easily shows that $f_{m,x}(2^l)$ is of the form

$$f_{m,x}(2^l) = F(\delta_0^2 2^{m/3};\ 2^{2m/3-2l}\delta_0^{-2}\tilde{B}_0(x, \delta, \sigma), \delta_0^{-2}2^{m/3-l}\tilde{B}_1(x, \delta, \sigma), 2^{l-m/3}, \delta;\ 2^{m-2l}, 2^l),$$

where F is a smooth function that is a Schwartz function with respect to the first variable, whose Schwartz norms are all uniformly bounded with respect to the other variables. Moreover, F is uniformly a classical symbol of order 0 in both of the last two variables. More precisely, we may estimate

$$\left|\partial_{\mu_1}^{\alpha_1}\partial_{\mu_2}^{\alpha_2}\partial_z^{\beta} F(z_1; z_2, z_3, z_4, z_5; \mu_1, \mu_2)\right| \le C_{N,\alpha_1,\alpha_2,\beta}(1 + |z_1|)^{-N}|\mu_1|^{-\alpha_1}|\mu_2|^{-\alpha_2},$$

for every $N \in \mathbb{N}$..

This clearly implies that $|f_{m,x}(2^l)| \lesssim (\delta_0^2 2^{m/3})^{-N}$ for every $N \in \mathbb{N}$. However, such an estimate is not sufficient in order to control the summation in l.

We therefore isolate the leading homogeneous term of order 0 of F with respect to the last two variables, which gives a smooth function

$$h(\delta_0^2 2^{m/3};\ 2^{2m/3-2l}\delta_0^{-2}\tilde{B}_0(x, \delta, \sigma), \delta_0^{-2}2^{m/3-l}\tilde{B}_1(x, \delta, \sigma), 2^{l-m/3}, \delta)$$

of its variables, and the remainder terms, which clearly can be estimated by a constant times $(\delta_0^2 2^{m/3})^{-N}((2^{m-2l})^{-1} + 2^{-l})$. The second factor allows us to sum in l, and then the first factor (choosing $N = 1$) leads again to an estimate of the form (5.46) for the contribution by the remainder terms.

In order to control the main term given by the function h, we shall again apply Lemma 2.7.

Let us here define a cuboid Q by the following set of restrictions, for suitable $R_1, \varepsilon_2 > 0$:

$$|y_1| = |2^{-2l}2^{2m/3}\delta_0^{-2}\tilde{B}_0(x, \delta, \sigma)| \le R_1, \quad |y_2| = |2^{-l}\delta_0^{-2}2^{m/3}\tilde{B}_1(x, \delta, \sigma)| \le \varepsilon_2,$$

$$|y_3| = |2^l 2^{-m/3}| \le \varepsilon_0, \quad |y_4| = |2^{-l}\delta_0^3 2^{m/2}| \le 1$$

(the last condition stems for the additional summation restriction in the definition of $\mu_{1+it}^{II,2}(x)$), and let us define $H_{m,\delta}(y_1, \dots, y_4) := h(\delta_0^2 2^{m/3};\ y_1, y_2, y_3, \delta)$. Then (choosing $N = 1$),

$$\|H_{m,\delta}\|_{C^1(Q)} \le C(\delta_0^2 2^{m/3})^{-1},$$

and thus Lemma 2.7 implies that the sum over the ls in the definition of $\mu_{1+it}^{II,2}(x)$ can be estimated by $C(\delta_0^2 2^{m/3})^{-1}$, so that the remaining sum in m can again be estimated by the expression in (5.46). This concludes the discussion of this subcase.

We are thus eventually reduced to those ls for which $2^{2m/3}2^{-l}|\tilde{B}_1(x,\delta,\sigma)| \sim \delta_0^2 2^{m/3} \gg 1$ and $2^{m-2l}|\tilde{B}_0(x,\delta,\sigma)| \lesssim \delta_0^2 2^{m/3}$. Assume more precisely that we consider here pairs (m,l) for which

$$\frac{1}{A}\delta_0^2 2^{m/3} \le 2^{2m/3}2^{-l}|\tilde{B}_1(x,\delta,\sigma)| \le A\delta_0^2 2^{m/3}, \tag{5.47}$$

where $A \gg 1$ is a fixed constant. In this situation, the phase Φ_5 will have only nondegenerate critical points of size 1 as a function of v or else none. The latter case can be treated as before, so assume that we do have a critical point v^c such that $|v^c| \sim 1$. Then we may apply the method of stationary phase in v in (5.43), which leads to the following estimate for $f_{m,x}(2^l)$:

$$|f_{m,x}(2^l)| \lesssim (\delta_0^2 2^{m/3})^{-1/2}.$$

But, given m, (5.47) means that we are summing over at most $\log A^2$ different ls, and thus the contribution of those $f_{m,x}(2^l)$ that we are considering here to the sum forming $\mu_{1+it}^{II,2}(x)$ can be estimated by

$$C \log A^2\, \delta_0^{1/3} \sum_{2^m > M_0^2 \delta_0^{-6}} 2^{m/18}(\delta_0^2 2^{m/3})^{-1/2} \lesssim 1.$$

Combining this estimate with the previous ones, we see that we can bound $|\mu_{1+it}^{II,2}(x)| \le C$, with a constant C that is independent of t, x, δ and σ. This concludes the proof of the estimate (5.39), hence of (5.36) and (5.33), finally of (4.53), and consequently of Proposition 4.2(c).

5.3 PROOF OF PROPOSITION 4.2(A), (B): COMPLEX INTERPOLATION

For the proofs of parts (a) and (b) of Proposition 4.2 we shall make use of similar interpolation schemes. A crucial result for part (a) will also be Lemma 2.9 on oscillatory double sums.

5.3.1 Estimate (4.51) in Proposition 4.2(a)

Recall that $\delta_0 = 2^{-j}$, that

$$\nu_{\delta,j}^V = \sum_{\lambda_1=2^{M+j}}^{2^{2j}} \sum_{\lambda_3=(2^{-M-j}\lambda_1)^2}^{2^{-M}\lambda_1} \nu_j^{(\lambda_1,\lambda_1,\lambda_3)}$$

(in this notation, summation is always meant to be over dyadic λ_js), and that, by (4.43), $\|\widehat{\nu_j^\lambda}\|_\infty \lesssim \lambda_1^{-1/2}\lambda_3^{-1/2}$. We therefore define an analytic family of measures for ζ in the strip $\Sigma = \{\zeta \in \mathbb{C} : 0 \le \operatorname{Re}\zeta \le 1\}$ by

$$\mu_\zeta(x) := \gamma(\zeta)\,\delta_0^{\zeta/3} \sum_{k_1=M+j}^{2j} \sum_{k_3=-2M+2k_1-2j}^{-M+k_1} 2^{(3-7\zeta)k_1/6}2^{(3-7\zeta)k_3/6}\nu_j^{(2^{k_1},2^{k_1},2^{k_3})},$$

where $\gamma(\zeta)$ is an entire function that will serve a similar role as the function $\gamma(z)$ in Subsection 5.2.1. Its precise definition will be given later (based on Remark 2.10). It will again be uniformly bounded on Σ, such that $\gamma(\theta_c) = \gamma(\frac{3}{7}) = 1$.

By T_ζ we denote the operator of convolution with $\widehat{\mu_\zeta}$. Observe that for $\zeta = \theta_c = \frac{3}{7}$, we have $\mu_{\theta_c} = \delta_0^{1/7} \nu_{\delta,j}^V$; hence, $T_{\theta_c} = 2^{-j/7} T_{\delta,j}^V$, so that, again by Stein's interpolation theorem, (4.51) will follow if we can prove the following estimates on the boundaries of the strip Σ :

$$\|\widehat{\mu_{it}}\|_\infty \le C \qquad \forall t \in \mathbb{R}, \tag{5.48}$$

$$\|\mu_{1+it}\|_\infty \le C \qquad \forall t \in \mathbb{R}. \tag{5.49}$$

As before, the first estimate (5.48) is an immediate consequence of the estimates (4.43), so let us concentrate on (5.49), that is, assume that $\zeta = 1 + it$, with $t \in \mathbb{R}$. We then have to prove that there is a constant C such that

$$|\mu_{1+it}(x)| \le C, \tag{5.50}$$

where C is independent of t, x, δ, and σ.

Let us introduce the measures $\mu_{\lambda_1,\lambda_3}$ given by

$$\mu_{\lambda_1,\lambda_3}(x) := (\lambda_1\lambda_3)^{-2/3} \nu_j^{(\lambda_1,\lambda_1,\lambda_3)}(x),$$

which allow us to rewrite

$$\mu_{1+it}(x) = \gamma(1+it)\, \delta_0^{(1+it)/3} \sum_{\lambda_1 = 2^M \delta_0^{-1}}^{\delta_0^{-2}} \ \sum_{\lambda_3 = 2^{-2M}(\delta_0\lambda_1)^2}^{2^{-M}\lambda_1} (\lambda_1\lambda_3)^{-7it/6} \mu_{\lambda_1,\lambda_3}(x). \tag{5.51}$$

Notice that according to Remark 4.3,

$$\begin{aligned} \mu_{\lambda_1,\lambda_3}(x) = \lambda_1^{4/3}\lambda_3^{1/3} \int & \check{\chi}_1(\lambda_1(x_1 - y_1))\, \check{\chi}_1(\lambda_1(x_2 - \delta_0 y_2 - y_1^2 \omega(\delta_1 y_1))) \\ & \times \check{\chi}_1(\lambda_3(x_3 - b_0(y,\delta)\, y_2^2 - \sigma y_1^n \beta(\delta_1 y_1)))\, \eta(y,\delta)\, dy, \end{aligned}$$

where η is supported where $y_1 \sim 1$ and $|y_2| \lesssim 1$. Assume first that $|x| \gg 1$. Since $\check{\chi}_1$ is rapidly decreasing, after scaling in y_1 by the factor $1/\lambda_1$, we then easily see that $|\mu_{\lambda_1,\lambda_3}(x)| \le C_N \lambda_1^{1/3} \lambda_3^{-N}$ for every $N \in \mathbb{N}$. Since $2^j \lesssim \lambda_1 \le 2^{2j}$ and $(2^{-j}\lambda_1)^2 \lesssim \lambda_3 \ll \lambda_1$ in the sum defining $\mu_{1+it}(x)$, this easily implies (5.50).

From now on, we may and shall therefore assume that $|x| \lesssim 1$.

By means of the change of variables $y_1 \mapsto x_1 - y_1/\lambda_1$, $y_2 \mapsto y_2/\lambda_3^{1/2}$ and Taylor expansion around x_1 we may rewrite

$$\mu_{\lambda_1,\lambda_3}(x) = \lambda_1^{1/3}\lambda_3^{-1/6}\tilde{\mu}_{\lambda_1,\lambda_3}(x),$$

with

$$\tilde{\mu}_{\lambda_1,\lambda_3}(x) := \iint \check{\chi}_1(y_1) F_\delta(\lambda_1, \lambda_3, x, y_1, y_2)\, dy_1\, dy_2, \tag{5.52}$$

where

$$\begin{aligned} F_\delta(\lambda_1, \lambda_3, x, y_1, y_2) := & \eta(x_1 - \lambda_1^{-1} y_1, \lambda_3^{-1/2} y_2, \delta)\, \check{\chi}_1(D - E y_2 + r_1(y_1)) \\ & \times \check{\chi}_1(A - y_2^2 b_0(x_1 - \lambda_1^{-1} y_1, \lambda_3^{-1/2} y_2, \delta) + \lambda_3 \lambda_1^{-1} r_2(y_1)). \end{aligned}$$

Here, the quantities $A = A(x, \lambda_3, \delta)$, $D = D(x, \lambda_1, \delta)$ and $E = E(\lambda_1, \lambda_3, \delta)$, given by

$$A := \lambda_3 Q_A(x), \quad D =: \lambda_1 Q_D(x), \quad E := \delta_0 \lambda_1 \lambda_3^{-1/2}$$
$$\text{with} \quad Q_A(x) := x_3 - \sigma x_1^n \beta(\delta_1 x_1), \; Q_D(x) := x_2 - x_1^2 \omega(\delta_1 x_1), \quad (5.53)$$

do not depend on y_1, y_2, and $r_i(y_1) = r_i(y_1; \lambda_1^{-1}, x_1, \delta)$, $i = 1, 2$, are smooth functions of y_1 (and λ_1^{-1} and x_1) satisfying estimates of the form

$$|r_i(y_1)| \le C|y_1|, \quad \left| \left(\frac{\partial}{\partial(\lambda_1^{-1})} \right)^l r_i(y_1; \lambda_1^{-1}, x_1, \delta) \right| \le C_l |y_1|^{l+1} \quad \text{for every} \quad l \ge 1. \tag{5.54}$$

Notice that we may here assume that $|y_1| \lesssim \lambda_1$, because of our assumption $|x| \lesssim 1$ and the support properties of η. It will also be important to observe that $E = \delta_0 \lambda_1 \lambda_3^{-1/2} \le 2^{M/2}$ for the index set of λ_1, λ_3 over which we sum in (5.51).

In order to verify (5.50), given x, we shall split the sum in (5.51) into three parts, according to whether $|A(x, \lambda_3, \delta)| \gg 1$, or $|A(x, \lambda_3, \delta)| \lesssim 1$ and $|D(x, \lambda_1, \delta)| \gg 1$, or $|A(x, \lambda_3, \delta)| \lesssim 1$ and $|D(x, \lambda_1, \delta)| \lesssim 1$.

1. The part where $|A| \gg 1$. Denote by $\mu^1_{1+it}(x)$ the contribution to $\mu_{1+it}(x)$ by the terms for which $|A(x, \lambda_3, \delta)| > K$, where $K \gg 1$ is a large constant. We claim that for every $\epsilon > 0$,

$$|\tilde{\mu}_{\lambda_1,\lambda_3}(x)| \lesssim |A|^{\epsilon - 1/2}, \qquad \text{if } |A| = |A(x, \lambda_3, \delta)| > K, \tag{5.55}$$

provided K is sufficiently large. This estimate will imply the right kind of estimate,

$$|\mu^1_{1+it}(x)| \lesssim \delta_0^{1/3} \sum_{\{\lambda_3 : 1 \le \lambda_3 \le \delta_0^{-2},\, \lambda_3 |Q_A(x)| \ge K\}} \; \sum_{\lambda_1 \le \delta_0^{-1} \lambda_3^{1/2}} \frac{\lambda_1^{1/3} \lambda_3^{-1/6}}{(\lambda_3 |Q_A(x)|)^{1/2-\epsilon}}$$
$$\lesssim \sum_{\{\lambda_3 : 1 \le \lambda_3 \le \delta_0^{-2},\, \lambda_3 |Q_A(x)| \ge K\}} \frac{1}{(\lambda_3 |Q_A(x)|)^{1/2-\epsilon}} \lesssim \frac{1}{K^{1/2-\epsilon}},$$

since we are summing over dyadic λ_3s.

In order to verify (5.55), observe first that if we apply the van der Corput–type estimate in Lemma 2.1(b) (with $M = 2$) to the integration in y_2 (making use of the last factor of F_δ), we obtain

$$\int |F_\delta(\lambda_1, \lambda_3, x, y_1, y_2)| \, dy_2 \le C,$$

where the constant C is independent of y_1, x, λ, and δ (recall that $|b_0| \sim 1$!). Let $\varepsilon > 0$. It follows in particular that the contribution of the region where $|y_1| \gtrsim |A|^\varepsilon$ to $\tilde{\mu}_{\lambda_1,\lambda_3}$ can be estimated by the right-hand side of (5.55) because of the rapidly decreasing factor $\check{\chi}_1(y_1)$ in the double integral defining $\tilde{\mu}_{\lambda_1,\lambda_3}(x)$.

Let us thus consider the part of $\tilde{\mu}_{\lambda_1,\lambda_3}(x)$ given by integrating over the region where $|y_1| \le C|A|^\varepsilon$, where C is a fixed positive number. Here, according to (5.54), we have $|r_2(y_1)| \lesssim |A|^\varepsilon$, and hence $|A + \lambda_3 \lambda_1^{-1} r_2(y_1)| \sim |A|$ if we choose, for instance, $\varepsilon < \frac{1}{2}$ and K sufficiently large.

Then an easy estimation for the y_2-integration leads to

$$\int \left| \check{\chi}_1(A - y_2^2 b_0(x_1 - \lambda_1^{-1} y_1, \lambda_3^{-1/2} y_2, \delta) + \lambda_3\lambda_1^{-1} r_2(y_1)) \right| dy_2 \lesssim |A|^{-1/2},$$

and integrating subsequently in y_1 over the region $|y_1| \leq C|A|^{\varepsilon}$, we again arrive at the right-hand side of (5.55).

2. The part where $|A| \lesssim 1$ and $|D| \gg 1$. Denote by $\mu^2_{1+it}(x)$ the contribution to $\mu_{1+it}(x)$ by the terms for which $|A(x, \lambda_3, \delta)| \leq K$ and $|D(x, \lambda_1, \delta)| > K$. We claim that here

$$|\tilde{\mu}_{\lambda_1,\lambda_3}(x)| \lesssim \frac{1}{|D|}, \qquad \text{if } |D| = |D(x, \lambda_1, \delta)| > K, \tag{5.56}$$

provided K is sufficiently large. It is again easy to see that this estimate will imply the right kind of estimate for $|\mu^2_{1+it}(x)|$ (just interchange the roles of A and D and of λ_1 and λ_3 in the arguments of the previous situation).

In order to prove (5.56), consider first the contribution to $\tilde{\mu}_{\lambda_1,\lambda_3}(x)$ given by the integration over the region where $|y_1| \geq C|D|^{\varepsilon}$, where C is a fixed positive number. Arguing in the same way as in the previous situation, we find that this part can be estimated by the right-hand side of (5.56).

Next, we consider the contribution to $\tilde{\mu}_{\lambda_1,\lambda_3}(x)$ given by the integration over the region where $|y_1| < C|D|^{\varepsilon}$ and $|y_2| \gg C|D|^{\varepsilon}$. According to (5.54), we then have that $|r_j(y_1)| \lesssim |D|^{\varepsilon}$, $j = 1, 2$, so that we may assume that $|A + \lambda_3\lambda_1^{-1} r_2(y_1)| \ll |D|^{\varepsilon}$; hence

$$|A - y_2^2 b_0(x_1 - \lambda_1^{-1} y_1, \lambda_3^{-1/2} y_2, \delta) + \lambda_3\lambda_1^{-1} r_2(y_1)| \gtrsim |D|^{2\varepsilon}.$$

This easily implies that this part of $\tilde{\mu}_{\lambda_1,\lambda_3}(x)$ can also be estimated by the right-hand side of (5.56).

What remains is the contribution by the region where $|y_1| < C|D|^{\varepsilon}$ and $|y_2| < C|D|^{\varepsilon}$ (with C sufficiently large, but fixed). Since $E \lesssim 1$, we here have that $|D - Ey_2 + r_1(y_1)| \gtrsim |D|$, and again we see that we can estimate using the right-hand side of (5.56).

3. The part where $|A| \lesssim 1$ and $|D| \lesssim 1$. Finally, denote by $\mu^3_{1+it}(x)$ the contribution to $\mu_{1+it}(x)$ by those terms for which $|A(x, \lambda_3, \delta)| \leq K$ and $|D(x, \lambda_1, \delta)| \leq K$. In this case, it is easily seen from formula (5.52) and (5.54) that

$$\tilde{\mu}_{\lambda_1,\lambda_3}(x) = \tilde{J}(A, D, E, \lambda_1^{-1}, \lambda_3^{-1/2}, \lambda_3\lambda_1^{-1}),$$

where $\tilde{J}$ is a smooth function of all its bounded variables; hence

$$\delta_0^{1/3} \mu_{\lambda_1,\lambda_3}(x) = E^{1/3} J(A, D, E^{1/3}, \lambda_1^{-1}, \lambda_3^{-1/2}, \lambda_3\lambda_1^{-1}),$$

where again J is a smooth function.

Let us write $\lambda_1 = 2^{m_1}$, $\lambda_3 = 2^{m_2}$, with $m_1, m_2 \in \mathbb{N}$. In combination with (5.51) we then see that $\delta_0^{-it/3} \mu_{1+it}(x)$ can be written in the form (2.19), with $(\alpha_1, \alpha_2) := (-\frac{7}{6}, -\frac{7}{6})$ and $M_1 = \delta_0^{-2}$, $M_2 := 2^{-M}\delta_0^{-2}$. The cuboid Q is defined by the following set of restrictions:

$$|y_1| = |\lambda_3 Q_A(x)| \le K, \quad |y_2| = \lambda_1 |Q_D(x)| \le K,$$
$$|y_3| = |E^{1/3}| = \lambda_1^{1/3}\lambda_3^{-1/6}\delta_0^{1/3} \le 2^{M/3}, \quad |y_4| = \lambda_1^{-1} \le 1,$$
$$|y_5| = \lambda_3^{-1/2} \le 1, \quad |y_6| = |\lambda_1^{-1}\lambda_3| \le 2^{-M},$$
$$|y_7| = |\lambda_1^2\lambda_3^{-1}\delta_0^2| \le 2^M, \quad |y_8| = |\lambda_1^{-1}\delta_0^{-1}| \le 2^{-M}.$$

The first three conditions arise from our assumptions $|A| \lesssim 1$, $|D| \lesssim 1$, and $|E| \lesssim 1$, and the last three arise from the restrictions on the summation indices in (5.51). Moreover, for the function H in Lemma 2.9, we my choose $H(y_1, \ldots, y_8) := y_3 J(y_1, \ldots, y_6)$. The corresponding vectors (β_1^k, β_2^k) are given by $(0, 1), (1, 0), (\frac{1}{3}, -\frac{1}{6}), (-1, 0), (0, -\frac{1}{2}), (2, -1), (-1, 1)$ and $(-1, 0)$. Therefore, if we choose for $\gamma(\zeta)$ the corresponding function $\gamma_{3/7}(\zeta)$ of Remark 2.10, then Lemma 2.9 shows that indeed $\mu_{1+it}^3(x)$ also satisfies the estimate (5.50).

This concludes the proof of Proposition 4.2 (a).

5.3.2 Estimate (4.52) in Proposition 4.2(b)

Recall that $\delta_0 = 2^{-j}$, that

$$\nu_{\delta,j}^{VI} := \sum_{\lambda_1=2^{M+j}}^{2^{2j}} \sum_{\lambda_3=2}^{(2^{-M-j}\lambda_1)^2} \nu_j^{(\lambda_1,\lambda_1,\lambda_3)},$$

and that, by (4.43), $\|\widehat{\nu_j^\lambda}\|_\infty \lesssim \lambda_1^{-1/2}\lambda_3^{-1/2}$. We therefore define an analytic family of measures for ζ in the strip $\Sigma = \{\zeta \in \mathbb{C} : 0 \le \operatorname{Re}\zeta \le 1\}$ by

$$\mu_\zeta(x) := \gamma(\zeta)\,\delta_0^{\zeta/3} \sum_{k_1=M+j}^{2j} \sum_{k_3=1}^{-2M+2k_1-2j} 2^{(3-7\zeta)k_1/6} 2^{(3-7\zeta)k_3/6} \nu_j^{(2^{k_1},2^{k_1},2^{k_3})},$$

where we shall put

$$\gamma(\zeta) := \frac{2^{7(1-z)/2} - 1}{3}.$$

By T_ζ we denote the operator of convolution with $\widehat{\mu_\zeta}$. Observe that for $\zeta = \theta_c = \frac{3}{7}$, we have $\mu_{\theta_c} = \delta_0^{1/7}\nu_{\delta,j}^{VI}$, hence, $T_{\theta_c} = 2^{-j/7} T_{\delta,j}^{VI}$, so that, arguing exactly as in the preceding subsection by means of Stein's interpolation theorem, (4.52) will follow if we can prove that there is constant C such that

$$|\mu_{1+it}(x)| \le C, \tag{5.57}$$

where C is independent of t, x, δ, and σ.

As before, we introduce the measures $\mu_{\lambda_1,\lambda_3}$ given by

$$\mu_{\lambda_1,\lambda_3}(x) := (\lambda_1\lambda_3)^{-2/3}\nu_j^{(\lambda_1,\lambda_1,\lambda_3)}(x),$$

which allow us to rewrite

$$\mu_{1+it}(x) = \gamma(1+it)\,\delta_0^{(1+it)/3} \sum_{\lambda_1=2^{M+j}}^{2^{2j}} \sum_{\lambda_3=2}^{(2^{-M-j}\lambda_1)^2} (\lambda_1\lambda_3)^{-7it/6}\mu_{\lambda_1,\lambda_3}(x). \tag{5.58}$$

Recall also that according to Remark 4.3,

$$\mu_{\lambda_1,\lambda_3}(x) = \lambda_1^{4/3}\lambda_3^{1/3}\int \check{\chi}_1(\lambda_1(x_1-y_1))\,\check{\chi}_1(\lambda_1(x_2-\delta_0 y_2 - y_1^2\omega(\delta_1 y_1)))$$

$$\times\check{\chi}_1(\lambda_3(x_3-b_0(y,\delta)\,y_2^2-\sigma y_1^n\beta(\delta_1 y_1))\,\eta(y,\delta)\,dy,$$

where η is supported where $y_1\sim 1$ and $|y_2|\lesssim 1$. Assume first that $|x|\gg 1$. If $|x_1|\gg 1$ or $|x_2|\gg 1$, this easily implies that $|\mu_{\lambda_1,\lambda_3}(x)|\le C_N\lambda_1^{-N}\le(\lambda_1\lambda_2)^{-N/2}$ for every $N\in\mathbb{N}$ because $\lambda_1\gg\lambda_3^{1/2}$. Thus (5.57) follows immediately.

And, if $|x_3|\gg 1$, we may estimate the last factor in the integrand by $C_N\lambda_3^{-N}$ and then easily obtain that $|\mu_{\lambda_1,\lambda_3}(x)|\le C_N\lambda_1^{4/3}\lambda_3^{1/3-N}\,\lambda_1^{-1}(\lambda_1\delta_0)^{-1}=2^j\lambda_1^{-2/3}\lambda_3^{1/3-N}$. Summing first over all $\lambda_1\gg 2^j\lambda_3^{1/2}$ and then over λ_3, we find that $|\mu_{1+it}(x)|\lesssim\delta_0^{1/3}2^{j/3}\lesssim 1$.

From now on, we may and shall therefore assume that $|x|\lesssim 1$.

By means of the change of variables $y_1\mapsto x_1-y_1/\lambda_1$, $y_2\mapsto y_2/(\delta_0\lambda_1)$ we rewrite

$$\mu_{\lambda_1,\lambda_3}(x)=\delta_0^{-1}\lambda_1^{-2/3}\lambda_3^{1/3}\,\tilde{\mu}_{\lambda_1,\lambda_3}(x),$$

with

$$\tilde{\mu}_{\lambda_1,\lambda_3}(x):=\iint\check{\chi}_1(y_1)\,\tilde{F}_\delta(\lambda_1,\lambda_3,x,y_1,y_2)\,dy_1\,dy_2, \tag{5.59}$$

where

$$\tilde{F}_\delta(\lambda_1,\lambda_3,x,y_1,y_2):=\eta(x_1-\lambda_1^{-1}y_1,\,\delta_0^{-1}\lambda_1^{-1}y_2,\delta)\,\check{\chi}_1(D-y_2+r_1(y_1))$$

$$\times\check{\chi}_1(A+y_2^2E\,b_0(x_1-\lambda_1^{-1}y_1,\delta_0^{-1}\lambda_1^{-1}y_2,\delta)+\lambda_3\lambda_1^{-1}r_2(y_1)).$$

The quantities

$$A:=\lambda_3 Q_A(x),\qquad D:=\lambda_1 Q_D(x),\qquad E:=\frac{\lambda_3}{(\delta_0\lambda_1)^2},$$

$$\text{with}\quad Q_A(x):=x_3-\sigma x_1^n\beta(\delta_1 x_1),\qquad Q_D(x):=x_2-x_1^2\omega(\delta_1 x_1), \tag{5.60}$$

appearing here again do not depend on y_1, y_2, and the functions $r_i(y_1)$ are as before (i.e., they are indeed smooth functions of y_1,λ_1^{-1},x_1, and δ, and again satisfy estimates of the form (5.54)). Notice that here we also have that $\lambda_3/\lambda_1\ll 1$. Recall also that we may assume that $|y_1|\lesssim\lambda_1$ because of our assumption $|x|\lesssim 1$ and the support properties of η, and that $\delta_0^{-1}\lambda_1^{-1}\ll 1$. Observe finally that our summation conditions imply that $E\ll 1$.

Notice also that the first factor $\check{\chi}_1(y_1)$ in (5.59) in combination with the second factor of F_δ clearly allow for a uniform estimate

$$|\tilde{\mu}_{\lambda_1,\lambda_3}(x)|\lesssim 1;\quad\text{hence,}\quad \delta_0^{1/3}|\mu_{\lambda_1,\lambda_3}(x)|\lesssim\left(\frac{\lambda_3^{1/2}}{\delta_0\lambda_1}\right)^{2/3}.$$

However, these estimate are not quite sufficient in order to prove estimate (5.57), so we need to improve on them. The second estimate suggests to introducing new dyadic summation variables λ_0, λ_4 in place of λ_1, λ_3 so that

$$\lambda_3 = \lambda_4^2 \quad \text{and} \quad \lambda_1 = \frac{\lambda_0\lambda_4}{\delta_0}, \tag{5.61}$$

so in these new variables we would have $\delta_0^{1/3}|\mu_{\lambda_1,\lambda_3}(x)| \lesssim \lambda_0^{-2/3}$.

More precisely, recalling that $\lambda_3 = 2^{k_3}$, we decompose the summation over k_3 in (5.58) into two arithmetic progressions by writing $k_3 = 2k_4 + i$, with $i \in \{0, 1\}$ fixed for each of these progressions. Since all of these sums can be treated in essentially the same way, let us assume for simplicity that $i = 0$, so that $k_3 = 2k_4$. Letting $\lambda_4 := 2^{k_4}$ and $\lambda_0 := 2^{k_0}$ and writing $k_1 := k_0 + k_4 + j$, we indeed obtain (5.61). Replacing without loss of generality the sum over the dyadic λ_3 in (5.58) by the sum over the corresponding arithmetic progression with $i = 0$, it is also easy to check that the summation restrictions $2^{M+j} \le \lambda_1 \le 2^{2j}$ and $2 \le \lambda_3 \le (2^{-M-j}\lambda_1)^2$ are equivalent to the conditions

$$2^M \le \lambda_0 \le (2\delta_0)^{-1}, \qquad 2 \le \lambda_4 \le (\delta_0\lambda_0)^{-1}.$$

We may thus estimate in (5.58)

$$|\mu_{1+it}(x)| \le \sum_{\lambda_0=2^M}^{(2\delta_0)^{-1}} \lambda_0^{-2/3} \left|\gamma(1+it) \sum_{\lambda_4=2}^{(\delta_0\lambda_0)^{-1}} \lambda_4^{-7it/2} \tilde{\mu}_{\lambda_0\lambda_4/\delta_0, \lambda_4^2}(x)\right|.$$

For λ_0 and x fixed, we let

$$f_{\lambda_0,x}(\lambda_4) := \tilde{\mu}_{\lambda_0\lambda_4/\delta_0, \lambda_4^2}(x),$$

$$\rho_{t,\lambda_0}(x) := \gamma(1+it) \sum_{\lambda_4=2}^{(\delta_0\lambda_0)^{-1}} \lambda_4^{-7it/2} f_{\lambda_0,x}(\lambda_4).$$

The previous estimate shows that in order to verify (5.57), it will suffice to prove the following uniform estimate: there exist constants $C > 0$ and $\epsilon \ge 0$ with $\epsilon < \frac{2}{3}$, so that for all x with $|x| \lesssim 1$ and δ sufficiently small, we have

$$|\rho_{t,\lambda_0}(x)| \le C\lambda_0^{\epsilon} \quad \text{for} \quad 2^M \le \lambda_0 \le (2\delta_0)^{-1}. \tag{5.62}$$

In order to prove this, observe that by (5.59),

$$f_{\lambda_0,x}(\lambda_4) = \iint \check{\chi}_1(y_1) F_\delta(\lambda_0, \lambda_4, x, y_1, y_2)\, dy_1\, dy_2, \tag{5.63}$$

where

$$\begin{aligned} F_\delta(\lambda_0, \lambda_4, x, y_1, y_2) := {} & \eta(x_1 - \delta_0(\lambda_0\lambda_4)^{-1}y_1, (\lambda_0\lambda_4)^{-1}y_2, \delta)\, \check{\chi}_1(D - y_2 + r_1(y_1)) \\ & \times \check{\chi}_1(A + y_2^2 E\, b_0(x_1 - \delta_0(\lambda_0\lambda_4)^{-1}y_1, (\lambda_0\lambda_4)^{-1}y_2, \delta) \\ & + \delta_0\lambda_4\lambda_0^{-1} r_2(y_1)) \end{aligned}$$

and

$$A = A(x, \lambda_4, \delta) = \lambda_4^2 Q_A(x), \quad D = D(x, \lambda_0, \lambda_4, \delta) = \frac{\lambda_0\lambda_4}{\delta_0} Q_D(x),$$

$$E = E(\lambda_0) = \frac{1}{\lambda_0^2},$$

$$\text{with} \quad Q_A(x) := x_3 - \sigma x_1^n \beta(\delta_1 x_1), \quad Q_D(x) := x_2 - x_1^2\omega(\delta_1 x_1). \tag{5.64}$$

Given x and λ_0, we shall split the summation in λ_4 into three parts, according to whether $|D| \gg 1$, $|D| \lesssim 1$ and $|A| \gg 1$ or $|D| \lesssim 1$ and $|A| \lesssim 1$.

1. The part where $|D| \gg 1$. Denote by $\rho^1_{t,\lambda_0}(x)$ the contribution to $\rho_{t,\lambda_0}(x)$ by the terms for which $|D| \gg 1$.

We first consider the contribution to $f_{\lambda_0,x}(\lambda_4)$ given by integrating in (5.63) over the region where $|y_1| \gtrsim |D|^\varepsilon$ (where $\varepsilon > 0$ is assumed to be sufficiently small). Here, the rapidly decaying first factor $\check{\chi}_1(y_1)$ in (5.59) leads to an improved estimate of this contribution of the order $|D|^{-N}$ for every $N \in \mathbb{N}$, which allows us to sum over the dyadic λ_4 for which $\lambda_4|\lambda_0 Q_D(x)/\delta_0| = |D| \gg 1$, and the contribution to $\rho_{t,\lambda_0}(x)$ is of order $O(1)$, which is stronger than what is needed in (5.62).

We may therefore restrict ourselves in the sequel to the region where $|y_1| \ll |D|^\varepsilon$. Observe that, because of (5.54), this implies in particular that $|r_i(y_1)| \ll |D|^\varepsilon$, $i = 1, 2$. By looking at the second factor in F_δ, we see that the contribution by the regions where, in addition, $|y_2| < |D|/2$ or $|y_2| > 3|D|/2$ is again of the order $|D|^{-N}$ for every $N \in \mathbb{N}$, and their contributions to $\rho^1_{t,\lambda_0}(x)$ are again admissible.

What remains is the region where $|y_1| \ll |D|^\varepsilon$ and $|D|/2 \le |y_2| \le 3|D|/2$. In addition, we may assume that y_2 and D have the same sign, since otherwise we can estimate as before. Let us therefore assume, for example, that $D > 0$ and that $D/2 \le y_2 \le 3D/2$.

The change of variables $y_2 \mapsto Dy_2$ then allows us to rewrite the corresponding contribution to $f_{\lambda_0,x}(\lambda_4)$ as

$$\tilde{f}_{\lambda_0,x}(\lambda_4) := D \int_{|y_1| \ll |D|^\varepsilon} \int_{1/2 \le y_2 \le 3/2} \check{\chi}_1(y_1) \tilde{F}_\delta(\lambda_0, \lambda_4, x, y_1, y_2)\, dy_2\, dy_1,$$

where here

$$\begin{aligned}\tilde{F}_\delta(\lambda_0, \lambda_4, x, y_1, y_2) :=\ & \eta(x_1 - \delta_0(\lambda_0\lambda_4)^{-1} y_1, (\lambda_0\lambda_4)^{-1} D y_2, \delta)\\ & \times \check{\chi}_1(D - Dy_2 + r_1(y_1))\\ & \times \check{\chi}_1(A + y_2^2 E D^2\, b_0(x_1 - \delta_0(\lambda_0\lambda_4)^{-1} y_1, (\lambda_0\lambda_4)^{-1} D y_2, \delta)\\ & + \delta_0 \lambda_4 \lambda_0^{-1} r_2(y_1)).\end{aligned}$$

Recall also that $|b_0| \sim 1$, and notice that, according to Remark 4.3, $|\partial_{y_2} b_0| \lesssim \delta_0\delta_2 \ll 1$ if ϕ is of type A, and similarly $|\partial_{y_2} b_0| \lesssim \delta_0\delta_2/\delta_1 = \delta_0\delta_1^{m-1} \ll 1$ if ϕ is of type D. In combination with the localization given by η, this shows that, given y_1, we may change variables from y_2 to $z := y_2^2 E D^2\, b_0(x_1 - \delta_0(\lambda_0\lambda_4)^{-1} y_1, (\lambda_0\lambda_4)^{-1} D y_2, \delta)$ and use the last factor of $\tilde{F}_\delta$ in order to estimate the integral in y_2 (respectively, z) by $C|ED^2|^{-1}$. Subsequently, we may estimate the integration with respect to y_1 by means of the factor $\check{\chi}_1(y_1)$ and find that

$$|\tilde{f}_{\lambda_0,x}(\lambda_4)| \le C \frac{D}{|ED^2|} = C \frac{1}{|ED|}.$$

Interpolating this with the trivial estimate $|\tilde{f}_{\lambda_0,x}(\lambda_4)| \le C$ leads to

$$|\tilde{f}_{\lambda_0,x}(\lambda_4)| \le C \frac{1}{|ED|^{\epsilon/2}} = C\lambda_0^\epsilon |D|^{-\epsilon/2},$$

where we chose $\epsilon > 0$ so that $\epsilon < \frac{2}{3}$. The factor $|D|^{-\epsilon/2}$ then allows us to sum in λ_4, and we see that altogether we arrive at the estimate $|\rho^1_{t,\lambda_0}(x)| \leq C\lambda_0^\epsilon$. This completes the proof of estimate (5.62) in this first case.

2. The part where $|D| \lesssim 1$ and $|A| \gg 1$. Denote by $\rho^2_{t,\lambda_0}(x)$ the contribution to $\rho_{t,\lambda_0}(x)$ by the terms for which $|D| \lesssim 1$ and $|A| \gg 1$. Arguing in a similar way as in the previous case, only with D replaced by A, we see that we may restrict ourselves to the regions where $|y_1| \lesssim |A|^\varepsilon$ and $|y_2| \lesssim |A|^\varepsilon$ (where $\varepsilon > 0$ is any fixed, positive constant). In the remaining regions, we can gain a factor $C_N|A|^{-N}$ in the estimate of $f_{\lambda_0,x}(\lambda_4)$ in a trivial way. But, if $|y_1| \lesssim |A|^\varepsilon$ and $|y_2| \lesssim |A|^\varepsilon$ and if $\varepsilon > 0$ is sufficiently small, then

$$\left|A + y_2^2 E\, b_0(x_1 - \delta_0(\lambda_0\lambda_4)^{-1}y_1, (\lambda_0\lambda_4)^{-1}y_2, \delta) + \delta_0\lambda_4\lambda_0^{-1}r_2(y_1)\right| \gtrsim |A|,$$

and thus we obtain an estimate of the same kind, that is,

$$|f_{\lambda_0,x}(\lambda_4)| \leq C_N|A|^{-N} \qquad \text{for every } N \in \mathbb{N}.$$

Summing over all dyadic λ_4 such that $\lambda_4^2|Q_A(x)| = |A| \gg 1$, this implies $|\rho^2_{t,\lambda_0}(x)| \leq C$.

3. The part where $|D| \lesssim 1$ and $|A| \lesssim 1$. Denote by $\rho^3_{t,\lambda_0}(x)$ the contribution to $\rho_{t,\lambda_0}(x)$ by the terms for which $|D| \leq K$ and $|A| \leq K$, where $K > 0$ is a sufficiently large constant. Observe that $\rho^3_{t,\lambda_0}(x)$ can again be estimated by means of Lemma 2.7. Indeed, the cuboid Q will here be defined by means of the conditions $|D| \leq K$, $|A| \leq K$ and $w_1 := \lambda_1^{-1}\delta_0^{-1} = \lambda_4^{-1}\lambda_0^{-1} \leq 2^{-M-1}$, $w_2 := \lambda_3/\lambda_1 = \lambda_4(\delta_0/\lambda_0) \leq 2^{-2M}$, $\lambda_4(\delta_0\lambda_0) \leq 1$ (compare also the properties of the functions $r_i(y_1)$), and if we define $M := 1/(\delta_0\lambda_0)$, $\alpha := -\frac{7}{2}$ and

$$\begin{aligned} &H_{x,\delta}(A, D, E, w_1, w_2) \\ &:= \iint \check{\chi}_1(y_1)\eta(x_1 - \delta_0 w_1 y_1,\ w_2 y_2, \delta)\check{\chi}_1(D - y_2 + r_1(y_1; \delta_0 w_1, x_1, \delta)) \\ &\times \check{\chi}_1(A + y_2^2 E\, b_0(x_1 - \delta_0 w_1 y_1, w_1 y_2, \delta) + w_2 r_2(y_1; \delta_0 w_1, x_1, \delta))\, dy_1\, dy_2, \end{aligned}$$

then (5.63) shows that $f_{\lambda_0,x}(\lambda_4) = H_{x,\delta}(A, D, E, w_1, w_2)$, and $\gamma(1+it)^{-1}\rho^3_{t,\lambda_0}(x)$ is an oscillatory sum of the form (2.15) (with summation index $l := k_4$). Moreover, one easily checks that

$$\|H_{x,\delta}\|_{C^1(Q)} \leq C,$$

with a constant C that does not depend on x and δ. Applying Lemma 2.7, we therefore obtain the estimate $|\rho^3_{t,\lambda_0}(x)| \leq C$. This completes the proof of estimate (5.62) and hence also the proof of Proposition 4.2 (b).

Chapter Six

The Case When $h_{\mathrm{lin}}(\phi) \geq 2$: Preparatory Results

We now turn to the case where $h_{\mathrm{lin}}(\phi) \geq 2$. In this case, our strategy will be as follows.

We first look at every edge γ_l of the Newton polyhedron $\mathcal{N}(\phi^a)$ of ϕ (when written in the adapted coordinates y given by (1.10), (1.11)), which lies completely within the closed half-space where $t_2 \geq t_1$. As explained in Chapter 1, each such edge is associated with a unique weight κ^l. Following a basic idea from [PS97], we then decompose the domain (3.1), whose contribution we still need to control and which is of the form $|y_2| \leq \varepsilon y_1^m$ in the adapted coordinates y, into subdomains of the form

$$D_l^a := \{(y_1, y_2) \in H^+ : \varepsilon_l y_1^{a_l} < |y_2| \leq N_l y_1^{a_l}\}$$

and intermediate domains

$$E_l^a := \{(y_1, y_2) \in H^+ : N_{l+1} y_1^{a_{l+1}} < |y_2| \leq \varepsilon_l y_1^{a_l}\},$$

with $\varepsilon_l > 0$ small and $N_l > 0$ large. Note that the domain D_l^a is homogeneous with respect to the dilations associated to the weight κ^l, whereas the domains E_l^a should be viewed as "transition domains" between two different homogeneities.

These domains will cover all of the domain (3.1), with the exception of a domain D_{pr}^a, which will be associated with the principal face $\pi(\phi^a)$ in a certain sense and whose study will create the most serious problems.

In Section 6.2 we shall deal with the Fourier restriction estimates for the pieces of the hypersurface S associated with the transition domains E_l (corresponding to the domains E_l^a in our original coordinates x), by means of decompositions into bidyadic rectangles and rescaling arguments, which will eventually allow us to reduce these estimates to surfaces and corresponding measures supported over squares of dimension 1×1 (in the coordinates y). The localizations to these bidyadic rectangles, which will indeed correspond to "curved" bidyadic boxes in our original coordinates x, will again be based on Littlewood-Paley theory, applied to the variables x_1 and x_3. Here, Condition (R) will be needed once more. After all these reductions to measures supported over squares of dimension 1×1, it will turn out that the corresponding hypersurfaces have nonvanishing curvature with respect to the first variable x_1, so that we can apply Greenleaf's restriction estimate, in fact for every p such that $p' \geq 6$. Since $p_c' \geq 2d + 2 \geq 6$, this allows us to cover, in particular, the range where $p' \geq p_c'$, and after undoing the rescalings that we have performed along the way, we shall verify that the Fourier restriction estimates that we obtain in this way for the contributions by the curved bidyadic boxes do sum absolutely. Notice that here the condition $d = h_{\text{lin}}(\phi) \geq 2$ is sufficient.

Next, in Section 6.3, we shall turn to the contributions by the domains D_l (i.e., the domains in our original coordinates x that correspond the domains D_l^a in the adapted coordinates y). For a better understanding of these domains, it may be useful to notice that for analytic ϕ, the domain D_l^a will contain all the (real) roots of ϕ^a belonging to the cluster $[l]$ if we choose ε_l sufficiently small and N_l sufficiently large—this is immediate from our discussions in Chapter 1. For the study of the contributions of these domains, we shall first perform a dyadic decomposition, followed by a rescaling, by means of the dilations associated to the weight κ^l. Again, we will be able to reduce to Fourier restriction estimates for "normalized measures" supported over squares of dimension 1×1, contained in the set where $y_1 \sim 1$. By covering such a unit square by a finite number of small squares, we may even reduce it to small squares of dimension $\varepsilon' \times \varepsilon'$. Assume that a given small square of this type is centered at a point $v = (1, c_0)$. Then we shall look at the κ^l-principal part $\phi^a_{\kappa^l}$ of ϕ^a and distinguish between three cases: the case where $\partial_2 \phi^a_{\kappa^l}(v) \neq 0$ (Case 1), the case where $\partial_2 \phi^a_{\kappa^l}(v) = 0$ and $\partial_1 \phi^a_{\kappa^l}(v) \neq 0$ (Case 2), and the case where $\partial_2 \phi^a_{\kappa^l}(v) = 0$ and $\partial_1 \phi^a_{\kappa^l}(v) = 0$ (Case 3). Each of these cases will be treated in a different way.

Case 1 turns out to be the easiest one, and it can be treated in a way that does not require the condition $h_{\text{lin}}(\phi) \geq 2$. The other two cases can be dealt with in a comparatively easy way, again by means of Greenleaf's restriction theorem. The reason for this is that under the assumption that the edge γ_l lies within the closed half plane bounded from below by the bisectrix, one can show that there is a good control of the multiplicity $B - 1$ of any real root of $\partial_2 \phi^a_{\kappa^l}$ in terms of the Newton distance d (compare (6.20) and (6.23)). We note that our arguments show that in Case 2 and Case 3, we do then get the desired Fourier restriction estimates for the normalized measures whenever $p' \geq 2d + 2$.

Such a strong control of the multiplicities $B - 1$ will no longer exist for the remaining domain D_{pr} corresponding to the domain D^a_{pr} in our original coordinate x. In our study of the domain D_{pr} in Section 6.5, we shall therefore devise a stopping-time algorithm that will allow us to decompose the domain D_{pr} into further subdomains of type $D_{(l)}$ and $E_{(l)}$, where each of the domains $D_{(l)}$ will be homogeneous in suitable "modified adapted" coordinates, and the domains $E_{(l)}$ can again be viewed as "transition domains." In the analytic case, when we are looking at our original coordinates x, our algorithm basically allows us in some sense to "zoom" into small, hornlike neighborhoods of every real root belonging to the cluster of roots whose leading part is given by the principal root jet ψ and thus "resolve" any possible branching of roots, so that in the end no further branching will take place. The new transition domains $E_{(l)}$ can be treated in the same way as in Section 6.2. As for the domains $D_{(l)}$, we shall again distinguish three cases, analogous to our approach in Section 6.3. However, our algorithm will be built in such a way that we will not stop whenever we arrive at Case 3, so that in the end only the Cases 1 and 2 will have to be dealt with. Case 1 can be treated as before, and thus what remains is Case 2. Now, under the assumption that $h_{\text{lin}}(\phi) \geq 5$, it turns out that Case 2 can be dealt with by means of a fibration of the given piece of surface into a family of curves, in combination with Drury's Fourier restriction theorem for curves with nonvanishing torsion (cf. Theorem 2 in [Dru85]). This will conclude our proof of Proposition 3.3 for that case.

Regretfully, that method breaks down when $h_{\rm lin}(\phi) < 5$, so that in the subsequent two chapters we shall devise an alternative approach for dealing with Case 2.

6.1 THE FIRST DOMAIN DECOMPOSITION

We begin by recalling that we are assuming that the original coordinates x are linearly adapted but not adapted to ϕ, so that we may assume from now on that $d = h_{\rm lin} \geq 2$. Notice that this implies that $h = h(\phi) > 2$. Moreover, based on Varchenko's algorithm, we can locally find an adapted coordinate system $y_1 = x_1$, $y_2 = x_2 - \psi(x_1)$ for the function ϕ near the origin. In these coordinates, ϕ is given by $\phi^a(y) := \phi(y_1, y_2 + \psi(y_1))$ (cf. (1.10), (1.11)). Observe next that the domain (3.1) in which the function ρ_1 is supported and that we still need to control in Proposition 3.3 can also be described as the subset of $\Omega \cap \overline{H^+}$ on which

$$|x_2 - \psi(x_1)| \leq \varepsilon x_1^m.$$

We also recall from Chapter 1 that the vertices of the Newton polyhedron $\mathcal{N}(\phi^a)$ of ϕ^a are assumed to be the points (A_l, B_l), $l = 0, \dots, n$, so that the Newton polyhedron $\mathcal{N}(\phi^a)$ is the convex hull of the set $\bigcup_l ((A_l, B_l) + \mathbb{R}^2_+)$, where $A_{l-1} < A_l$ for every $l \geq 1$. Moreover, $L_l := \{(t_1, t_2) \in \mathbb{R}^2 : \kappa_1^l t_1 + \kappa_2^l t_2 = 1\}$ denotes the line passing through the points (A_{l-1}, B_{l-1}) and (A_l, B_l), and $a_l = \kappa_2^l / \kappa_1^l$. The a_l can be identified as the distinct leading exponents of the set of all roots of ϕ^a in the case where ϕ^a is analytic (cf. Remark 1.8), and the cluster of those roots whose leading exponent in their Puiseux series expansion is given by a_l is associated with the edge $\gamma_l = [(A_{l-1}, B_{l-1}), (A_l, B_l)]$ of $\mathcal{N}(\phi^a)$.

Following essentially the discussion in Section 8.2 of [IKM10], we choose the integer $l_0 \geq 1$ in such a way that

$$0 =: a_0 < \cdots < a_{l_0 - 1} \leq m < a_{l_0} < \cdots < a_l < a_{l+1} < \cdots < a_n < a_{n+1} := \infty.$$

Then the vertex (A_{l_0-1}, B_{l_0-1}) lies strictly above the bisectrix, that is, $A_{l_0-1} < B_{l_0-1}$, since the original coordinates x were assumed to be nonadapted (compare Section 3 of [IKM10]).

Following in a slightly modified way the discussion in Section 3 of [IKM10], we single out a particular edge by fixing the corresponding index $l_{\rm pr} \geq l_0$, through the distinction of the following.

Cases:

(a) If the principal face $\pi(\phi^a)$ of $\mathcal{N}(\phi^a)$ is a compact edge, we choose $l_{\rm pr}$ so that the edge $\gamma_{l_{\rm pr}} = [(A_{l_{\rm pr}-1}, B_{l_{\rm pr}-1}), (A_{l_{\rm pr}}, B_{l_{\rm pr}})]$ is the principal face $\pi(\phi^a)$ of the Newton polyhedron of ϕ^a.

(b) If $\pi(\phi^a)$ is the vertex (h, h), we choose $l_{\rm pr}$ so that $(h, h) = (A_{l_{\rm pr}-1}, B_{l_{\rm pr}-1})$. Then (h, h) is the right endpoint of the compact edge $\gamma_{l_{\rm pr}-1}$.

(c) If the principal face $\pi(\phi^a)$ is unbounded, that is, a half line given by $t_1 \geq A$ and $t_2 = h := B$, with $A < B$, then we distinguish two subcases:

(c1) If the point (A, B) is the right endpoint of a compact edge of $\mathcal{N}(\phi^a)$, then we again choose $l_{\rm pr}$ so that this edge is given by $\gamma_{l_{\rm pr}-1}$.

(c2) Otherwise, (A, B) is the only vertex of $\mathcal{N}(\phi^a)$, that is, $\mathcal{N}(\phi^a) = (A, B) + \mathbb{R}^2_+$.

We also let

$$a := \begin{cases} a_{l_{\rm pr}} & \text{in Case (a);} \\ a_{l_{\rm pr}-1} & \text{in Case (b) and Case (c1);} \\ m & \text{in Case (c2).} \end{cases} \tag{6.1}$$

In order to also prove Proposition 3.3 in the case where $h_{\rm lin}(\phi) \geq 2$, following [IKM10] and [IM11b] we shall decompose the preceding domain, in which ρ_1 is supported in cases (a)–(c1), into various subdomains, given by the intersections of the domains listed here with the neighborhood Ω of the origin: the domains

$$D_l := \{(x_1, x_2) \cap H^+ : \varepsilon_l x_1^{a_l} < |x_2 - \psi(x_1)| \leq N_l x_1^{a_l}\}, \quad l = l_0, \dots, l_{\rm pr} - 1,$$

which correspond to the κ^l-homogeneous domains

$$D_l^a := \{(y_1, y_2) \in H^+ : \varepsilon_l y_1^{a_l} < |y_2| \leq N_l y_1^{a_l}\}$$

in our adapted coordinates y, and intermediate “transition” domains

$$E_l := \{(x_1, x_2) \in H^+ : N_{l+1} x_1^{a_{l+1}} < |x_2 - \psi(x_1)| \leq \varepsilon_l x_1^{a_l}\},$$

where $l = l_0, \dots, l_{\rm pr} - 1$ in Case (a) and $l = l_0, \dots, l_{\rm pr} - 2$ in all other cases, as well as the “first” transition domain

$$E_{l_0-1} := \{(x_1, x_2) \in H^+ : N_{l_0} x_1^{a_{l_0}} < |x_2 - \psi(x_1)| \leq \varepsilon_{l_0} x_1^m\}.$$

In our adapted coordinates y, these transition domains correspond to the domains

$$E_l^a := \{(y_1, y_2) \in H^+ : N_{l+1} y_1^{a_{l+1}} < |y_2| \leq \varepsilon_l y_1^{a_l}\},$$

respectively,

$$E_{l_0-1}^a := \{(y_1, y_2) \in H^+ : N_{l_0} y_1^{a_{l_0}} < |y_2| \leq \varepsilon_{l_0} y_1^m\}.$$

Here, κ^l again denotes the weight associated to the edge γ_l of $\mathcal{N}(\phi^a)$ (cf. Chapter 1). Moreover, the $\varepsilon_l > 0$ are small and the $N_l > 0$ are large parameters, which will be fixed later. We remark that the domain E_{l_0-1} can be written like E_l, with $l = l_0 - 1$, if we replace, with some slight abuse of notation, a_{l_0-1} by m and κ_{l_0-1} by κ. We shall make use of this unified way of describing E_l in the sequel.

In our to avoid possible confusion, we note that here the superscript a will always refer to the representation of the given domain in adapted coordinates and is not related to the number a defined in (6.1).

What will remain after removing these domains is a domain of the form

$$D_{\rm pr} := \begin{cases} \{(x_1, x_2) \in H^+ : |x_2 - \psi(x_1)| \leq N x_1^a\} & \text{in Case (a);} \\ \{(x_1, x_2) \in H^+ : |x_2 - \psi(x_1)| \leq \varepsilon x_1^a\}, & \text{in all other cases,} \end{cases} \tag{6.2}$$

where N is sufficiently large and ε is sufficiently small. Notice here that if $a_{l_0} = \infty$, then we must be in Case (c1), with $a = m$, and so the domain $|x_2 - \psi(x_1)| \leq \varepsilon x_1^m$ does agree with $D_{\rm pr}$.

In Cases (c1) and (c2), we shall furthermore regard the domains

$$E_{l_{\rm pr}-1} := D_{\rm pr} = \{(x_1, x_2) \in H^+ : |x_2 - \psi(x_1)| \leq \varepsilon x_1^a\}$$

as "generalized" transition domains. Notice that in Case (c2), this domain will cover the domain in (3.1), since here $a = m$, so that the proof of Proposition 3.3 will be complete once we have handled all these transition domains in the next section. In a similar way, the discussion of Case (c1) will be complete once we have handled the domains E_l and D_l. This will eventually reduce our problem to studying the domain $D_{\rm pr}$ in Cases (a) and (b).

In order to simplify the exposition, in the sequel we shall always regard the domains that will appear as subdomains of the half plane H^+ on which $x_1 > 0$, usually without further mentioning.

6.2 RESTRICTION ESTIMATES IN THE TRANSITION DOMAINS E_l WHEN $h_{\rm lin}(\phi) \geq 2$

Following a standard approach, we would like to study the contributions of the domains E_l by means of a decomposition of the corresponding y-domains E_l^a into dyadic rectangles. These rectangles correspond to a kind of "curved boxes" in the original coordinates x, so that we cannot achieve the localization to them by means of Littlewood-Paley decompositions in the variables x_1 and x_2. However, the following lemma shows that this localization can, nevertheless, be induced by means of Littlewood-Paley decompositions in the variables x_1 and x_3.

We shall formulate this lemma for a general smooth, finite-type function Φ with $\Phi(0,0) = 0$ and $\nabla\Phi(0,0) = 0$ in place of ϕ^a, since it will be applied not only to ϕ^a. However, we shall keep the notation introduced for ϕ^a, denoting, for instance, by (A_l, B_l), $l = 0, \dots, n$, the vertices of the Newton polyhedron of Φ, by κ^l the weight associated to the edge $\gamma_l = [(A_{l-1}, B_{l-1}), (A_l, B_l)]$, and so on.

Lemma 6.1. *For $l \geq l_0$, let $[(A_{l-1}, B_{l-1}), (A_l, B_l)]$ and $[(A_l, B_l), (A_{l+1}, B_{l+1})]$ be two subsequent compact edges of $\mathcal{N}(\Phi)$, with common vertex (A_l, B_l), and associated weights κ^l and κ^{l+1}. Recall also that $a_l = \kappa_2^l/\kappa_1^l < a_{l+1} = \kappa_2^{l+1}/\kappa_1^{l+1}$. For a given $M > 0$ and $\delta > 0$ sufficiently small, consider the domain*

$$E^a := \{(y_1, y_2) : 0 < y_1 < \delta,\ 2^M y_1^{a_{l+1}} < |y_2| \leq 2^{-M} y_1^{a_l}\}.$$

Then the following holds true:

(a) There is a constant $C > 0$ such that

$$\Phi(y) = c_{A_l,B_l} y_1^{A_l} y_2^{B_l}\Big(1 + O(\delta^C + 2^{-M})\Big) \quad on\ E^a, \tag{6.3}$$

where c_{A_l,B_l} denotes the Taylor coefficient of Φ corresponding to (A_l, B_l). More precisely, $\Phi(y) = c_{A_l,B_l} y_1^{A_l} y_2^{B_l}(1 + g(y))$, where $|g^{(\beta)}(y)| \leq C_\beta(\delta^C + 2^{-M})$ $|y_1^{-\beta_1} y_2^{-\beta_2}|$ for every multi-index $\beta \in \mathbb{N}^2$.

(b) For $M, j \in \mathbb{N}$ sufficiently large, the following conditions are equivalent:

(i) $y_1 \sim 2^{-j}$, $(y_1, y_2) \in E^a$, *and* $2^{A_l j + B_l k}\Phi(y) \sim 1$;

(ii) $y_1 \sim 2^{-j}$, $y_2 \sim 2^{-k}$ *and* $a_l j + M \leq k \leq a_{l+1} j - M$.

Moreover, if we set $\phi_{j,k}(x) := 2^{A_l j + B_l k}\Phi(2^{-j}x_1, 2^{-k}x_2)$, *then under the previous conditions we have that* $\phi_{j,k}(x) = c_{A_l,B_l} x_1^{A_l} x_2^{B_l}(1 + O(2^{-Cj} + 2^{-M}))$ *on the set where* $x_1 \sim 1$, $|x_2| \sim 1$, *in the sense of the* C^∞*-topology.*

The statements in (a) and (b) remain valid also in the case where $l = l_0 - 1$.

Proof. When Φ is analytic, these results have essentially been proven in Section 8.3 of [IKM10], at least implicitly. We shall here give an elementary proof that also works for smooth functions Φ.

We begin with the case where $l > l_0$. Notice first that (b) is an immediate consequence of (a). In order to prove (a), let us denote by Φ_N the Taylor polynomial of degree N of Φ centered at the origin. Since $(\Phi - \Phi_N)(y_1, y_2) = O(|y_1|^N + |y_2|^N)$, it is easily seen that $y_1^{-A_l} y_2^{-B_l}(\Phi - \Phi_N)(y_1, y_2) = O(2^{-B_l M})$ on E^a, provided N is sufficiently large and δ is small. It therefore suffices to prove (6.3) for Φ_N in place of Φ.

If $\Phi(y_1, y_2) \sim \sum_{\alpha_1,\alpha_2=0}^\infty c_{\alpha_1,\alpha_2} y_1^{\alpha_1} y_2^{\alpha_2}$ is the Taylor series of Φ centered at the origin, then we decompose the polynomial Φ_N as $\Phi_N = P^+ + P^-$, where

$$P^+(y_1, y_2) := \sum_{\alpha_1+\alpha_2 \leq N, \alpha_2 > B_l} c_{\alpha_1,\alpha_2} y_1^{\alpha_1} y_2^{\alpha_2},$$

$$P^-(y_1, y_2) := \sum_{\alpha_1+\alpha_2 \leq N, \alpha_2 \leq B_l} c_{\alpha_1,\alpha_2} y_1^{\alpha_1} y_2^{\alpha_2}.$$

Let (α_1, α_2) be one of the multi-indices appearing in P^- and assume it is different from (A_l, B_l). Let $(y_1, y_2) \in E^a$, and assume, for notational convenience, that $y_2 > 0$. Since clearly $A_l, B_l > 0$, we have

$$\frac{y_1^{\alpha_1} y_2^{\alpha_2}}{y_1^{A_l} y_2^{B_l}} = y_1^{\alpha_1 - A_l} y_2^{\alpha_2 - B_l} \leq y_1^{\alpha_1 - A_l}\left(2^M y_1^{a_{l+1}}\right)^{\alpha_2 - B_l} = 2^{(\alpha_2 - B_l)M} y_1^{\alpha_1 + a_{l+1}\alpha_2 - (A_l + a_{l+1}B_l)}.$$

It is easy to see that $A_l + a_{l+1}B_l = A_{l+1} + a_{l+1}B_{l+1}$, so that

$$\frac{y_1^{\alpha_1} y_2^{\alpha_2}}{y_1^{A_l} y_2^{B_l}} \leq 2^{(\alpha_2 - B_l)M} y_1^{\alpha_1 + a_{l+1}\alpha_2 - (A_{l+1} + a_{l+1}B_{l+1})}. \tag{6.4}$$

But, since γ_{l+1} is an edge of $\mathcal{N}(\Phi)$, we have that $\kappa_1^{l+1}\alpha_1 + \kappa_2^{l+1}\alpha_2 \geq 1$, that is, $\alpha_1 + a_{l+1}\alpha_2 \geq (\kappa_1^{l+1})^{-1}$, whereas $A_{l+1} + a_{l+1}B_{l+1} = (\kappa_1^{l+1})^{-1}$. Thus, (6.4) implies that $y_1^{\alpha_1} y_2^{\alpha_2} \leq 2^{(\alpha_2 - B_l)M} y_1^{A_l} y_2^{B_l}$, so that $y_1^{\alpha_1} y_2^{\alpha_2} \leq 2^{-M} y_1^{A_l} y_2^{B_l}$ when $\alpha_2 < B_l$. And, when $\alpha_2 = B_l$, then (α_1, α_2) lies in the interior of $\mathcal{N}(\Phi)$, so that $\alpha_1 + a_{l+1}\alpha_2 - (A_{l+1} + a_{l+1}B_{l+1}) > 0$, hence $y_1^{\alpha_1} y_2^{\alpha_2} \leq \delta^C y_1^{A_l} y_2^{B_l}$ for some positive constant C.

The estimates of the derivatives of $g(y) = \Phi(y)/c_{A_l,B_l} y_1^{A_l} y_2^{B_l} - 1$ follow in a very similar way.

The terms in P^+ can be estimated analogously, making use here of the estimates $y_2 \leq 2^{-M} y_1^{a_l}$ and $\kappa_1^l\alpha_1 + \kappa_2^l\alpha_2 \geq 1$. This proves (a).

Finally, if $l = l_0$, exactly the same arguments work if we redefine a_{l_0-1} to be m and κ_{l_0-1} to be κ since $\kappa_2/\kappa_1 = m$. Q.E.D.

A similar result also applies to the generalized transition domains $E_{l_{\rm pr}-1}$ arising in Cases (c1) and (c2), provided we can factor the root $y_2 = 0$ to its given order, which applies in particular when Φ is real analytic (some easy examples show that it may be false otherwise). Recall that in these cases, the principal face of $\mathcal{N}(\phi^a)$ is an unbounded half line with left endpoint (A, B). More generally, we have the following result.

Lemma 6.2. *Assume that (A, B) is a vertex of $\mathcal{N}(\Phi)$ such that the unbounded horizontal half line with left endpoint (A, B) is a face of $\mathcal{N}(\Phi)$, and assume in addition that Φ factors as $\Phi(y_1, y_2) = y_2^B \Upsilon(y_1, y_2)$, with a smooth function Υ. Moreover, let $L_\kappa := \{(t_1, t_2) \in \mathbb{R}^2 : \kappa_1 t_1 + \kappa_2 t_2 = 1\}$ be a nonhorizontal supporting line for $\mathcal{N}(\Phi)$ (i.e., $\kappa_1 > 0$) passing through (A, B), and let $a := \kappa_2/\kappa_1$. We then let*

$$E^a := \{(y_1, y_2) : 0 < y_1 < \delta,\ |y_2| \leq 2^{-M} y_1^a\}.$$

Then the following hold true:

(a) There is a constant $C > 0$ such that

$$\Phi(y) = c_{A,B}\, y_1^A y_2^B \Big(1 + O(\delta^C + 2^{-M})\Big) \quad \text{on } E^a,$$

where $c_{A,B}$ denotes the Taylor coefficient of Φ corresponding to (A, B). More precisely, $\Phi(y) = c_{A,B} y_1^A y_2^B (1 + g(y))$, where $|g^{(\beta)}(y)| \leq C_\beta(\delta^C + 2^{-M})|y_1^{-\beta_1} y_2^{-\beta_2}|$ for every multi-index $\beta \in \mathbb{N}^2$.

(b) For $M, j \in \mathbb{N}$ sufficiently large, the following conditions are equivalent:

(i) $y_1 \sim 2^{-j}$, $(y_1, y_2) \in E^a$, and $2^{Aj+Bk}\Phi(y) \sim 1$;

(ii) $y_1 \sim 2^{-j}$, $y_2 \sim 2^{-k}$, and $aj + M \leq k$.

Moreover, if we set $\phi_{j,k}(x) := 2^{Aj+Bk}\Phi(2^{-j}x_1, 2^{-k}x_2)$, then under the previous conditions we have that $\phi_{j,k}(x) = c_{A,B} x_1^A x_2^B (1 + O(2^{-Cj} + 2^{-M}))$ on the set where $x_1 \sim 1$, $|x_2| \sim 1$, in the sense of the C^∞-topology.

Proof. It suffices again to prove (a).

By our assumption, $\Phi(y_1, y_2) = y_2^B \Upsilon(y_1, y_2)$, so that $\Phi(y)/y_1^A y_2^B = \Upsilon(y)/y_1^A$. Approximating Υ by its Taylor polynomial of sufficiently high degree, we again see that we may reduce to the case where Υ, hence Φ, is a polynomial. Then let (α_1, α_2) be any point different from (A, B) in its Taylor support. Since $\alpha_2 \geq B$, assuming again that $y_2 > 0$, we see that

$$\frac{y_1^{\alpha_1} y_2^{\alpha_2}}{y_1^A y_2^B} = y_1^{\alpha_1 - A} y_2^{\alpha_2 - B} \leq y_1^{\alpha_1 - A} \left(2^{-M} y_1^a\right)^{\alpha_2 - B} = 2^{-(\alpha_2 - B)M} y_1^{\alpha_1 + a\alpha_2 - (A + aB)}.$$

Moreover, clearly $\alpha_1 + a\alpha_2 \geq A + aB$, and $\alpha_1 + a\alpha_2 > A + aB$ when $\alpha_2 = B$. We can thus argue in a very similar way as in the proof of Lemma 6.1 to finish the proof. Q.E.D.

Let us now fix $l \in \{l_0 - 1, \dots, l_{\rm pr} - 1\}$, and consider the corresponding (generalized) transition domain E_l from Section 6.1, which can be written as

$$E_l = \{(x_1, x_2) : Nx_1^{a_{l+1}} < |x_2 - \psi(x_1)| \le \varepsilon x_1^{a_l}\},$$

where, with some slight abuse of notation, we have again redefined $a_{l_0-1} := m$ and let $a_{l_{\rm pr}} := \infty$ in Cases (c1) and (c2), so that $x_1^{a_{l_{\rm pr}}} := 0$, by definition.

Following [IKM10], we shall localize to the domain E_l by means of a cutoff function

$$\tau_l(x_1, x_2) := \chi_0\Big(\frac{x_2 - \psi(x_1)}{\varepsilon x_1^{a_l}}\Big)(1 - \chi_0)\Big(\frac{x_2 - \psi(x_1)}{Nx_1^{a_{l+1}}}\Big),$$

with $\varepsilon = \varepsilon_l$ and $N = N_l$ and where $\chi_0 \in C_0^\infty(\mathbb{R})$ is again supported in $[-1, 1]$ and $\chi_0 \equiv 1$ on $[-\frac{1}{2}, \frac{1}{2}]$ (actually, χ_0 may depend on l). In Case (c), when $l = l_{\rm pr} - 1$ and $a_{l_{\rm pr}} = \infty$, the second factor has to be interpreted as 1, that is,

$$\tau_{l_{\rm pr}-1}(x_1, x_2) = \chi_0\Big(\frac{x_2 - \psi(x_1)}{\varepsilon x_1^{a}}\Big).$$

Recall that ϕ is assumed to satisfy Condition (R).

Proposition 6.3. *Let $l \in \{l_0 - 1, \dots, l_{\rm pr} - 1\}$. Then, if $\varepsilon > 0$ is chosen sufficiently small and $N > 0$ sufficiently large,*

$$\Big(\int_S |\widehat{f}|^2 \, d\mu^{\tau_l}\Big)^{1/2} \le C_p \|f\|_{L^p(\mathbb{R}^3)}, \qquad f \in \mathcal{S}(\mathbb{R}^3),$$

whenever $p' \ge p_c'$.

Proof. Consider partitions of unity $\sum_j \chi_j(s) = 1$ and $\sum_k \tilde{\chi}_{j,k}(s) = 1$ on $\mathbb{R} \setminus \{0\}$ with $\chi, \tilde{\chi} \in C_0^\infty(\mathbb{R})$ supported in $[-2, -\frac{1}{2}] \cup [\frac{1}{2}, 2]$ (respectively, $[-2^{B_l}, -2^{-B_l}] \cup [2^{-B_l}, 2^{B_l}]$), where $\chi_j(s) := \chi(2^j s)$ and, for j fixed, $\tilde{\chi}_{j,k}(s) := \chi(2^{A_l j + B_l k} s)$, and let

$$\chi_{j,k}(x_1, x_2, x_3) := \chi_j(x_1)\tilde{\chi}_{j,k}(x_3) = \chi(2^j x_1)\tilde{\chi}(2^{A_l j + B_l k} x_3), \qquad j, k \in \mathbb{Z}.$$

Notice here that $B_l > B_{l+1} \ge 0$. We next let $\mu_{j,k} := \chi_{j,k}\mu^{\tau_l}$ and assume that μ has sufficiently small support near the origin. Then clearly $\mu_{j,k} = 0$ unless $j \ge j_0$, where $j_0 > 0$ is a large number, which we shall choose in a suitable way later. But then, according to Lemma 6.1, we may assume in addition that

$$a_l j + M \le k \le a_{l+1} j - M, \tag{6.5}$$

where M is a large number. Indeed, we may choose $N := 2^M$ and $\varepsilon := 2^{-M}$, and then Lemma 6.1 (b) shows that $\mu_{j,k} = 0$ for all pairs (j, k) not satisfying (6.5). Notice that this also implies that $k \ge k_0$ for some large number k_0. Observe also that the measure $\mu_{j,k}$ is supported over a "curved box" given by $x_1 \sim 2^{-j}$ and $|x_2 - \psi(x_1)| \sim 2^{-k}$. This shows that the localization that we have achieved by means of the cutoff function $\chi_{j,k}$ is very similar to the localization that we could have imposed by means of the cutoff function $\chi(2^j x_1)\chi(2^k(x_2 - \psi(x_1)))$.

Then, applying again Littlewood-Paley theory, now in the variables x_1 and x_3, and interpolating with the trivial $L^1 \to L^\infty$ estimate for the Fourier transform, we see that in order to prove Proposition 6.3, it suffices to prove uniform restriction estimates for the measures $\mu_{j,k}$ at the critical exponent, that is,

$$\int_S |\widehat{f}|^2 \, d\mu_{j,k} \leq C \|f\|^2_{L^{p_c}(\mathbb{R}^3)}, \qquad \text{when } (j,k) \text{ satisfies (6.5) and } j \geq j_0, \tag{6.6}$$

provided M and j_0 are chosen sufficiently large.

We introduce the normalized measures $\nu_{j,k}$ given by

$$\langle \nu_{j,k}, f \rangle := \int f\big(x_1,\, 2^{mj-k}x_2 + x_1^m \omega(2^{-j}x_1),\, \phi_{j,k}(x_1,x_2)\big)\, a_{j,k}(x)\, dx,$$

where

$$a_{j,k}(x) = \eta\big(2^{-j}x_1,\, 2^{-k}x_2 + \psi(2^{-j}x_1)\big)\, \chi_0\left(2^{a_l j+M-k}\frac{x_2}{x_1^{a_l}}\right)$$
$$\times (1-\chi_0)\left(2^{a_{l+1}j-M-k}\frac{x_2}{x_1^{a_{l+1}}}\right) \chi(x_1)\tilde{\chi}\left(\phi_{j,k}(x_1,x_2)\right).$$

Here, according to Lemma 6.1, the functions $\phi_{j,k}$ satisfy

$$\phi_{j,k}(x_1,x_2) = c x_1^{A_l} x_2^{B_l} + O(2^{-M}) \quad \text{in} \quad C^\infty$$

on domains where $x_1 \sim 1$, $|x_2| \sim 1$, and the amplitude $a_{j,k}$ in the preceding integral is supported in such a domain.

Observe that

$$\langle \mu_{j,k}, f \rangle = 2^{-j-k} \int f\big(2^{-j}y_1,\, 2^{-mj}y_2,\, 2^{-(A_l j+B_l k)}y_3\big)\, d\nu_{j,k}(y), \tag{6.7}$$

which follows easily by means of a change to adapted coordinates in the integral defining the measure $\mu_{j,k}$ and scaling in x_1 by the factor 2^{-j} and in x_2 by the factor 2^{-k}.

We observe that the measure $\nu_{j,k}$ is supported on the surface given by

$$S_{j,k} := \{(x_1,\, 2^{mj-k}x_2 + x_1^m\omega(2^{-j}x_1),\, \phi_{j,k}(x_1,x_2)) : x_1 \sim 1 \sim x_2\},$$

which is a small perturbation of the limiting surface

$$S_\infty := \{(x_1,\, x_1^m \omega(0),\, c x_1^{A_l} x_2^{B_l}) : x_1 \sim 1 \sim x_2\}$$

since $mj - k \leq a_l j - k \leq -M$ because of (6.5). Notice also that $|\partial(c x_1^{A_l} x_2^{B_l})/\partial x_2| \sim 1$ since $B_l \geq 1$. This shows that S_∞, and hence also $S_{j,k}$ (for j and M sufficiently large), is a smooth hypersurface with one nonvanishing principal curvature (with respect to x_1) of size ~ 1. This implies that

$$|\widehat{\nu_{j,k}}(\xi)| \leq C(1+|\xi|)^{-1/2}$$

uniformly in j and k.

Moreover, the total variations of the measures $\nu_{j,k}$ are uniformly bounded, that is, $\sup_{j,k} \|\nu_{j,k}\|_1 < \infty$.

We may thus apply again Greenleaf's Theorem 1.1 in order to prove that a uniform estimate

$$\int |\widehat{f}|^2 \, d\nu_{j,k} \leq C \|f\|^2_{L^p(\mathbb{R}^3)} \tag{6.8}$$

holds true whenever $p' \geq 6$, with a constant C that is independent of j, k. Since $p'_c \geq 2d + 2 \geq 6$, this holds in particular for $p = p_c$. Rescaling this estimate by means of (6.7), this implies that

$$\int |\widehat{f}|^2 \, d\mu_{j,k} \leq C 2^{-j-k+2[(m+1+A_l)j+B_l k]/p'_c} \|f\|^2_{L^{p_c}(\mathbb{R}^3)}. \tag{6.9}$$

But, we may write k in the form $k = \theta a_l j + (1-\theta) a_{l+1} j + \tilde{M}$ with $0 \leq \theta \leq 1$ and $|\tilde{M}| \leq M$, and then

$$-j - k + 2\frac{(m+1+A_l)j + B_l k}{p'_c} = -j\theta \left[1 + a_l - 2\frac{m+1+A_l+a_l B_l}{p'_c}\right]$$

$$-j(1-\theta)\left[1 + a_{l+1} - 2\frac{m+1+A_l+a_{l+1}B_l}{p'_c}\right] + \left(-1 + 2\frac{B_l}{p'_c}\right)\tilde{M}.$$

Recall next that by the definitions of the notion of r-height and of the critical exponent p'_c, we have $p'_c \geq 2(h_l + 1)$ whenever $l \geq l_0$. And, (1.17) shows that

$$h_l + 1 = \frac{1 + (1+m)\kappa_1^l}{|\kappa^l|} = \frac{m + 1 + 1/\kappa_1^l}{1 + a_l}. \tag{6.10}$$

Moreover, we have seen in the proof of Lemma 6.1 that $A_l + a_l B_l = 1/\kappa_1^l$, so that

$$2(h_l + 1) = 2\frac{m + 1 + A_l + a_l B_l}{1 + a_l}.$$

We thus find that $1 + a_l - 2(m + 1 + A_l + a_l B_l)/p'_c \geq 0$. Arguing in a similar way for $l+1$ in place of l, by using that $p'_c \geq 2(h_{l+1} + 1)$ and $A_l + a_{l+1} B_l = 1/\kappa_1^{l+1}$, we also see that $1 + a_{l+1} - 2(m + 1 + A_l + a_{l+1} B_l)/p'_c \geq 0$.

Consequently, the exponent on the right-hand side of the estimate (6.9) is uniformly bounded from above, which verifies the claimed estimate (6.6).

Assume next that $l = l_0 - 1$. Observe that in this case, by following Varchenko's algorithm, one observes that the left endpoint (A_{l_0-1}, B_{l_0-1}) of the edge $[(A_{l_0-1}, B_{l_0-1}), (A_{l_0}, B_{l_0})]$ of the Newton polyhedron of ϕ^a belongs also to the Newton polyhedron of ϕ and lies on the principal line $L = L_\kappa$ of $\mathcal{N}(\phi)$, whose slope is the reciprocal of $\kappa_2/\kappa_1 = m$. Thus, if we formally replace h_{l_0-1} by d in the previous argument (compare also Remark 1.9(a)), it is easily seen that the previous argument works in exactly the same way.

What remains to be considered are the generalized transition domains $E_{l_{\mathrm{pr}}-1}$ in Cases (c1) and (c2). Observe that in this case, our Condition (R) allows us to assume that $\Phi := \phi^a$ satisfies the factorization hypothesis of Lemma 6.2, possibly after modifying ψ by adding a suitable flat function.

We may therefore argue in a similar way as before, by applying Lemma 6.2 in place of Lemma 6.1, and obtain the estimate

$$\int_S |\widehat{f}|^2 \, d\mu_{j,k} \leq C 2^{-j-k+2[(m+1+A)j+Bk]/p'_c} \|f\|^2_{L^{p_c}(\mathbb{R}^3)}, \tag{6.11}$$

where here $B = h$ is the height of ϕ and where now we may assume only that

$$a_l j + M \le k.$$

Observe next that $-1 + 2B/p'_c \le 0$ since, by (1.18), we have $p'_c = 2(h^r + 1) \ge 2h = 2B$. We may thus estimate the exponent in (6.11) by

$$\begin{aligned} -j - k + 2\frac{(m+1+A)j + Bk}{p'_c} &\le -j\Big[a + 1 - 2\frac{m+1+A+aB}{p'_c}\Big] + \Big(-1 + \frac{2B}{p'_c}\Big)M \\ &\le -j\frac{a+1}{p'_c}\Big[p'_c - 2\frac{m+1+A+aB}{a+1}\Big]. \end{aligned}$$

And, in Case (c1), arguing as before we see that $2(m+1+A+aB)/(a+1) = 2(h_{l_{\rm pr}} + 1) \le p'_c$.

Finally, in Case (c2), we have $m = a$. Moreover, the point (A, B) lies on the principal line L of $\mathcal{N}(\phi)$, so that $\kappa_1 A + \kappa_2 B = 1$, that is, $A + aB = 1/\kappa_1$. This shows that

$$2\frac{m+1+A+aB}{a+1} = 2\Big(1 + \frac{1}{\kappa_1 + \kappa_2}\Big) = 2(1+d) \le p'_c.$$

We thus see that the uniform estimate (6.6) is valid also for the generalized transition domains. Q.E.D.

6.3 RESTRICTION ESTIMATES IN THE DOMAINS D_l, $l < l_{\rm pr}$, WHEN $h_{\rm lin}(\phi) \ge 2$

We shall now consider the domains D_l, $l = l_0, \dots, l_{\rm pr} - 1$, which are homogeneous in the adapted coordinates. Recall that for such l we have $a_l > m$. Again following [IKM10] we can localize to these domains by means of cutoff functions

$$\rho_l(x_1, x_2) := \chi_0\Big(\frac{x_2 - \psi(x_1)}{N x_1^{a_l}}\Big) - \chi_0\Big(\frac{x_2 - \psi(x_1)}{\varepsilon x_1^{a_l}}\Big), \qquad l = l_0, \dots, l_{\rm pr} - 1,$$

with $\varepsilon = \varepsilon_l$ and $N = N_l$ and where χ_0 is as in the previous section. Recall that such domains do appear only in Cases (a), (b) and (c1).

PROPOSITION 6.4. *Let $h_{\rm lin}(\phi) \ge 2$, and assume that $l < l_{\rm pr}$. Then, if $\varepsilon > 0$ is chosen sufficiently small and $N > 0$ sufficiently large,*

$$\Big(\int_S |\widehat{f}|^2 \, d\mu^{\rho_l}\Big)^{1/2} \le C_p \|f\|_{L^p(\mathbb{R}^3)}, \qquad f \in \mathcal{S}(\mathbb{R}^3),$$

whenever $p' \ge p'_c$.

Proof. Similarly to the proof of Proposition 3.1, we denote by $\{\delta_r\}_{r>0}$ the dilations associated to the weight κ^l, that is, $\delta_r y := (r^{\kappa_1^l} y_1, r^{\kappa_2^l} y_2)$, where by y we again denote our adapted coordinates. Recall that the κ^l-principal part $\phi^a_{\kappa^l}$ of ϕ^a is homogeneous of degree one with respect to these dilations and that we are interested in a κ^l-homogeneous (for small dilations) domain of the form $D_l^a = \{(y_1, y_2) : 0 < y_1 < \delta,\ \varepsilon y_1^{a_l} < |y_2| \le N y_1^{a_l}\}$ with respect to the y-coordinates, where $\delta > 0$ can still be chosen as small as we please.

We shall prove that, given any real number c_0 with $\varepsilon \le |c_0| \le N$, there is some $\varepsilon' > 0$ such that the desired restriction estimate holds true on the domain $D(c_0)$ in x-coordinates corresponding to the homogeneous domain

$$D^a(c_0) := \{(y_1, y_2) : 0 < y_1 < \delta,\ |y_2 - c_0 y_1^{a_l}| \le \varepsilon' y_1^{a_l}\}$$

in y-coordinates. Since we can cover D_l^a by a finite number of such narrow domains, this will imply Proposition 6.4.

We can essentially localize to a domain $D(c_0)$ by means of a cutoff function

$$\rho_{(c_0)}(x_1, x_2) := \chi_0\Big(\frac{x_2 - \psi(x_1) - c_0 x_1^{a_l}}{\varepsilon' x_1^{a_l}}\Big).$$

Let us again fix a suitable smooth cutoff function $\chi \ge 0$ on $\mathbb{R}^2$ supported in an annulus $\mathcal{A} := \{x \in \mathbb{R}^2 : \frac{1}{2} \le |y| \le R\}$ such that the functions $\chi_k^a := \chi \circ \delta_{2^k}$ form a partition of unity. In the original coordinates x, these correspond to the functions $\chi_k(x) := \chi_k^a(x_1, x_2 - \psi(x_1))$. We then decompose the measure $\mu^{\rho_{(c_0)}}$ dyadically as

$$\mu^{\rho_{(c_0)}} = \sum_{k \ge k_0} \mu_k, \tag{6.12}$$

where $\mu_k := \mu^{\chi_k \rho_{(c_0)}}$. Notice that by choosing the support of η sufficiently small, we can choose $k_0 \in \mathbb{N}$ as large as we need. It is also important to observe that this decomposition can essentially be achieved by means of a dyadic decomposition with respect to the variable x_1, which again allows us to apply Littlewood-Paley theory!

Moreover, changing to adapted coordinates in the integral defining μ_k and scaling by $\delta_{2^{-k}}$, we find that

$$\langle \mu_k, f\rangle = 2^{-k|\kappa^l|} \int f\big(2^{-\kappa_1^l k} x_1,\ 2^{-\kappa_2^l k} x_2 + 2^{-m\kappa_1^l k} x_1^m \omega(2^{-\kappa_1^l k} x_1),\ 2^{-k}\phi_k(x)\big)$$
$$\times\ \eta(\delta_{2^{-k}} x)\chi(x)\ \chi_0\Big(\frac{x_2 - c_0 x_1^{a_l}}{\varepsilon' x_1^{a_l}}\Big)\, dx,$$

where

$$\phi_k(x) := 2^k \phi^a(\delta_{2^{-k}} x) = \phi_{\kappa^l}^a(x) + \text{error terms of order } O(2^{-\delta k}) \tag{6.13}$$

with respect to the C^∞ topology (and $\delta > 0$).

We consider the corresponding normalized measure ν_k given by

$$\langle \nu_k, f\rangle := \int f\Big(x_1,\ 2^{(m\kappa_1^l - \kappa_2^l)k} x_2 + x_1^m \omega(2^{-\kappa_1^l k} x_1),\ \phi_k(x)\Big)\ \tilde{\eta}(x)\, dx,$$

with amplitude $\tilde{\eta}(x) := \eta(\delta_{2^{-k}} x)\chi(x)\chi_0((x_2 - c_0 x_1^{a_l})/(\varepsilon' x_1^{a_l}))$.

Observe that the support of the integrand is contained in the thin neighborhood

$$U(v) := \mathcal{A} \cap \{(x_1, x_2) : |x_2 - c_0 x_1^{a_l}| \le 2\varepsilon' x_1^{a_l}\}$$

of $v = v(c_0) := (1, c_0)$ and that the measure ν_k is supported on the hypersurface

$$S_k := \{g_k(x_1, x_2) := \big(x_1,\ 2^{(m\kappa_1^l - \kappa_2^l)k} x_2 + x_1^m \omega(2^{-\kappa_1^l k} x_1),\ \phi_k(x_1, x_2)\big) : (x_1, x_2) \in U(v)\},$$

which, for k sufficiently large, is a small perturbation of the limiting variety

$$S_\infty := \{g_\infty(x_1, x_2) := \big(x_1,\ \omega(0)x_1^m,\ \phi^a_{\kappa^l}(x)\big) : (x_1, x_2) \in U(v)\}$$

since $m\kappa_1^l - \kappa_2^l < a_l\kappa_1^l - \kappa_2^l = 0$ and since ϕ^k tends to $\phi^a_{\kappa^l}$ because of (6.13). The corresponding limiting measure will be denoted by ν_∞.

By Littlewood-Paley theory (applied to the variable x_1) and interpolation, in order to prove the desired restriction estimates for the measure $\mu^{\rho(c_0)}$, it suffices again to prove uniform restriction estimates for the measures μ_k, that is,

$$\Big(\int |\widehat{f}|^2\, d\mu_k\Big)^{1/2} \leq C\, \|f\|_{L^{p_c}}, \tag{6.14}$$

with a constant C not depending on $k \geq k_0$. We shall obtain these by first proving restriction estimates for the measures ν_k.

Indeed, we shall prove that for ε' sufficiently small, the estimate

$$\Big(\int |\widehat{f}|^2\, d\nu_k\Big)^{1/2} \leq C\|f\|_{L^{p_c}} \tag{6.15}$$

holds true, with a constant C which does not depend on k. Then, after rescaling, estimate (6.15) implies the following estimate for μ_k:

$$\Big(\int |\widehat{f}|^2\, d\mu_k\Big)^{1/2} \leq C\, 2^{-k\left(|\kappa^l|/2 - (\kappa_1^l(1+m)+1)/p_c'\right)}\, \|f\|_{L^{p_c}}. \tag{6.16}$$

But, by (1.17) (respectively, (6.10)) we have that

$$\frac{|\kappa^l|}{2} - \frac{\kappa_1^l(1+m)+1}{p_c'} = \frac{|\kappa^l|}{2}\Big(1 - \frac{2(h_l+1)}{p_c'}\Big),$$

where, by definition, $p_c' \geq 2(h_l + 1)$. This shows that the exponent on the right-hand side of (6.16) is less than or equal to zero, which verifies (6.14).

We turn to the proof of (6.15). Recall that $v = (1, c_0)$. Depending on the behavior of $\phi^a_{\kappa^l}$ near v, we shall distinguish among three cases.

Case 1. $\boldsymbol{\partial_2\phi^a_{\kappa^l}(v) \neq 0}$. This assumption implies that we may use $y_2 := \phi^a_{\kappa^l}(x_1, x_2)$ in place of x_2 as a new coordinate for S_∞ (which thus is a hypersurface, too) and then also for S_k, in place of x_2, provided ε' is chosen small enough and k, sufficiently large. Since $x_1 \sim 1$ on $U(v)$, this then shows that S_k is a hypersurface with one nonvanishing principal curvature. Therefore, we can again apply Theorem 1.1 and obtain that for $p' \geq 6$ and k sufficiently large, the estimate

$$\Big(\int |\widehat{f}|^2\, d\nu_k\Big)^{1/2} \leq C_p\|f\|_{L^p} \tag{6.17}$$

holds true, with a constant C_p that does not depend on k. This applies in particular to $p = p_c$ since $p_c' \geq 2d + 2 \geq 6$, and we obtain (6.15).

Assume next that $\partial_2\phi^a_{\kappa^l}(v) = 0$. Then $v = (1, c_0)$ is a real root of $\partial_2\phi^a_{\kappa^l}$ of multiplicity, say, $B - 1 \geq 1$, so that a Taylor expansion with respect to x_2 around c_0 and homogeneity show that

$$\partial_2\phi^a_{\kappa^l}(x_1, x_2) = (x_2 - c_0 x_1^{a_l})^{B-1}\tilde{Q}(x_1, x_2),$$

where $\tilde{Q}$ is a κ^l-homogenous fractionally smooth function (polynomial in x_2) such that $\tilde{Q}(v) \neq 0$. Integrating in x_2 and again making use of the κ^l-homogeneity of $\phi^a_{\kappa^l}$, we find that

$$\phi^a_{\kappa^l}(x_1, x_2) = (x_2 - c_0 x_1^{a_l})^B x_2^D Q(x_1, x_2) + c_1 x_1^{1/\kappa_1^l}, \tag{6.18}$$

where $B \geq 2$, $D \in \mathbb{N}$, and $c_1 \in \mathbb{R}$ and where Q is again a κ^l-homogenous fractionally smooth function satisfying $Q(1, c_0) \neq 0$ and $Q(1, 0) \neq 0$ (recall that $c_0 \neq 0$).

On the other hand, since $l < l_{\text{pr}}$, the edge $\gamma_l = [(A_{l-1}, B_{l-1}), (A_l, B_l)]$ lies above the bisectrix (but notice that if $l = l_{\text{pr}} - 1$, then in Case (b) its right endpoint will lie on the bisectrix). If we then factor

$$\phi^a_{\kappa^l}(x_1, x_2) = x_2^{B_l} Q_0(x_1, x_2),$$

where $Q_0(x_1, x_2)$ is a κ^l-homogenous fractionally smooth function such that $Q_0(1, 0) \neq 0$, then this implies that

$$B_l \geq h > d,$$

where again $d = d(\phi)$. Moreover, since $a_l > m \geq 2$, so that the edge γ_l is less steep than the line L (which intersects the bisectrix at (d, d)), we have $1/\kappa_2 > 1/\kappa_2^l$.

We claim that this implies that

$$\deg_{x_2} Q_0 < \frac{d}{2}, \tag{6.19}$$

where $\deg_{x_2}$ denotes the degree with respect to the variable x_2. Indeed, since $\phi^a_{\kappa^l}$ is κ^l-homogenous of degree one, we have $(B_l + \deg_{x_2} Q_0)\kappa_2^l \leq 1$; hence

$$\deg_{x_2} Q_0 \leq \frac{1}{\kappa_2^l} - B_l < \frac{1}{\kappa_2^l} - d < \frac{1}{\kappa_2} - \frac{1}{\kappa_2 + \kappa_1} = \frac{\kappa_1}{(\kappa_2 + \kappa_1)\kappa_2} = \frac{d}{m} \leq \frac{d}{2}.$$

We then need to distinguish two cases, which behave differently.

Case 2. $\partial_2\phi^a_{\kappa^l}(v) = 0$ and $\partial_1\phi^a_{\kappa^l}(v) \neq 0$. Then clearly $c_1 \neq 0$, and this implies that $D = 0$, for otherwise the point $(1/\kappa_1^l, 0)$ would belong to the edge γ_l, which would contradict our assumption that $l < l_{\text{pr}}$. Observe also that

$$\partial_2\phi^a_{\kappa^l}(x) = x_2^{B_l - 1}\left(B_l Q_0(x_1, x_2) + x_2\partial_2 Q_0(x_1, x_2)\right),$$

where the degree with respect to x_2 of the polynomial in parentheses is bounded by $\deg_{x_2} Q_0$. This obviously implies that $B - 1 \leq \deg_{x_2} Q_0 < d/2$; hence

$$B < \frac{d}{2} + 1. \tag{6.20}$$

Observe that this implies that $d > 2$ since $B \geq 2$.

Let us perform a bidyadic frequency domain decomposition by putting

$$\widehat{\nu_k^{\lambda_0,\lambda_3}}(\xi_1,\xi_2,\xi_3) := \tilde{\chi}_1\left(\frac{(\xi_1,\xi_2)}{\lambda_0}\right)\chi_1\left(\frac{\xi_3}{\lambda_3}\right)\widehat{\nu_k}(\xi_1,\xi_2,\xi_3),$$

where $\tilde{\chi}_1$ and χ_1 are smooth bump functions on $\mathbb{R}^2$, respectively, $\mathbb{R}$, supported where $|(\xi_1,\xi_2)| \sim 1$, respectively, $|\xi_3| \sim 1$, and where $\lambda_0 \geq 1$ and $\lambda_3 \geq 1$ are dyadic numbers. We claim that the measures $\nu_k^{\lambda_0,\lambda_3}$ satisfy the following estimates, uniformly in $k \geq k_0$, provided k_0 is sufficiently large and ε' sufficiently small:

$$\|\widehat{\nu_k^{\lambda_0,\lambda_3}}\|_\infty \leq C \begin{cases} \lambda_0^{-1/3}\lambda_3^{-1/B}, & \text{if } \lambda_0 \lesssim \lambda_3, \\ \lambda_0^{-1/2}\lambda_3^{-1/B}, & \text{if } \lambda_0 \gg \lambda_3; \end{cases} \tag{6.21}$$

$$\|\nu_k^{\lambda_0,\lambda_3}\|_\infty \leq C\lambda_0\lambda_3^{1-1/B}. \tag{6.22}$$

Indeed,

$$\widehat{\nu_k^{\lambda_0,\lambda_3}}(\xi) = \tilde{\chi}_1\left(\frac{(\xi_1,\xi_2)}{\lambda_0}\right)\chi_1\left(\frac{\xi_3}{\lambda_3}\right)\int e^{-i[\xi_1x_1+\xi_2(2^{(m\kappa_1^l-\kappa_2^l)k}x_2+x_1^m\omega(2^{-\kappa_1^l k}x_1))+\xi_3\phi_k(x)]} \\ \times\tilde{\eta}(x)\,dx,$$

which, in the limit as $k \to \infty$, simplifies as

$$\widehat{\nu_\infty^{\lambda_0,\lambda_3}}(\xi) = \tilde{\chi}_1\Big(\frac{(\xi_1,\xi_2)}{\lambda_0}\Big)\chi_1\Big(\frac{\xi_3}{\lambda_3}\Big)\int e^{-i[\xi_1x_1+\xi_2\omega(0)x_1^m+\xi_3\phi^a_{\kappa^l}(x)]}\,\tilde{\eta}(x)\,dx.$$

Now, if $|\xi_3| \gg |(\xi_1,\xi_2)|$, then we may first apply the method of stationary phase to the integration in x_1, and, subsequently, the van der Corput–type Lemma 2.1 to the integration in x_2 (with $M = B$) and obtain that $|\widehat{\nu_\infty^{\lambda_0,\lambda_3}}(\xi)| \lesssim |\xi_3|^{-1/2-1/B}$ (compare (6.18)). Next, if $|\xi_3| \sim |(\xi_1,\xi_2)|$, then the worst case will arise when $|\xi_1| \sim |\xi_2| \sim |\xi_3|$ (in all other cases we may still apply the method of stationary phase to the integration in x_1 and obtain better estimates). But, if $|\xi_1| \sim |\xi_2| \sim |\xi_3|$, then we may argue in a similar way as in Chapter 5, making use of a suitable analogue of Lemma 5.2 and the "Airy" Lemma 2.2 (with $B = 3$), in order to first control the integration in x_1 and subsequently apply the van der Corput–type Lemma 2.1 to the integration in x_2 (with $M = B$) in order to see that $|\widehat{\nu_\infty^{\lambda_0,\lambda_3}}(\xi)| \lesssim |\xi|^{-1/3-1/B}$. Finally, if $|\xi_3| \ll |(\xi_1,\xi_2)|$, then we may either apply the method of stationary phase to the x_1-integration (in case that there is a critical point) or integrate by parts in x_1 and subsequently apply again the van der Corput–type Lemma 2.1 to the integration in x_2 (with $M = B$) in order to obtain $|\widehat{\nu_\infty^{\lambda_0,\lambda_3}}(\xi)| \lesssim |(\xi_1,\xi_2)|^{-1/2}|\xi_3|^{-1/B}$. These arguments are stable under small perturbations of the phase, and we thus obtain (6.21).

In order to verify (6.22), observe that

$$\nu_\infty^{\lambda_0,\lambda_3}(x_1,x_2,x_3) = \lambda_0^2\lambda_3\int(\mathcal{F}^{-1}\tilde{\chi}_1)\big(\lambda_0(x_1-y_1),\lambda_0(x_2-\omega(0)y_1^m)\big) \\ \times(\mathcal{F}^{-1}\chi_1)\big(\lambda_3(x_3-\phi^a_{\kappa^l}(y_1,y_2))\big)\,\tilde{\eta}(y)\,dy_1\,dy_2,$$

where $\mathcal{F}$ denotes the Fourier transform, and thus we may estimate

$$|\nu_\infty^{\lambda_0,\lambda_3}(x_1,x_2,x_3)| \le \lambda_0^2\lambda_3 \int \rho(\lambda_0 x_1 - \lambda_0 y_1)\, \rho\big(\lambda_3 x_3 - \lambda_3 \phi_{\kappa^l}^a(y_1,y_2)\big) \\ \times |\tilde{\eta}|(y_1,y_2)\, dy_1\, dy_2,$$

where ρ and η_1 are suitable, nonnegative Schwartz functions and η_1 localizes again to $U(v)$.

However, since $|\partial_2^B \phi_{\kappa^l}^a(y_1,y_2))| \simeq 1$ on the domain of integration, we may first apply the van der Corput–type estimate of Theorem 2.1 (b) to the integration in y_2 and obtain the estimate

$$\int \rho(\lambda_0 x_1 - \lambda_0 y_1)\, \rho\big(\lambda_3 x_3 - \lambda_3\phi_{\kappa^l}^a(y_1,y_2)\big)\, |\tilde{\eta}|(y_1,y_2)\, dy_1 dy_2 \le C\lambda_3^{-1/B} \\ \times \int \rho(\lambda_0 x_1 - \lambda_0 y_1)\, dy_1,$$

with a constant C that does not depend on λ_3. The subsequent integration in y_1 then allows to gain an additional factor of order $O(\lambda_0^{-1})$. Observing that our argument is again stable under small perturbations, we thus obtain (6.22).

Assume first that $\lambda_0 \lesssim \lambda_3$. Interpolating the estimates from (6.21) and (6.22), we then find that if $\theta = 2/p'$; then

$$\|T_k^{\lambda_0,\lambda_3}\|_{p\to p'} \lesssim \lambda_0^{(4\theta-1)/3}\lambda_3^{\theta-1/B}.$$

In the case where $\theta \le \frac{1}{4}$, we obtain

$$\sum_{\{\lambda_0\ge 1:\lambda_0\lesssim\lambda_3\}} \|T_k^{\lambda_0,\lambda_3}\|_{p\to p'} \lesssim (1+\log_2\lambda_3)\lambda_3^{\theta-1/B}.$$

But, if $p' \ge p'_c$, then $p' \ge 2d+2$; hence $\theta \le 1/(d+1)$. Moreover, by (6.20), we have $d+1 > 2B-1$, so that

$$\theta - \frac{1}{B} < \frac{1}{2B-1} - \frac{1}{B} = \frac{1-B}{B(2B-1)} < 0$$

since $B \ge 2$. Thus we can also sum in λ_3, that is,

$$\sum_{\lambda_0\ge 1,\lambda_3\ge 1} \|T_k^{\lambda_0,\lambda_3}\|_{p\to p'} \le C_p$$

with a constant C_p that does not depend on k for k sufficiently large. This proves, estimate (6.17), in fact, even for $p' \ge 2d+2$. Notice here that the frequency regions where $|(\xi_1,\xi_2)| \lesssim 1$ or $|\xi_3| \lesssim 1$, which we have ignored so far, can be treated in a similar way as the contributions given by $\lambda_0 \sim 1$ or $\lambda_3 \sim 1$.

And, if $\theta > \frac{1}{4}$, then we obtain

$$\sum_{\{\lambda_0\ge 1:\lambda_0\lesssim\lambda_3\}} \|T_k^{\lambda_0,\lambda_3}\|_{p\to p'} \lesssim \lambda_3^{(4\theta-1)/3}\lambda_3^{\theta-1/B} = \lambda_3^{7\theta/3-1/B-1/3}.$$

But,

$$\frac{7}{3}\theta - \frac{1}{B} - \frac{1}{3} < \frac{7}{3}\frac{1}{2B-1} - \frac{1}{B} - \frac{1}{3} = -\frac{2}{3}\frac{B^2 - B - 3/2}{B(2B-1)} < 0$$

since $B \ge 2$, and again we can sum in λ_3.

Assume next that $\lambda_0 \gg \lambda_3$. Interpolating the estimates from (6.21) and (6.22), we now find that if $\theta = 2/p'$, then

$$\|T_k^{\lambda_0,\lambda_3}\|_{p\to p'} \lesssim \lambda_0^{(3\theta-1)/2}\lambda_3^{\theta-1/B}.$$

Since $\theta \le 1/(d+1) < \frac{1}{3}$, we see that

$$\sum_{\{\lambda_0:\lambda_0\gg\lambda_3\}} \|T_k^{\lambda_0,\lambda_3}\|_{p\to p'} \lesssim \lambda_3^{(3\theta-1)/2}\lambda_3^{\theta-1/B} = \lambda_3^{5\theta/2-1/B-1/2}.$$

But,

$$\frac{5}{2}\theta - \frac{1}{B} - \frac{1}{2} < \frac{5}{2}\frac{1}{2B-1} - \frac{1}{B} - \frac{1}{2} = -\frac{B^2 - B - 1}{B(2B-1)} < 0$$

since $B \ge 2$, and again we can sum in λ_3.

We have thus shown that (6.17) holds true in Case 2 whenever $p' \ge 2d+1$, which proves, in particular, (6.15).

Case 3. $\partial_2\phi^a_{\kappa^l}(v) = 0$ and $\partial_1\phi^a_{\kappa^l}(v) = 0$. In this case, we have $c_1 = 0$, which implies that $D = B_l$, and thus $Q_0(x_1, x_2) = (x_2 - c_0 x_1^{a_l})^B Q(x_1, x_2)$. By (6.19) this clearly implies that

$$B < \frac{d}{2}. \tag{6.23}$$

We can now apply similar arguments as in Case 2, but it will suffice to perform a coarser dyadic frequency domain decomposition by putting

$$\widehat{\nu_k^\lambda}(\xi) := \chi_1\left(\frac{\xi}{\lambda}\right)\widehat{\nu_k}(\xi),$$

where χ_1 is a smooth bump function supported where $|\xi| \sim 1$ and where $\lambda \ge 1$ is again a dyadic number. We claim that in this case we have

$$\|\widehat{\nu_k^\lambda}\|_\infty \le C\lambda^{-1/B}, \tag{6.24}$$

$$\|\nu_k^\lambda\|_\infty \le C\lambda^{2-1/B}. \tag{6.25}$$

Indeed, the latter estimate follows by essentially the same arguments as in Case 2 if we put $\lambda_0 = \lambda_3 = \lambda$. Moreover, if $|\xi_3| \ge c|(\xi_1, \xi_2)|$, then an application of van der Corput's lemma to the integration in x_2 yields $|\widehat{\nu_\infty^\lambda}(\xi)| \lesssim |\xi_3|^{-1/B}$ (compare (6.18)), and if $|\xi_3| \ll |(\xi_1, \xi_2)|$, then we may apply van der Corput's lemma to the x_1-integration and obtain $|\widehat{\nu_\infty^\lambda}(\xi)| \lesssim |(\xi_1, \xi_2)|^{-1/2}$. Since $B \ge 2$ and because van der Corput's estimates are stable under small perturbations, we thus obtain (6.24).

Interpolating the estimates (6.24) and (6.25), it is again easily seen that we can sum the corresponding estimates over all dyadic $\lambda \geq 1$ and obtain the L^p-L^2 restriction estimate

$$\Big(\int |\widehat{f}|^2 \, d\nu_k\Big)^{1/2} \leq C_p \|f\|_{L^p}$$

uniformly in k for k sufficiently large, provided $p' > 4B$.

But, by (6.23) we have $4B < 2d < 2d + 2 \leq p'_c$, and thus we have also verified (6.15) in this case. Q.E.D.

In combination with Proposition 6.3, we immediately obtain the following.

Corollary 6.5. *The restriction estimate in Proposition 3.3 holds true in Case (c), that is, when the principal face of the Newton polyhedron of ϕ^a is unbounded.*

The following remark will already indicate the approach that we shall use in the next section in order to treat the contribution by the remaining domain $D_{\rm pr}$ when $h_{\rm lin} \geq 5$.

Remark 6.6. *When $h_{\rm lin} \geq 5$, then Case 2, where $\partial_2 \phi^a_{\kappa^l}(v) = 0$, and $\partial_1 \phi^a_{\kappa^l}(v) \neq 0$, could in fact be handled more easily by means of Drury's Fourier restriction theorem for curves with nonvanishing torsion (cf. Theorem 2 in* [Dru85]*). This approach will allow us to treat the analogous case also for the remaining domain $D_{\rm pr}$, provided $h_{\rm lin} \geq 5$, since it does not require the condition $B < d/2 + 1$, which may not hold true in $D_{\rm pr}$.*

Indeed, recall that in Case 2 we have $c_1 \neq 0$ in (6.18). Moreover,

$$2 \leq m < a_l = \frac{\kappa_2^l}{\kappa_1^l} < \frac{1}{\kappa_1^l}, \tag{6.26}$$

since $\kappa_1^l A_{l-1} + \kappa_2^l B_{l-1} = 1$ with $B_{l-1} \geq h > 1$, so that $\kappa_2^l < 1$. Observe next that the mapping $F : (x_1, c) \mapsto (x_1, c x_1^{a_l})$ provides local smooth coordinates (x_1, c) near $v = (1, c_0)$, since the Jacobian J_F of F at the point $(1, c_0)$ is given by $J_F(1, c_0) = 1$. We may, therefore, fiber the variety S_∞ into the family of curves

$$\gamma_c(x_1) := g_\infty(F(x_1, c)) = (x_1, \omega(0) x_1^m, \phi^a_{\kappa^l}(F(x_1, c)), \qquad c \in V(c_0),$$

where $V(c_0)$ is a sufficiently small neighborhood of c_0, provided ε' is chosen sufficiently small. But, (6.26) implies that the curve $\gamma_{c_0}(x_1) = (x_1, \omega(0) x_1^m, c_1 x_1^{1/\kappa_1^l})$ has nonvanishing torsion near v_1, since $v_1 \neq 0$, and so the same is true for the curves γ_c when c is sufficiently close to c_0.

If, in a similar way, we fiber the surface S_k into the family of curves

$$\gamma_c^k(x_1) := g_k(F(x_1, c)), \qquad c \in V(c_0),$$

then for k sufficiently large and $V(c_0)$ sufficiently small, these curves will have nonvanishing torsion uniformly bounded from above and below, and the measure ν_k will decompose into the direct integral

$$\langle \nu_k, f \rangle = \iint f(\gamma_c^k(x_1)) \, \tilde{\eta}(x_1, c) \, dx_1 \, dc = \int_{V(c_0)} \int_{W(v_1)} f \, d\Gamma_c \, dc,$$

where $\tilde{\eta}$ is a smooth function with compact support in $W(v_1) \times V(c_0)$ and $W(v_1)$ is a sufficiently small neighborhood of v_1. Here, $d\Gamma_c$ is a measure which has a smooth density with respect to the arc length measure on the curve γ_c^k.

We may thus apply Drury's Fourier restriction theorem for curves with nonvanishing torsion (cf. Theorem 2 in [Dru85]) to the measures $d\Gamma_c$ and obtain

$$\Big(\int_{W(v_1)} |\hat{f}|^2 \, d\Gamma_c\Big)^{1/2} \leq C_p \|f\|_{L^p(\mathbb{R}^3)},$$

provided $p' > 7$ and $2 \leq p'/6$, that is, if $p' \geq 12$. The constant C_p will then be independent of c provided the neighborhoods $V(c_0)$ and $W(v_1)$ are sufficiently small and k is sufficiently large. But, if $h_{\rm lin} \geq 5$, then we do have $p'_c \geq 2(h_{\rm lin} + 1) \geq 12$, so that we also obtain estimate (6.15) in this way.

6.4 RESTRICTION ESTIMATES IN THE DOMAIN $D_{\rm pr}$ WHEN $h_{\rm lin}(\phi) \geq 5$

What remains to be studied is the piece of the surface S corresponding to the domain $D_{\rm pr}$, in Cases (a) and (b), that is,

$$D_{\rm pr} := \begin{cases} \{(x_1, x_2) : |x_2 - \psi(x_1)| \leq N x_1^a\}, & \text{in Case (a)}, \\ \{(x_1, x_2) : |x_2 - \psi(x_1)| \leq \varepsilon x_1^a\}, & \text{in Case (b)}, \end{cases}$$

where $N > 0$ can be any given number in Case (a) and where ε may be assumed to be sufficiently small in Case (b) (cf. (6.2)). Recall that this domain is associated to the edge $\gamma_{l_{\rm pr}}$ of $\mathcal{N}(\phi^a)$, whose (modulus of) slope is given by $1/a$ (cf. (6.1)). Our goal will to prove the following.

Proposition 6.7. *Assume that $h_{\rm lin}(\phi) \geq 5$ and that we are in Case (a) or (b). Then for any given $N > 0$ in Case (a), respectively, every sufficiently small ε in Case (b), we have*

$$\Big(\int_{D_{\rm pr}} |\widehat{f}|^2 \, d\mu^{\rho_{l_{\rm pr}}}\Big)^{1/2} \leq C_p \|f\|_{L^p(\mathbb{R}^3)}, \qquad f \in \mathcal{S}(\mathbb{R}^3),$$

whenever $p' \geq p'_c$.

Within the domain $D_{\rm pr}$, the upper bound $B < d/2$ for the multiplicity B of real roots that we used in our discussion of Case 3 in the previous section, as well as the condition $B < d/2 + 1$ that we used in our discussion of Case 2, will in general no longer be satisfied. Even the weaker condition $B < h^r(\phi)/2$, which would still suffice for the previous argument in Case 3, may fail, as the following examples shows.

Examples 6.8. (a) The first example,

$$\phi(x_1, x_2) := (x_2 - x_1^2 - x_1^3)(x_2 - x_1^2 - x_1^4)^3,$$

deals with Case 3. Here, $\phi_{\rm pr}(x_1, x_2) = (x_2 - x_1^2)^4$, the multiplicity of the root x_1^2 satisfies $4 > d(\phi) = \frac{8}{3}$, so that the coordinates (x_1, x_2) are not adapted to ϕ.

Adapted coordinates are given by $y_1 := x_1$, $y_2 := x_2 - x_1^2$, and in these coordinates ϕ is given by

$$\phi^a(y_1, y_2) = (y_2 - y_1^3)(y_2 - y_1^4)^3.$$

$\mathcal{N}(\phi^a)$ has three vertices, $(A_0, B_0) := (0, 4)$, $(A_1, B_1) := (3, 3)$, and $(A_2, B_2) := (0, 15)$, with corresponding edges $\gamma_1 := [(0, 4), (3, 3)]$ and $\gamma_2 := [(3, 3), (0, 15)]$ and associated weights $\kappa^1 := (\frac{1}{12}, \frac{1}{4})$ and $\kappa^2 := (\frac{1}{15}, \frac{4}{15})$. Moreover, one easily computes by means of (1.17) that $h_1 = \frac{11}{4}$ and $h_2 = \frac{13}{5}$. We thus see that $h^r(\phi) = h_1 = \frac{11}{4}$. The multiplicity of the root $r_1(y_1) := y_1^3$ associated to the first edge γ_1 lying above the bisectrix is $1 < \frac{8}{3}/2$ and thus satisfies the condition (6.23), whereas the root $r_2(y_1) := y_1^4$ of multiplicity $B = 3$ associated with the edge γ_2 below the bisectrix does not even satisfy $B < h^r(\phi)$, since $3 > \frac{11}{4}$.

(b) The second example,

$$\phi(x_1, x_2) := (x_2 - x_1^2 - x_1^3)(x_2 - x_1^2 - x_1^4)^3 + x_1^{20},$$

dealing with Case 2, is a small modification of the previous one. Here, arguments similar to those in the first example show that

$$\phi^a(y_1, y_2) = (y_2 - y_1^3)(y_2 - y_1^4)^3 + y_1^{20}.$$

Again $d = d(\phi) = \frac{8}{3}$, $\mathcal{N}(\phi^a)$ has the same edges as before, and $h^r = h^r(\phi) = h_1 = \frac{11}{4}$. The multiplicity of the root $r_2(y_1) = y_1^4$ of $\partial_2\phi^a_{\kappa^{l_{\mathrm{pr}}}}$ associated to the edge γ_2 lying below the bisectrix is given by $B - 1 = 2$, whereas $d/2 + 1 = \frac{7}{3} < 3 = B$.

Following our approach from the previous section, we shall consider small, $\kappa^{l_{\mathrm{pr}}}$-homogeneous neighborhoods of a given point $v = (1, c_0)$ and distinguish between the following three cases:

Case 1. $\partial_2\phi^a_{\kappa^{l_{\mathrm{pr}}}}(v) \neq 0$.

Case 2. $\partial_2\phi^a_{\kappa^{l_{\mathrm{pr}}}}(v) = 0$ and $\partial_1\phi^a_{\kappa^{l_{\mathrm{pr}}}}(v) \neq 0$.

Case 3. $\nabla\phi^a_{\kappa^{l_{\mathrm{pr}}}}(v) = 0$.

In Case 1 we can argue as in the corresponding case in Section 6.3, since our arguments in that case did not make use of the assumption $l > l_{\mathrm{pr}}$.

Observe next that if we are in Case 3, then by homogeneity we shall also have $\phi^a_{\kappa^{l_{\mathrm{pr}}}}(v) = 0$, so that v is a root of $\phi^a_{\kappa^{l_{\mathrm{pr}}}}$ of order two or higher. Notice that in the analytic setting, multiple roots of $\phi^a_{\kappa^{l_{\mathrm{pr}}}}$ will correspond to possibly branching roots of ϕ^a. In order to overcome the problem of an insufficient control of the multiplicity of this root that we saw in Example 6.8(a), we shall perform a suitable change of coordinates, which will allow for a localization to small, horn-shaped neighborhoods of branches of these roots corresponding to subclusters of roots. A similar idea is underlying Varchenko's algorithm for the construction of adapted coordinates. Then we shall again distinguish between the corresponding three cases arising in the new coordinates and, if, necessary, iterate this procedure until no further branching of roots will take place. The details of this algorithm, which will allow us to deal with the problems related to Case 3, will be outlined in the next subsection.

Finally, the problems related to Case 2 will be handled in this chapter in an easy way by means of Drury's Fourier restriction theorem for curves with nonvanishing torsion, as indicated in Remark 6.6. This will work when $h_{\mathrm{lin}} \geq 5$, but in order to handle the general case where $h_{\mathrm{lin}} \geq 2$, substantially more refined methods will be required. These will be developed in the next chapters.

The study of the domain D_{pr} will thus require finer decompositions into further transition and homogeneous domains (with respect to further weights). These will be devised by means of an iteration scheme, somewhat resembling Varchenko's algorithm. If ϕ is analytic, then the effect of any further step of the stopping time algorithm that we shall devise can be interpreted as a localization from a given domain containing one cluster of roots to a finite number of smaller, horn-shaped subdomains, each of them containing either exactly one subcluster of the given cluster of roots of ϕ ("homogeneous domains") or none ("transition domains; compare Remark 1.8).

We also note that Varchenko's algorithm shows that the principal root jet ψ is actually a polynomial

$$\psi(x_1) = cx_1^m + \cdots + c_{\mathrm{pr}} x_1^a$$

of degree $\leq a = a_{l_{\mathrm{pr}}}$ in Cases (a) and (b) (cf. [IM11a]). Notice that in Case (a), we will have $c_{\mathrm{pr}} = 0$, and the same holds true in Case (b) if a is not an integer.

6.5 REFINED DOMAIN DECOMPOSITION OF D_{pr}: THE STOPPING-TIME ALGORITHM

6.5.1 First step of the algorithm

Let us begin with Case (a), where $D_{\mathrm{pr}} \cap \Omega = \{(x_1, x_2) : 0 < x_1 < \delta, |x_2 - \psi(x_1)| \leq N x_1^a\}$, with a possibly large constant $N > 0$. We then put $D_{(1)} := D_{\mathrm{pr}} \cap \Omega$, $\phi^{(1)} := \phi^a$, $\psi^{(1)} := \psi$ and $a_{(1)} := a, \kappa^{(1)} := \kappa^{l_{\mathrm{pr}}}$, so that $D_{(1)}$ can be rewritten as

$$D_{(1)} = \{(x_1, x_2) : 0 < x_1 < \delta, |x_2 - \psi^{(1)}(x_1)| \leq N x_1^{a_{(1)}}\}.$$

As in the discussion of the domains D_l in the previous chapter, we can cover the domain $D_{(1)}$ by finitely many narrow domains of the form

$$D_{(1)}(c_0) := \{(x_1, x_2) : 0 < x_1 < \delta, |x_2 - \psi(x_1) - c_0 x_1^{a_{(1)}}| \leq \varepsilon x_1^{a_{(1)}}\},$$

where $\varepsilon > 0$ can be chosen as small as we need and where $0 \leq |c_0| \leq N$. Fix any of these domains, and again put $v := (1, c_0)$.

We distinguish again between the cases where $\partial_2 \phi^{(1)}_{\kappa^{(1)}}(v) \neq 0$ (Case 1), $\partial_2 \phi^{(1)}_{\kappa^{(1)}}(v) = 0$ and $\partial_1 \phi^{(1)}_{\kappa^{(1)}}(v) \neq 0$ (Case 2), and $\nabla \phi^{(1)}_{\kappa^{(1)}}(v) = 0$ (Case 3).

Now, in Case 1, we can argue as in the corresponding case in Section 6.3, since our arguments in that case did not make use of the condition $l > l_{\mathrm{pr}}$.

In Case 2, the argument given in Section 6.3 may fail, since it made use of the estimate $B < d/2$, which here no longer may hold true. However, as explained in Remark 6.6, if $h_{\rm lin} \geq 5$, we may use the alternative argument based on Drury's restriction estimate for curves in this case.

If Case 3 does not appear for any choice of c_0, then we stop our algorithm and are done.

Otherwise, assume Case 3 applies to c_0, so that $c_0 x_1^{a_{(1)}}$ is a root of $\phi^{(1)}_{\kappa^{(1)}}$, say of multiplicity $M_1 \geq 2$. In this case, we define new coordinates y by putting

$$y_1 := x_1 \qquad \text{and} \qquad y_2 := x_2 - \psi^{(2)}(x_1), \tag{6.27}$$

where

$$\psi^{(2)}(x_1) := \psi(x_1) + c_0 x_1^{a_{(1)}}.$$

We denote by $x = s_{(2)}(y)$ the corresponding change of coordinates, which in general is a fractional shear only, since the exponent $a_{(1)} = a$ may be noninteger (but rational). In these coordinates (y_1, y_2), ϕ is given by $\phi^{(2)} := \phi \circ s_{(2)}$, and the domain $D_{(1)}(c_0)$ becomes the domain

$$D'^{a}_{(1)} := \{(y_1, y_2) : 0 < y_1 < \delta, \ |y_2| \leq \varepsilon y_1^{a_{(1)}}\},$$

which is still $\kappa^{(1)}$ homogeneous.

Let us see to which extent the Newton polyhedra of $\phi^{(1)}$ and $\phi^{(2)}$ will differ.

Claim 1. The Newton polyhedra of $\phi^{(1)}$ and $\phi^{(2)}$ agree in the region above the bisectrix. In particular, the line $\Delta^{(m)}$ intersects the boundary of the augmented Newton polyhedron $\mathcal{N}^r(\phi^{(1)}) = \mathcal{N}^r(\phi^a)$ at the same point as the augmented Newton polyhedron $\mathcal{N}^r(\phi^{(2)})$ of $\phi^{(2)}$, so that we can use the modified "adapted" coordinates (6.27) in place of our earlier adapted coordinates to compute the r-height of ϕ.

To see this, observe that $\phi^{(2)}(x_1, x_2) = \phi^{(1)}(x_1, x_2 + c_0 x_1^{a_{(1)}})$, where the exponent $a_{(1)}$ is just the reciprocal of the (modulus of the) slope of the line containing the principal face of the Newton polyhedron of $\phi^{(1)} = \phi^a$. This implies that the edges of $\mathcal{N}(\phi^{(1)})$ and $\mathcal{N}(\phi^{(2)})$ that lie strictly above the bisectrix are the same (compare corresponding discussions in [IM11a] or in Chapter 9). Moreover, if $\gamma_{(1)} = [(A_{(0)}, B_{(0)}), (A_{(1)}, B_{(1)})] = [(A_{l_{\rm pr}-1}, B_{l_{\rm pr}-1}), (A_{l_{\rm pr}}, B_{l_{\rm pr}})]$ is the principal face of $\mathcal{N}(\phi^{(1)})$, then it is easy to see that the principal face of $\mathcal{N}(\phi^{(2)})$ is given by the edge $\gamma'_{(1)} := [(A_{(0)}, B_{(0)}), (A'_{(1)}, B'_{(1)})]$, where

$$A'_{(1)} := A_{(1)} + a_{(1)}(B_{(1)} - M_1), \qquad B'_{(1)} = M_1$$

(write $\phi^{(1)}_{\kappa^{(1)}}$ in the normal form (1.15) and use that $c_0 x_1^{a_{(1)}}$ is a root of multiplicity M_1 of $\phi^{(1)}_{\kappa^{(1)}}$). Observe also that $M_1 \leq h$, because ϕ^a is in adapted coordinates. We thus see that the right endpoint of $\gamma'_{(1)}$ still lies on or below the bisectrix. This proves the claim.

Our considerations show that it suffices to study the contributions of narrow domains of the form

$$D'_{(1)} = \{(x_1, x_2) : 0 < x_1 < \delta, |x_2 - \psi^{(2)}(x_1)| \leq \varepsilon x_1^{a_{(1)}}\}$$

in place of $D_{(1)}$ (these actually depend on the choice of the real root of $\phi^{(1)}_{\kappa^{(1)}}$—this corresponds to a "fine splitting" of roots of ϕ in the case where ϕ is analytic).

Case A. $\mathcal{N}(\phi^{(2)}) \subset \{(t_1, t_2) : t_2 \geq B'_{(1)} = M_1\}$. In this case, we again stop our algorithm.

Case B. $\mathcal{N}(\phi^{(2)})$ contains a point below the horizontal line given by $t_2{=}B'_{(1)}{=}M_1$.

Then $\mathcal{N}(\phi^{(2)})$ will contain a further compact edge

$$\gamma_{(2)} = [(A'_{(1)}, B'_{(1)}), (A_{(2)}, B_{(2)})],$$

so that $(A'_{(1)}, B'_{(1)})$ is a vertex at which the edges $\gamma'_{(1)}$ and $\gamma_{(2)}$ meet. Determine the weight $\kappa^{(2)}$ by requiring that $\gamma_{(2)}$ lies on the line

$$\kappa_1^{(2)} t_1 + \kappa_2^{(2)} t_2 = 1,$$

and put $a_{(2)} := \kappa_2^{(2)}/\kappa_1^{(2)}$. Then clearly $a_{(1)} < a_{(2)}$.

Next, we decompose the domain $D'_{(1)}$ into the domains

$$E_{(1)} := \{(x_1, x_2) : 0 < x_1 < \delta,\ N x_1^{a_{(2)}} < |x_2 - \psi^{(2)}(x_1)| \leq \varepsilon x_1^{a_{(1)}}\}$$

and

$$D_{(2)} := \{(x_1, x_2) : 0 < x_1 < \delta,\ |x_2 - \psi^{(2)}(x_1)| \leq N x_1^{a_{(2)}}\},$$

where $N > 0$ will be a sufficiently large constant.

The contributions by the transition domain $E_{(1)}$ can be estimated in exactly the same way as we did for the domains E_l in Section 6.2. Indeed, notice that our arguments for the domains E_l did apply to any $l \geq l_0$ as long as $B_l \geq 1$, so that this statement is immediate when $c_0 = 0$, where the coordinates y in (6.27) do agree with our original adapted coordinates. When $c_0 \neq 0$, there are two minor twists in the arguments needed: first, observe that Lemma 6.1 remains valid for $\Phi = \phi^{(2)}$ and the domain

$$E^a_{(1)} := \{(y_1, y_2) : 0 < y_1 < \delta,\ 2^M y_1^{a_{(2)}} < |y_2| \leq 2^{-M} y_1^{a_{(1)}}\}$$

corresponding to the domain $E_{(1)}$ in the coordinates (6.27) when $\varepsilon = 2^{-M}$ and $N = 2^M$. The fact that $a_{(2)}$ may be noninteger, but rational, say $a_{(2)} = p/q$, with $p, q \in \mathbb{N}$, requires only minor changes of the proof: just consider the Taylor expansion of the smooth function $\Phi(y_1^q, y_2)$. Second, if we define, in analogy with h_l in (1.17), the corresponding quantity associated to the edges $\gamma'_{(1)}$ and $\gamma_{(2)}$ of $\mathcal{N}(\phi^{(2)})$ by

$$h_{(1)} := \frac{1 + m\kappa_1^{(1)} - \kappa_2^{(1)}}{\kappa_1^{(1)} + \kappa_2^{(1)}} = h_{l_{\text{pr}}} \quad \text{and} \quad h_{(2)} := \frac{1 + m\kappa_1^{(2)} - \kappa_2^{(2)}}{\kappa_1^{(2)} + \kappa_2^{(2)}},$$

then Claim 1 shows that $\max\{h_{(1)}, h_{(2)}\} \leq h^r(\phi)$, which replaces the condition $\max\{h_l, h_{l+1}\} \leq h^r(\phi)$ that was needed in the proof of Proposition 6.3.

6.5.2 Further steps of the algorithm

We are thus left with the domains $D_{(2)}$, which formally look exactly like $D_{(1)}$, only with $\psi^{(1)}$ replaced by $\psi^{(2)}$ and $a_{(1)}$ replaced by $a_{(2)}$. This allows us to iterate our first step of the algorithm that led from $D_{(1)}$ to $D_{(2)}$, producing in this way nested sequences of domains,

$$D_{\text{pr}} = D_{(1)} \supset D_{(2)} \supset \cdots \supset D_{(l)} \supset D_{(l+1)} \supset \cdots,$$

of the form

$$D_{(l)} := \{(x_1, x_2) : 0 < x_1 < \delta, \ |x_2 - \psi^{(l)}(x_1)| \le N x_1^{a_{(l)}}\},$$

where $N = N_l$ may be large and where the functions $\psi^{(l)}$ are of the form

$$\psi^{(l)}(x_1) = \psi(x_1) + \sum_{j=1}^{l-1} c_{j-1} x_1^{a_{(j)}},$$

with real coefficients c_j, and where the exponents $a_{(j)}$ form a strictly increasing sequence

$$a = a_{(1)} < a_{(2)} < \cdots < a_{(l)} < a_{(l+1)} < \cdots$$

of rational numbers.

Moreover, each of the domains $D_{(l)}$ will be covered by a finite number of narrow domains of the form

$$D_{(l)}(c_l) := \{(x_1, x_2) : 0 < x_1 < \delta, \ |x_2 - \psi^{(l)}(x_1) - c_l x_1^{a_{(l)}}| \le \varepsilon x_1^{a_{(l)}}\},$$

with $|c_l| \le N_l$, for every given $\varepsilon = \varepsilon_l > 0$. Notice that these domains can be rewritten as domains $D'_{(l)}$ of the form

$$D'_{(l)} = \{(x_1, x_2) : 0 < x_1 < \delta, |x_2 - \psi^{(l+1)}(x_1)| \le \varepsilon x_1^{a_{(l)}}\}; \qquad (6.28)$$

here $\varepsilon > 0$ can be chosen as small as we please. These, in turn, will decompose as

$$D'_{(l)} = E_{(l)} \cup D_{(l+1)},$$

where $E_{(l)}$ is a transition domain of the form

$$E_{(l)} := \{(x_1, x_2) : 0 < x_1 < \delta, \ N x_1^{a_{(l+1)}} < |x_2 - \psi^{(l+1)}(x_1)| \le \varepsilon x_1^{a_{(l)}}\}.$$

Putting

$$\phi^{(l)}(x_1, x_2) := \phi(x_1, x_2 + \psi^{(l)}(x_1)),$$

one finds that the Newton polyhedron $\mathcal{N}(\phi^{(l+1)})$ agrees with that one of $\phi^a = \phi^{(1)}$ in the region above the bisectrix, and it will have subsequent "edges"

$$\begin{aligned}\gamma'_{(1)} &= [(A_{(0)}, B_{(0)}), (A'_{(1)}, B'_{(1)})], \gamma'_{(2)} = [(A'_{(1)}, B'_{(1)}), (A'_{(2)}, B'_{(2)})], \ldots, \\ \gamma'_{(l)} &= [(A'_{(l-1)}, B'_{(l-1)}), (A'_{(l)}, B'_{(l)})], \gamma_{(l+1)} = [(A'_{(l)}, B'_{(l)}), (A_{(l+1)}, B_{(l+1)})],\end{aligned}$$

crossing or lying below the bisectrix, at least (possibly more). In fact, it is possible that some of these "edges" degenerate and become a single point (we then shall still speak of an edge, with a slight abuse of notation). The edge with index l will lie on a line

$$L^{(l)} := \{(t_1, t_2) \in \mathbb{R}^2 : \kappa_1^{(l)} t_1 + \kappa_2^{(l)} t_2 = 1\},$$

where $a_{(l)} = \kappa_2^{(l)}/\kappa_1^{(l)}$. Moreover, $c_{l-1}x_1^{a_{(l)}}$ is any real root of the $\kappa^{(l)}$-homogeneous polynomial $\phi^{(l)}_{\kappa^{(l)}}$, of multiplicity $M_l \geq 2$. Notice that when ϕ is real analytic, then this just means that $\psi^{(l)}$ is a leading term of a root of ϕ belonging to the cluster of roots defined by ψ. Our algorithm thus follows any possible "fine splitting" of the roots belonging to this cluster, and the domains $D_{(l)}$, and so on, depend on the branches of these roots that we choose along the way.

By our construction, we see that $M_l = B'_{(l)}$, which shows that the sequence of multiplicities is decreasing, that is,

$$M_1 \geq M_2 \geq \cdots \geq M_l \geq M_{l+1} \geq \cdots . \tag{6.29}$$

Observe also that the transition domains $E_{(l)}$ can be handled by the same reasoning that we had applied to $E_{(1)}$.

When will our algorithm stop? Clearly, this will happen at step l, when $\phi^{(l)}_{\kappa^{(l)}}$ has no real root, so that only Case 1 and Case 2 will arise at this step. In that case, we do obtain the desired Fourier restriction estimate for the piece of surface corresponding to $D_{(l)}$, just by the same reasoning that we applied in Section 6.3. Otherwise, we shall also stop our algorithm in step l when

$$\mathcal{N}(\phi^{(l+1)}) \subset \{(t_1, t_2) : t_2 \geq B'_{(l)} = M_l\}.$$

In this situation, the domain that still needs to be understood is the domain $D'_{(l)}$ given by (6.28).

Notice that in this case, by Condition (R) there exists a function $\tilde{\psi}^{(l+1)} \sim \psi^{(l+1)}$ such that ϕ can be factored as

$$\phi(x_1, x_2) = (x_2 - \tilde{\psi}^{(l+1)}(x_1))^{M_l}\tilde{\phi}(x_1, x_2), \tag{6.30}$$

with a fractionally smooth function $\tilde{\phi}$. This means that Lemma 6.2 (respectively, its immediate extension to fractionally smooth functions) applies to the function $\Phi(y_1, y_2) := \phi(y_1, y_2 + \tilde{\psi}^{(l+1)}(y_1))$, and since the domain $D'_{(l)}$ can be regarded as a generalized transition domain, like the domains $E_{l_{\rm pr}-1}$ that appeared when the principal face of $\mathcal{N}(\phi^a)$ was an unbounded horizontal edge, we can argue in the same way as we did for the domains $E_{l_{\rm pr}-1}$ in Section 6.2 to derive the required restriction estimates for the piece of S corresponding to $D'_{(l)}$.

There is, finally, the possibility that our algorithm does not terminate. In this case, (6.29) shows that the sequence of integers M_l will eventually become constant. We then choose L minimal so that $M_l = M_L$ for all $l \geq L$. Note that, by our construction, $M_L \geq 2$. For every $l \geq L+1$, the point $(A, B) := (A_{(L)}, B_{(L)}) = (A_L, M_L)$ will be a vertex of $\mathcal{N}(\phi^{(l)})$ that is contained in the line $L^{(l)}$, whose slope $1/a_{(l)}$ tends to zero as $l \to \infty$, and $\mathcal{N}(\phi^{(l)})$ is contained in the half-plane bounded by $L^{(l)}$ from below.

Notice also that there is a fixed rational number $1/q$, with q an integer, such that every $a_{(l)}$ is a multiple of $1/q$. This can be proven in the same way as the corresponding statement in [IKM10] on page. 240.

We can thus apply a classical theorem of E. Borel [Bo95] (see also [H90], Theorem 1.2.6) in a way similar to [IM11a] in order to show that there is a smooth function h of x_1 whose Taylor series expansion is given by the formal series

$$h(x_1) \sim \psi(x_1^q) + \sum_{j=1}^{\infty} c_{j-1} x_1^{q a_{(j)}}.$$

If we put $\psi^{(\infty)}(x_1) := h(x_1^{1/q})$ and set $\phi^{(\infty)}(y_1, y_2) := \phi(y_1, y_2 + \psi^{(\infty)}(y_1))$, then it is easily seen by means of a straightforward adaption of the proof of Theorem 5.1 in [IM11a] that $\mathcal{N}(\phi^{(\infty)}) \subset \{(t_1, t_2) : t_2 \geq B\}$. Therefore, Condition (R) in Theorem 1.14 guarantess that, possibly after adding a flat function to $\psi^{(\infty)}$, we may assume that ϕ factors as $\phi(x_1, x_2) = (x_2 - \psi^{(\infty)}(x_1))^B \tilde{\phi}(x_1, x_2)$, which means that the analogue of (6.30) holds true. We can thus argue as before to complete also this case and hence also the discussion of the Case (a), in which the principal face of $\mathcal{N}(\phi^a)$ is a compact edge.

Finally, in Case (b), where the principal face of $\mathcal{N}(\phi^a)$ is a vertex, we have that $D_{\mathrm{pr}} = \{(x_1, x_2) : |x_2 - \psi(x_1)| \leq \varepsilon x_1^a\}$, which corresponds to the domain $D'_{(1)}$ in the discussion of Case (a). This means that we can just drop the initial step of the algorithm described before and from then on may proceed as in Case (a).

We have thus established the desired restriction estimates for the piece of the surface S corresponding to the remaining domain D_{pr}, which completes the proof of Proposition 6.7 and hence also of Theorem 1.14, in the case where $h_{\mathrm{lin}}(\phi) \geq 5$.

Chapter Seven

How to Go beyond the Case $h_{\text{lin}}(\phi) \geq 5$

So far, we have been able to cover the cases where $h_{\text{lin}}(\phi) < 2$ or $h_{\text{lin}}(\phi) \geq 5$. In the latter case, we had made use of Drury's restriction estimate for curves, but this approach is indeed limited to the case where $h_{\text{lin}}(\phi) \geq 5$. In order also to cover the case where $2 \leq h_{\text{lin}}(\phi) < 5$, quite different and substantially more involved arguments are needed, which will be developed in the sequel. It turns out that these arguments will work whenever $h_{\text{lin}}(\phi) \geq 2$.

In Section 7.1 we shall briefly remind the reader of the regions for which we still need to prove restriction estimates, namely, the narrow domains of the type $D'_{(l)}$, which in the modified adapted coordinates $y_1 := x_1$ and $y_2 := x_2 - \psi^{(l+1)}(x_1)$ are of the form $D'^a_{(l)} := \{(y_1, y_2) : 0 < y_1 < \varepsilon, |y_2| \leq \varepsilon y_1^{a_{(l)}}\}$. Since we are left with the Case 2 situation, we will assume that $\partial_2 \phi^{(l+1)}_{\kappa^{(l)}}(1,0) = 0$ and $\partial_1 \phi^{(l+1)}_{\kappa^{(l)}}(1,0) \neq 0$. Again, we shall decompose these domains dyadically, and Littlewood-Paley theory allows us to reduce everything to proving uniform restriction estimates for the dyadic constituents μ_k of the part of the measure μ corresponding to the domain $D'_{(l)}$ (cf. Proposition 7.2).

It will turn out that the treatment of the domains of type $D'_{(1)}$, which are associated with the principal face of the Newton polyhedron of ϕ^a, is by far more involved than the treatment of the remaining domains of type $D'_{(l)}$ with $l \geq 2$, which are associated with edges of the Newton polyhedra lying below the bisectrix. We shall therefore mainly concentrate on the domains of type $D'_{(1)}$ and only briefly indicate the modifications also needed to cover the domains of type $D'_{(l)}$ with $l \geq 2$ in Section 7.10.

In Section 7.2, we shall basically follow our approach from earlier chapters and first reduce our restriction estimates for the measures μ_k to corresponding uniform estimates for families of normalized measures ν_δ (compare (7.8)) associated with phase functions of the form

$$\phi_\delta(x) = x_2^B\, b(x_1, x_2, \delta) + x_1^n \alpha(\delta_1 x_1) + \sum_{j=1}^{B-1} \delta_{j+2}\, x_2^j x_1^{n_j} \alpha_j(\delta_1 x_1).$$

These phase functions will depend on small perturbation parameters $\delta = (\delta_0, \ldots, \delta_{B+1})$, where each δ_i is a fractional power of 2^{-k}.

In some situations, it is possible to remove the term of order $j = B - 1$ in the last sum by means of another smooth change of coordinates of nonlinear shear type, as shown in Section 7.3. This will turn out to be quite useful, in particular for the discussion of the case where $B = 3$ (which will create the most serious difficulties among all), for instance in our discussion of the operators $T_\delta^{III_1}$ in Section 7.6, as well as in Chapter 8.

Next, in Section 7.4, we shall establish lower bounds for the r-height $h^r = h^r(\phi)$. These will turn out to be very useful in the subsequent sections, since we will be able to prove in many situations that for our normalized measures ν_δ, uniform L^p-L^2 Fourier restriction estimates will hold true for wider L^p-ranges of the form $p' \geq 2h^r_b + 2$ than the range $p' \geq 2h^r + 2$, where h^r_b is one of the lower bounds that we have found for h^r.

Indeed, these restriction estimates will be established in Section 7.5, where we shall follow to some extent the approach that we had devised in Chapter 4. Again, we perform an additional spectral localization to dyadic frequency domains given by $|\xi_i| \sim \lambda_i$, $i = 1, 2, 3$, and denote by ν_δ^λ the corresponding complex measures. These should be compared with the measures ν_j^λ from Section 4.1, where δ_0 will now take over the role that 2^{-j} played in Chapter 4. Again, we will distinguish various cases, depending on the relative sizes of the quantities $\lambda_1, \lambda_2, \lambda_3$, and δ_0, in a way analogous to what we did in Section 4.1. The most difficult situation will arise when $\lambda_1 \sim \lambda_2 \sim \lambda_3$, and the study of this case will therefore be initiated only later in Section 7.7. In all other situations, we can obtain the desired Fourier restriction estimates for the measures ν_δ^λ in a comparatively easy way, by making use of our lower bounds for h^r, with the exception of a few cases in which interpolation arguments are needed in order to cover the endpoint $p = p_c$. In some cases, this can be done by means of the real interpolation method due to Bak and Seeger, more precisely by means of Proposition 2.6 (compare Section 7.6), and in others we need again to apply complex interpolation. In all these exceptional situations, we shall have $m = 2$ and $B = 2$ or $B = 3$.

In Sections 7.7 and 7.8, where we assume that $\lambda_1 \sim \lambda_2 \sim \lambda_3$, we will again be able to handle most cases in a relatively easy way, with the exception of a few cases. One of these exceptional situations concerns the case where $B = 5$, but it turns out that this case can still be dealt with in an easy way by means of a slight improvement of our previous methods (see Section 7.8). As already mentioned, the most difficult situations will arrive when $m = 2$ and $B = 3$. An important tool for the study of this situation will be provided by Proposition 7.9, which gives a description of the augmented Newton polyhedron of $\tilde{\phi}^a$ under the additional assumption that $3 < h^r + 1 \leq 3.5$. Only when $h^r + 1$ lies within this range, our lower bounds for h^r are insufficient in order to obtain the desired estimates in the chapter, but the precise value for h^r provided by Proposition 7.9 will work.

What remains in the end are a few exceptional cases, which will be compiled in Section 7.9. In these remaining cases, we have in particular that $m = 2$ and either $B = 3$ or $B = 4$.

Their treatment will be rather involved, and it will form the content of Chapter 8.

7.1 THE CASE WHEN $h_{\text{lin}}(\phi) \geq 2$: REMINDER OF THE OPEN CASES

Assume from now on that $h_{\text{lin}}(\phi) \geq 2$. We recall the following two cases from Chapter 6:

(a) The principal face $\pi(\phi^a)$ of the Newton polyhedron $\mathcal{N}(\phi^a)$ of ϕ^a is a compact edge, lying on the line L^a, which we call the *principal line* of $\mathcal{N}(\phi^a)$;

(b) $\pi(\phi^a)$ is the vertex (h, h).

What has remained open in our preceding discussion is the study of the piece of the surface S corresponding to the domain D_{pr} containing the principal root jet ψ in Cases (a) and (b), that is,

$$D_{\mathrm{pr}} \cap \Omega := \begin{cases} \{(x_1, x_2) : 0 < x_1 < \varepsilon,\ |x_2 - \psi(x_1)| \leq N x_1^a\}, & \text{in Case (a)}, \\ \{(x_1, x_2) : 0 < x_1 < \varepsilon,\ |x_2 - \psi(x_1)| \leq \varepsilon x_1^a\}, & \text{in Case (b)}, \end{cases}$$

when $2 \leq h_{\mathrm{lin}}(\phi) < 5$. Indeed, we shall develop an approach that will work whenever $h_{\mathrm{lin}}(\phi) \geq 2$. Our goal will thus be to prove the following extension of Proposition 6.7 to the case where $h_{\mathrm{lin}}(\phi) \geq 2$.

Proposition 7.1. *Assume that $h_{\mathrm{lin}}(\phi) \geq 2$, and that we are in Case (a) or (b). Then for any given $N > 0$ in Case (a), respectively, every sufficiently small ε in Case (b), we have*

$$\left(\int_{D_{\mathrm{pr}}} |\widehat{f}|^2 \, d\mu^{\rho_{l_{\mathrm{pr}}}} \right)^{1/2} \leq C_p \|f\|_{L^p(\mathbb{R}^3)}, \qquad f \in \mathcal{S}(\mathbb{R}^3),$$

whenever $p' \geq p'_c$.

In order to prove this proposition, we follow the domain decomposition algorithm for the domain D_{pr} developed in Section 6.5 and begin with Case (a). In this case, this algorithm led to a finite family of subdomains $E_{(l)}$ (so-called transition domains), and domains $D'_{(l)}, l \geq 1$ of the form

$$D'_{(l)} := \{(x_1, x_2) : 0 < x_1 < \delta,\ |x_2 - \psi^{(l+1)}(x_1)| \leq \varepsilon x_1^{a_{(l)}}\},$$

where the functions $\psi^{(l)}$ are of the form

$$\psi^{(l)}(x_1) = \psi(x_1) + \sum_{j=1}^{l-1} c_{j-1} x_1^{a_{(j)}},$$

with real coefficients c_j, and where the exponents $a_{(j)}$ form a strictly increasing sequence

$$a = a_{(1)} < a_{(2)} < \cdots a_{(l)} < a_{(l+1)} < \cdots$$

of rational numbers. Moreover, in the modified adapted coordinates given by

$$y_1 := x_1, \quad y_2 := x_2 - \psi^{(l+1)}(x_1),$$

the function ϕ is given by

$$\phi^{(l+1)}(y_1, y_2) := \phi(y_1, y_2 + \psi^{(l+1)}(y_1)).$$

Notice that we have $\psi^{(1)} = \psi$ and $\phi^{(1)} = \phi^a$. Recall also that the domains and modified adapted coordinates that we encounter here do depend on the choice of the real numbers c_l that we make in every step of our algorithm.

Moreover, the domain $D'_{(l)}$ is associated with an "edge" $\gamma'_{(l)} = [(A'_{(l-1)}, B'_{(l-1)}), (A'_{(l)}, B'_{(l)})]$ (which is, indeed, an edge or can degenerate to a single point) of the Newton polyhedron of $\phi^{(l+1)}$ in the following way.

The edge with index l will lie on a line

$$L_{(l)} := \{(t_1, t_2) \in \mathbb{R}^2 : \kappa_1^{(l)} t_1 + \kappa_2^{(l)} t_2 = 1\}$$

of "slope" $1/a_{(l)}$, where $a_{(l)} = \kappa_2^{(l)}/\kappa_1^{(l)}$. Introduce corresponding "$\kappa^{(l)}$-dilations" $\delta_r = \delta_r^{\kappa^{(l)}}$ by putting $\delta_r(y_1, y_2) = (r^{\kappa_1^{(l)}} y_1, r^{\kappa_2^{(l)}} y_2)$, $r > 0$. Then the domain

$$D'^{a}_{(l)} := \{(y_1, y_2) : 0 < y_1 < \varepsilon,\ |y_2| \le \varepsilon y_1^{a_{(l)}}\},$$

which represents the domain $D'_{(l)}$ in the coordinates (y_1, y_2), is invariant under these dilations for $r \le 1$, and the Newton diagram of the $\kappa^{(l)}$-principal part $\phi^{(l+1)}_{\kappa^{(l)}}$ of $\phi^{(l+1)}$ consists exactly of the edge $\gamma'_{(l)}$.

Recall also that the first edge $\gamma'_{(1)}$ agrees with the principal face $\pi(\phi^{(2)})$ of $\phi^{(2)}$ and lies on the principal line L^a of the Newton polyhedron of ϕ^a, and it intersects the bisectrix Δ, whereas for $l \ge 2$ the edge $\gamma'_{(l)}$ will lie within the closed half-space below the bisectrix (this case will turn out to be easier).

Moreover, the Newton polyhedra of ϕ^a and of $\phi^{(l)}$ do agree in the closed half space above the bisectrix.

Now, as we have seen in Section 6.5, what remained to be controlled is the Case 2 situation for any given narrow domain $D'_{(l)} = D_{(l)}(c_l)$, that is, the situation where $\partial_2 \phi^{(l)}_{\kappa^{(l)}}(1, c_l) = 0$ and $\partial_1 \phi^{(l)}_{\kappa^{(l)}}(1, c_l) \ne 0$. But, observe that if we define our modified adapted coordinates $y_1 := x_1,\ y_2 := x_2 - \psi^{(l+1)}(x_1)$ in such a way that $\psi^{(l+1)}(x_1) := \psi^{(l)}(x_1) + c_l x_1^{a_{(l)}}$, with c_l as in the definition of the domain $D_{(l)}(c_l)$, then in these coordinates y the point $(1, c_l)$ (given in the coordinates associated to $\phi^{(l)}$) corresponds to the point $v = (1, 0)$, and we may rewrite the previous Case 2 condition as follows:

$$\partial_2 \phi^{(l+1)}_{\kappa^{(l)}}(v) = 0 \quad \text{and} \quad \partial_1 \phi^{(l+1)}_{\kappa^{(l)}}(v) \ne 0.$$

Only in Case 2, we had made use of the assumption $h_{\text{lin}}(\phi) \ge 5$, so that we may concentrate in the sequel on this case.

Notice also that our decomposition algorithm worked as well in Case (b), only we had to skip the first step of the algorithm. We shall therefore study the Fourier restriction estimates for the pieces of the surface S corresponding to the domains $D'_{(l)}$ and shall begin with the most difficult case $l = 1$, that is, the domain $D'_{(1)}$ in Case (a).

In the last section we shall describe the minor modifications needed to also treat the domains $D'_{(l)}$ for $l \ge 2$, which will then also cover Case (b) at the same time.

Observe finally that we can localize to the domain $D'_{(l)}$ by means of a cutoff function

$$\rho_{(l)}(x_1, x_2) := \chi_0\left(\frac{x_2 - \psi^{(l+1)}(x_1)}{\varepsilon x_1^{a_{(l)}}}\right),$$

where $\chi_0 \in \mathcal{D}(\mathbb{R})$. Let us again fix a suitable smooth cutoff function $\chi \ge 0$ on $\mathbb{R}^2$ supported in an annulus $\mathcal{A} := \{x \in \mathbb{R}^2 : \frac{1}{2} \le |y| \le R\}$ such that the functions

$\chi_k^a := \chi \circ \delta_{2^k}$ form a partition of unity. Here, $\delta_r = \delta_r^{\kappa^{(l)}}$ denote the dilations associated with the weight $\kappa^{(l)}$. In the original coordinates x, these correspond to the functions $\chi_k(x) := \chi_k^a(x_1, x_2 - \psi^{(l+1)}(x_1))$. We then decompose the measure $\mu^{\rho_{(l)}}$ dyadically as

$$\mu^{\rho_{(l)}} = \sum_{k \geq k_0} \mu_k, \tag{7.1}$$

where

$$\mu_k := \mu_k^{(l)} := \mu^{\chi_k \rho_{(l)}}.$$

Notice that by choosing the support of η sufficiently small, we can choose $k_0 \in \mathbb{N}$ as large as we need. It is also important to observe that this decomposition can essentially be achieved by means of a dyadic decomposition with respect to the variable x_1, which again allows to apply Littlewood-Paley theory.

Moreover, changing to modified adapted coordinates in the integral defining μ_k and scaling by $\delta_{2^{-k}}$, we find that

$$\langle \mu_k, f \rangle = 2^{-k|\kappa^{(l)}|} \int f(2^{-\kappa_1^{(l)} k} x_1, 2^{-\kappa_2^{(l)} k} x_2 + 2^{-m\kappa_1^{(l)} k} x_1^m \omega(2^{-\kappa_1^{(l)} k} x_1), 2^{-k} \phi_k(x)) \times \eta(x)\, dx, \tag{7.2}$$

where $\omega = \omega^{(l)}$ is given by

$$\psi^{(l+1)}(x_1) = x_1^m \omega(x_1),$$

so that $\omega(0) \neq 0$, $\eta = \eta_k^{(l)}$ is a smooth function supported where $x_1 \sim 1$, $|x_2| < \varepsilon$ (for some small $\varepsilon > 0$), whose derivatives are uniformly bounded in k and where

$$\phi_k(x) = \phi_k^{(l+1)}(x) := 2^k \phi^{(l+1)}(\delta_{2^{-k}} x) = \phi_{\kappa^{(l)}}^{(l+1)}(x) + \text{error terms of order } O(2^{-\delta k})$$

with respect to the C^∞ topology (and $\delta > 0$).

In order to prove Proposition 7.1, we still need to prove the following.

Proposition 7.2. *Assume that $h_{\text{lin}} \geq 2$, that we are in Case 2, that is, $\partial_2 \phi_{\kappa^{(l)}}^{(l+1)}(1,0) = 0$ and $\partial_1 \phi_{\kappa^{(l)}}^{(l+1)}(1,0) \neq 0$, and recall that $p_c' = 2h^r + 2$. When $\varepsilon > 0$ is sufficiently small and $k_0 \in \mathbb{N}$ is sufficiently large, then for every $l \geq 1$,*

$$\left(\int |\widehat{f}|^2 \, d\mu_k \right)^{1/2} \leq C_{p_c} \|f\|_{L^{p_c}(\mathbb{R}^3)}, \qquad f \in \mathcal{S}(\mathbb{R}^3),\ k \geq k_0,$$

where the constant C_p is independent of k.

7.2 RESTRICTION ESTIMATES FOR THE DOMAINS $D'_{(1)}$: REDUCTION TO NORMALIZED MEASURES ν_δ

Let us assume that we are in Case (a), where the principal face $\pi(\phi^a)$ is a compact edge. In the enumeration of edges γ_l of the Newton polyhedron associated with ϕ^a in Chapter 6, this edge corresponds to the index $l = l_{\text{pr}}$, that is,

$$\pi(\phi^a) = \gamma_{l_{\text{pr}}}.$$

Here, the weight $\kappa^{(1)}$ is the principal weight $\kappa^{l_{\rm pr}}$ from Chapter 6, and the line $L_{(1)}$ is the principal line $L^a = L_{l_{\rm pr}}$ of the Newton polyhedron of ϕ^a. We then put

$$\tilde{\kappa} := \kappa^{(1)}, \quad \text{so that} \quad a = \frac{\tilde{\kappa}_2}{\tilde{\kappa}_1}, \; \phi^a_{\tilde{\kappa}} = \phi^a_{\rm pr}.$$

In particular, $h_{l_{\rm pr}} + 1$ is the second coordinate of the point of intersection of the line

$$\Delta^{(m)} = \{(t, t+m+1) : t \in \mathbb{R}\}$$

with the line L^a, and according to the identity (1.17), we have

$$h_{l_{\rm pr}} + 1 = \frac{1 + (m+1)\tilde{\kappa}_1}{|\tilde{\kappa}|}. \tag{7.3}$$

The domain $D'_{(1)}$ that we have to study is then of the form

$$D'_{(1)} = \{(x_1, x_2) : 0 < x_1 < \varepsilon, \; |x_2 - \psi(x_1) - c_0 x_1^a| \leq \varepsilon x_1^a\},$$

where $\psi(x_1) + c_0 x_1^a = \psi^{(2)}(x_1)$. Moreover,

$$\phi^{(2)}(x_1, x_2) = \phi^a(x_1, x_2 + c_0 x_1^a) =: \tilde{\phi}^a(x_1, x_2), \tag{7.4}$$

so that $\tilde{\phi}^a$ represents ϕ in the modified adapted coordinates

$$y_1 := x_1, \quad y_2 := x_2 - \psi(x_1) - c_0 x_1^a, \tag{7.5}$$

compared to the adapted coordinates given by $y_1 := x_1$, $y_2 := x_2 - \psi(x_1)$, in which ϕ is represented by ϕ^a.

Notice that the exponent a may be noninteger (but rational), so that $\psi^{(2)}$ is, in general, only fractionally smooth, that is, a smooth function of x_2 and some fractional power of x_1 only. The same applies to every $\psi^{(l)}$ with $l \geq 2$, whereas ϕ^a is still smooth; that is, when we express ϕ in our adapted coordinates, we still get a smooth function, whereas when we pass to modified adapted coordinates, we may only get fractionally smooth functions.

We shall write D^a for the domain $D'^a_{(1)}$, that is,

$$D^a := \{(y_1, y_2) : 0 < y_1 < \varepsilon, \; |y_2| < \varepsilon y_1^a\},$$

so that D^a represents our domain $D'_{(1)}$ in our modified adapted coordinates, in which ϕ is represented by $\tilde{\phi}^a$.

We assume that we are in Case 2, so that $\partial_2 \tilde{\phi}^a{}_{\tilde{\kappa}}(1, 0) = 0$ and $\partial_1 \tilde{\phi}^a{}_{\tilde{\kappa}}(1, 0) \neq 0$.

Next we choose $B \geq 2$ to be minimal so that $\partial_2^B \tilde{\phi}^a_{\tilde{\kappa}}(1, 0) \neq 0$. Since $\tilde{\phi}^a_{\tilde{\kappa}}$ is $\tilde{\kappa}$-homogeneous, the principal part of $\tilde{\phi}^a$ is then of the form (cf. (6.18))

$$\tilde{\phi}^a_{\tilde{\kappa}}(y_1, y_2) = y_2^B Q(y_1, y_2) + c_1 y_1^n, \quad c_1 \neq 0, \; Q(1, 0) \neq 0, \tag{7.6}$$

where Q is a $\tilde{\kappa}$-homogeneous smooth function. Note that n is rational but not necessarily an integer since we are in modified adapted coordinates.

Observe also that this implies that we may write

$$\tilde{\phi}^a(y_1, y_2) = y_2^B b_B(y_1, y_2) + y_1^n \alpha(y_1) + \sum_{j=1}^{B-1} y_2^j b_j(y_1), \tag{7.7}$$

with fractionally smooth functions b_B, α and b_j such that $\alpha(0) \neq 0$ and

$$b_B(y_1, y_2) = Q(y_1, y_2) + \text{ terms of } \tilde{\kappa}\text{-degree strictly bigger than that of } Q.$$

Moreover, the functions $b_1, \dots, b_{B-1}$ of y_1 are either flat or of "finite type," that is, $b_j(y_1) = y_1^{n_j}\alpha_j(y_1)$, with rational exponents $n_j > 0$ and fractionally smooth functions α_j such that $\alpha_j(0) \neq 0$.

For convenience, we shall also write $b_j(y_1) = y_1^{n_j}\alpha_j(y_1)$ when b_j is flat, keeping in mind that in this case we may choose $n_j \in \mathbb{N}$ as large as we please (but $\alpha_j(0) = 0$).

Notice that then, for $j = 1, \dots, B-1$, $y_2^j b_j(y_1)$ consists of terms of $\tilde{\kappa}$-degree strictly bigger than 1.

Recall that the Newton diagram of the $\tilde{\kappa}$-principal part $\tilde{\phi}^a_{\tilde{\kappa}}$ is the line segment $\gamma'_{(1)} = [(A'_{(0)}, B'_{(0)}), (A'_{(1)}, B'_{(1)})]$, which consequently must contain a point of the Newton support of $\tilde{\phi}^a$, with the second coordinate given by B and none below that point, with the exception of $(n, 0)$. It then follows easily that the following relations hold true:

$$\tilde{\kappa}_2 < 1, \quad 2 \leq m < \frac{1}{\tilde{\kappa}_1} = n, \quad B \leq B'_{(0)}.$$

Actually, since $\tilde{\kappa}_2 B'_{(0)} \leq 1$, we even have

$$\tilde{\kappa}_2 B \leq 1, \quad m < a = \frac{\tilde{\kappa}_2}{\tilde{\kappa}_1} \leq \frac{n}{B}.$$

As in Section 6.3, we define normalized measures ν_k corresponding to the μ_k by putting

$$\langle \nu_k, f \rangle := \int f\left(x_1,\, 2^{(m\tilde{\kappa}_1 - \tilde{\kappa}_2)k} x_2 + x_1^m \omega(2^{-k\tilde{\kappa}_1} x_1),\, \phi_k(x)\right) \eta(x)\, dx,$$

where η is again a smooth function with $\operatorname{supp} \eta \subset \{x_1 \sim 1,\, |x_2| < \varepsilon\}$ (for some small $\varepsilon > 0$), and where by (7.7) the function

$$\phi_k(x_1, x_2) := 2^k \tilde{\phi}^a(2^{-\tilde{\kappa}_1 k} x_1, 2^{-\tilde{\kappa}_2 k} x_2)$$

is of the form

$$\phi_k(x_1, x_2) = x_2^B(Q(x) + O(2^{-\varepsilon' k})) + x_1^n \alpha(2^{-\tilde{\kappa}_1 k} x_1) + \sum_{j=1}^{B-1} x_2^j\, 2^{(1 - j\tilde{\kappa}_2)k}\, b_j(2^{-\tilde{\kappa}_1 k} x_1)$$

for some $\varepsilon' > 0$. Observe that

$$2^{(1 - j\tilde{\kappa}_2)k}\, b_j(2^{-\tilde{\kappa}_1 k} x_1) = x_1^{n_j} 2^{-(j\tilde{\kappa}_2 + n_j\tilde{\kappa}_1 - 1)k} \alpha_j(2^{-\tilde{\kappa}_1 k} x_1),$$

where $(j\tilde{\kappa}_2 + n_j\tilde{\kappa}_1 - 1) > 0$.

In the sequel, we shall rewrite ν_k as ν_δ by putting

$$\langle \nu_\delta, f\rangle := \int f\left(x_1,\ \delta_0 x_2 + x_1^m \omega(\delta_1 x_1),\ \phi_\delta(x)\right) \eta(x)\, dx, \tag{7.8}$$

where ϕ_δ is of the form

$$\phi_\delta(x) := x_2^B\, b(x_1, x_2, \delta) + x_1^n \alpha(\delta_1 x_1) + r(x_1, x_2, \delta), \tag{7.9}$$

with

$$r(x_1, x_2, \delta) := \sum_{j=1}^{B-1} \delta_{j+2}\, x_2^j x_1^{n_j} \alpha_j(\delta_1 x_1), \tag{7.10}$$

and $\delta = (\delta_0,\ \delta_1,\ \delta_2,\ \delta_3, \ldots, \delta_{B+1})$ is given by

$$\delta := (2^{-k(\tilde\kappa_2 - m\tilde\kappa_1)},\ 2^{-k\tilde\kappa_1},\ 2^{-k\tilde\kappa_2},\ 2^{-(n_1\tilde\kappa_1+\tilde\kappa_2-1)k}, \ldots, 2^{-(n_{B-1}\tilde\kappa_1+(B-1)\tilde\kappa_2-1)k}). \tag{7.11}$$

Recall that $\alpha(0) \neq 0$ and that either $\alpha_j(0) \neq 0$, and then n_j is fixed (the type of the finite-type function b_j) or $\alpha_j(0) = 0$, and then we may assume that n_j is as large as we please.

Observe that the components of δ can be viewed as small perturbation parameters, since $\delta \to 0$ as $k \to \infty$, and that every δ_j is a power of δ_0,

$$\delta_j = \delta_0^{q_j}, \qquad j = 1, \ldots, B+1,$$

with positive exponents $q_j > 0$ that are fixed rational numbers, except for those $j \geq 3$ for which $\alpha_{j-2}(0) = 0$, for which we may choose the exponents q_j as large as we please.

Moreover, $b(x_1, x_2, \delta)$ is a smooth function of all three arguments, and

$$b(x_1, x_2, 0) = Q(x_1, x_2). \tag{7.12}$$

For δ sufficiently small, this implies in particular that $b(x_1, x_2, \delta) \neq 0$ when $x_1 \sim 1$ and $|x_2| < \varepsilon$.

Assume we can prove that for δ sufficiently small, we have

$$\left(\int |\hat f|^2 d\nu_\delta\right)^{1/2} \leq C\|f\|_{L^{p_c}}, \tag{7.13}$$

with C independent of δ. Then straightforward rescaling by means of the $\tilde\kappa-$ dilations leads to the estimate

$$\left(\int |\hat f|^2 d\mu_k\right)^{1/2} \leq C 2^{-k|\tilde\kappa|(1-2(h_{l_{\rm pr}}+1)/p_c')/2}\|f\|_{L^{p_c}}, \tag{7.14}$$

where $p_c' \geq 2(h_{l_{pr}} + 1)$ (compare (6.16) and the subsequent identity). So, our goal is to verify (7.13).

Observe also that the $\tilde\kappa$-principal parts of $\tilde\phi^a$ and ϕ_δ do agree.

7.3 REMOVAL OF THE TERM $y_2^{B-1} b_{B-1}(y_1)$ IN (7.7)

In the case where the left endpoint of the principal face $\pi(\tilde\phi^a)$ of the Newton polyhedron of $\tilde\phi^a$ (which agrees with the one for ϕ^a) is of the form (A, B), it is possible to remove the term $y_2^{B-1}\, b_{B-1}(y_1)$ in our formula (7.7) for $\tilde{\phi}^a$ by means of an additional smooth change of coordinates (again of nonlinear shear type). The new coordinates turn out to be smooth and can be used as alternative adapted coordinates. This will become important in some situations—in particular, when $B = 3$.

Proposition 7.3. *Assume that the left endpoint of the principal face $\pi(\tilde{\phi}^a)$ of $\mathcal{N}(\tilde{\phi}^a)$ is of the form (A, B), where $\tilde{\phi}^a$ and B are as in (7.7). Then the numbers $a = \tilde{\kappa}_2/\tilde{\kappa}_2$, A, and n are integers, and $\tilde{\phi}^a$ is smooth. Moreover, there exists an additional, smooth change of coordinates of the form $z_1 = x_1, z_2 = x_2 - (\psi^{(2)}(x_1) + \rho(x_1))$, which allows us to remove the term of order $j = B - 1$ in (7.7), that is, after applying this change of coordinates and denoting the corresponding coordinates again by (y_1, y_2), we may assume that*

$$\tilde{\phi}^a(y_1, y_2) = y_2^B\, b_B(y_1, y_2) + y_1^n \alpha(y_1) + \sum_{j=1}^{B-2} y_2^j\, b_j(y_1). \tag{7.15}$$

Proof. In view of our definition of B (cf. (7.6)), we see that our assumptions imply that the $\tilde{\kappa}$- homogeneous function Q in (7.6) must be of the form $Q(y_1, y_2) = c_0 y_1^A$, with $c_0 \neq 0$. Let us assume without loss of generality that $c_0 = 1$, so that by (7.6)

$$\tilde{\phi}^a_{\tilde{\kappa}}(y) = y_1^A y_2^B + c_1 y_1^n, \qquad c_1 \neq 0. \tag{7.16}$$

But then $\phi^a_{\tilde{\kappa}}$ will be given by

$$\phi^a_{\tilde{\kappa}}(x) = x_1^A (x_2 - c_0 x_1^a)^B + c_1 x_1^n$$

(compare (7.4)). This must be a polynomial in (x_1, x_2), since ϕ^a is smooth. Expanding $(x_2 - c_0 x_1^a)^B$, it is clear that A and a must be integers. But then, $\tilde{\phi}^a$ is also necessarily smooth, which implies that n must be an integer, too.

Observe next that (7.16) also implies that $\partial_1^A \partial_2^{B-1} \tilde{\phi}^a_{\tilde{\kappa}}(y_1, 0) \equiv 0$ and $\partial_1^A \partial_2^B \tilde{\phi}^a_{\tilde{\kappa}}(y_1, 0) \equiv A!B! \neq 0$ for $|y_1| < \varepsilon$. Moreover, since (A, B) is a vertex of $\mathcal{N}(\tilde{\phi}^a)$, we also have $\partial_1^A \partial_2^B \tilde{\phi}^a(0, 0) = A!B! \neq 0$, whereas $\partial_1^A \partial_2^{B-1} \tilde{\phi}^a(0, 0) = 0$. Recall also that $a \geq 2$.

We can now argue in a similar way as in the proof of Proposition 2.11: if $A = 0$, then the implicit function theorem shows that, for ε sufficiently small, there is a smooth function $\rho(y_1)$, $-\varepsilon < y_1 < \varepsilon$, such that

$$\partial_2^{B-1} \tilde{\phi}^a(y_1, \rho(y_1)) \equiv 0,$$

and comparing $\tilde{\kappa}$-principal parts, it is easy to see that the $\tilde{\kappa}$-principal part of ρ has $\tilde{\kappa}$-degree strictly bigger than the degree of y_1^a. Thus, if we perform the further change of coordinates $(z_1, z_2) := (y_1, y_2 - \rho(y_1))$, in which $\phi(x_1, x_2)$ is represented, say, by $\tilde{\phi}(z_1, z_2)$, it is easily seen that the Newton polyhedra of $\tilde{\phi}^a$ and $\tilde{\phi}$ as well as their $\tilde{\kappa}$-principal parts are the same (cf. similar arguments in [IM11a]). Replacing $\psi^{(2)}(y_1)$ by $\psi^{(2)}(y_1) + \rho(y_1)$ and modifying $\omega(y_1)$ accordingly, we then find that in the corresponding modified adapted coordinates $z_1 = x_1, z_2 = x_2 - (\psi^{(2)}(x_1) + \rho(x_1))$, the function $\tilde{\phi}$ satisfies

$$\partial_2^{B-1} \tilde{\phi}(z_1, 0) \equiv 0. \tag{7.17}$$

On the other hand, like $\tilde{\phi}^a$, it must be of the form

$$\tilde{\phi}(z_1, z_2) = z_2^B\, b_B(z_1, z_2) + z_1^n \alpha(z_1) + \sum_{j=1}^{B-1} z_2^j\, b_j(z_1),$$

where $b_B(z_1, z_2) = 1 +$ terms of $\tilde{\kappa}$-degree strictly bigger than 0, and then (7.17) implies that $b_{B-1}(z_1) \equiv 0$.

Assume next that $A \geq 1$. Then, by (7.16), $\partial_2^{B-1}\tilde{\phi}^a(y_1, y_2) = c_2 y_1^A y_2 + R(y_1, y_2)$, with $c_2 := B!$ and where R consists of terms of $\tilde{\kappa}$-degree strictly bigger than the degree of $y_1^A y_2$. Consider the equation

$$\partial_2^{B-1}\tilde{\phi}^a(y_1, y_2) = c_2 y_1^A y_2 + R(y_1, y_2) = 0. \tag{7.18}$$

It is easily seen that the line given by $t_1 + at_2 = C$, where $C := A + a \in \mathbb{N}$, is a supporting line to the Newton polyhedron $\mathcal{N}(\partial_2^{B-1}\phi^a)$, which intersects $\mathcal{N}(\partial_2^{B-1}\phi^a)$ in exactly one point, namely, the point $(A, 1)$. Moreover, R is a smooth function consisting of terms $c_{j_1, j_2} y_1^{j_1} y_2^{j_2}$ for which $j_1 + aj_2 \geq 1 + C$ (more precisely, the Newton polyhedron of R is contained in the half space where $t_1 + at_2 \geq 1 + C$).

In analogy to the proof of the last part of Proposition 2.11, we write, for $y_1 \neq 0$, $y_2 = y_1^a z$. Then (7.18) is equivalent to the equation

$$c_2 y_1^C z + R(y_1, y_1^a z) = 0. \tag{7.19}$$

Clearly, $R(y_1, y_1^a z)$ is a smooth function, and in view of our previous observation we may factor it as $R(y_1, y_1^a z) = y_1^{C+1} g(y_1, z)$, with a smooth function $g(y_1, z)$. Thus, for $y_1 \neq 0$, equation (7.19) is equivalent to the equation

$$c_2 z + y_1 g(y_1, z) = 0.$$

Now we can again apply the implicit function theorem to conclude that near the origin this equation has a unique smooth solution

$$z = \psi_1(y_1), \quad \text{with} \quad \psi_1(0) = 0.$$

In particular, we find that

$$c_2 y_1^C \psi_1(y_1) + R(y_1, y_1^a \psi_1(y_1)) \equiv 0.$$

Thus, putting $\rho(y_1) := y_1^a \psi_1(y_1)$, we see that ρ is smooth, $\rho(0) = 0$, and

$$\partial_2^{B-1}\tilde{\phi}^a(y_1, \rho(y_1)) \equiv 0.$$

Notice also that the $\tilde{\kappa}$-principal part of ρ clearly has $\tilde{\kappa}$-degree strictly bigger than the degree of y_1^a. From now on we may thus argue as in the previous case where we had $A = 0$ and change coordinates from (y_1, y_2) to $(z_1, z_2) := (y_1, y_2 - \rho(y_1))$, with the effect that in these new coordinates, we again have (7.17).

On the other hand, like $\tilde{\phi}^a$, $\tilde{\phi}$ must here be of the form

$$\tilde{\phi}(z_1, z_2) = z_2^B\, b_B(z_1, z_2) + z_1^n \alpha(z_1) + \sum_{j=1}^{B-1} z_2^j\, b_j(z_1),$$

where $b_B(z_1, z_2) = z_1^A +$ terms of $\tilde{\kappa}$-degree strictly bigger than $\tilde{\kappa}_1 A$, and thus (7.18) implies that $b_{B-1}(z_1) \equiv 0$, and we obtain (7.15). Q.E.D.

7.4 LOWER BOUNDS FOR $h^r(\phi)$

Recall that $B \geq 2$ and $d \geq 2$. We shall often use the interpolation parameter

$$\theta_c := \frac{2}{p_c'} = \frac{1}{h^r + 1}.$$

Since, by definition, $h^r \geq d$, the second assumption implies that

$$\theta_c \leq \frac{1}{3}. \tag{7.20}$$

We first derive some useful estimates from the following for $p_c' = 2(h^r + 1)$. We let $H := 1/\tilde{\kappa}_2$, so that

$$n = \frac{1}{\tilde{\kappa}_1}, \qquad H = \frac{1}{\tilde{\kappa}_2}. \tag{7.21}$$

Note that H is rational but not necessarily entire. We next define

$$\tilde{h}^r := \frac{mH}{m+1}, \; \tilde{p}_c' := 2(\tilde{h}^r + 1), \; \tilde{\theta}_c := \frac{2}{\tilde{p}_c'} = \frac{m+1}{mH + m + 1} \leq \tilde{\theta}_B := \frac{m+1}{mB + m + 1}.$$

We also let $\tilde{p}_B' := 2/\tilde{\theta}_B \leq \tilde{p}_c'$ and

$$p_H' := \frac{12H}{3 + H}, \qquad \theta_H := \frac{2}{p_H'} = \frac{1}{2H} + \frac{1}{6}$$

and define p_B', θ_B accordingly, with H replaced by B.

Lemma 7.4. (a) *We have $p_c' > \tilde{p}_c'$, unless $h^r = \tilde{h}^r = d$ and $h^r + 1 \geq H$. In the latter case, $p_c' = \tilde{p}_c' = 2(d+1)$.*

(b) *If $m \geq 3$ and $H \geq 2$ or $m = 2$ and $H \geq 3$, then*

$$\tilde{p}_c' \geq p_H' \geq p_B',$$

where the inequality $\tilde{p}_c' \geq p_H'$ is even strict unless $m = 2$ and $H = 3$.

Proof. (a) The Newton polyhedron $\mathcal{N}(\tilde{\phi}^a)$ of $\tilde{\phi}^a$ is contained in the closed half space bounded from below by the principal line L^a of $\tilde{\phi}^a$, which passes through the points $(0, H)$ and $(n, 0)$. Moreover, it is known that the principal line L of ϕ is a supporting line to $\mathcal{N}(\tilde{\phi}^a)$ (this follows from Varchenko's algorithm), and it has slope $1/m$. It is therefore parallel to the line $\tilde{L}$ passing through the points $(0, H)$ and $(mH, 0)$ and lies "above" $\tilde{L}$ (see Figure 7.1). Thus the second coordinate $d + 1$ of the point of intersection of L with $\Delta^{(m)}$ is greater than or equal to the second coordinate t_2 of the point of intersection (t_1, t_2) of $\tilde{L}$ with $\Delta^{(m)}$, so that $h^r \geq d \geq t_2 - 1$.

But, the point (t_1, t_2) is determined by the equations $t_2 = m + 1 + t_1$ and $t_2 = H - t_1/m$, so that $t_2 = (mH + m + 1)/(m + 1) = \tilde{h}^r + 1$. This shows that $h^r \geq \tilde{h}^r$; hence, $p_c' \geq \tilde{p}_c'$.

Notice also that $d + 1 > t_2$; hence, $p_c' > \tilde{p}_c'$, unless $L = \tilde{L}$.

So, assume that $L = \tilde{L}$.

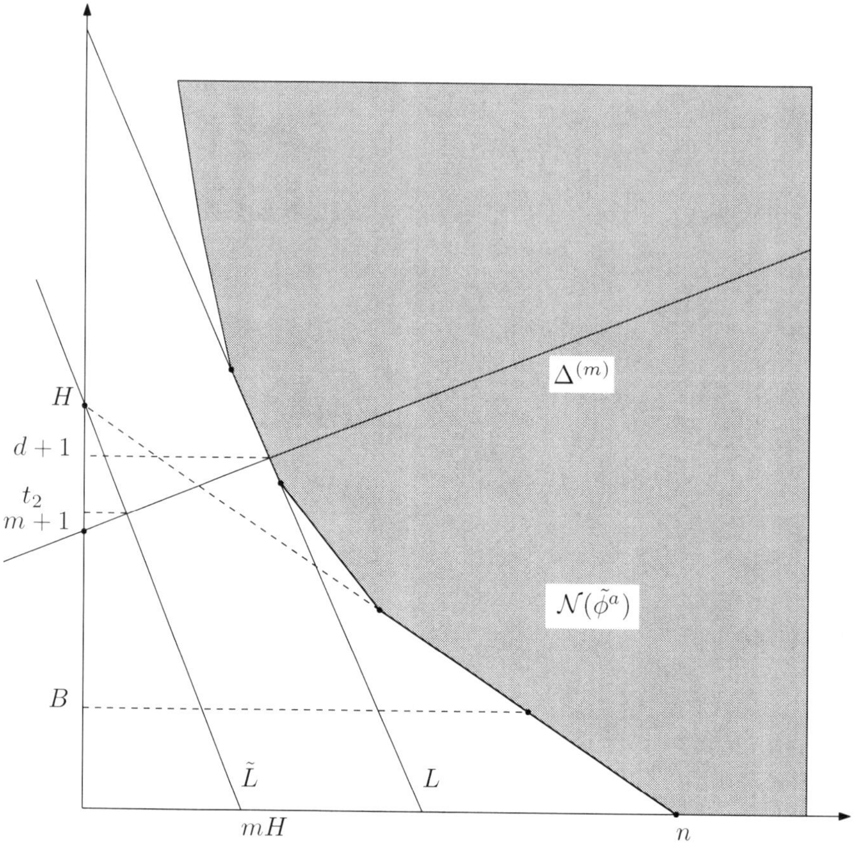

Figure 7.1

Then $d = 1/(1/H + 1/mH) = \tilde{h}^r$, and the principal face $\pi(\tilde{\phi}^a)$ of $\mathcal{N}(\tilde{\phi}^a)$ must be the edge $[(0, H), (n, 0)]$ (see Figure 7.1). Thus, if $h^r + 1 \geq H$, then clearly $h^r = d = \tilde{h}^r$ and $p'_c = \tilde{p}'_c$. And, if $h^r + 1 < H$, then we see that $h^r + 1$ is the second coordinate of the point of intersection of $\Delta^{(m)}$ with $\pi(\tilde{\phi}^a)$, and thus

$$\tilde{h}^r + 1 < h^r + 1 = h_{l_{\mathrm{pr}}} + 1 = \frac{1 + (m+1)\tilde{\kappa}_1}{|\tilde{\kappa}|}$$

(cf. (7.3)).

(b) The inequality $\tilde{p}'_c \geq p'_H$ is equivalent to

$$mH^2 - (2m+5)H + 3m + 3 \geq 0,$$

so that the remaining statements are elementary to check. Q.E.D.

The following corollary is a straightforward consequence of the definition of θ_H and Lemma 7.4.

COROLLARY 7.5. (a) *If $m \geq 3$ and $H \geq 2$ or $m = 2$ and $H \geq 3$, then $\theta_c < \theta_B$ unless $m = 2$ and $H = B = 3$ (where $\theta_c = \theta_B = \frac{1}{3}$).*

(b) *If $h^r + 1 \leq B$, then $\theta_c < \tilde{\theta}_c$ unless $B = H = h^r + 1 = d + 1$, where $\theta_c = \tilde{\theta}_c$.*

(c) *If $H \geq 3$, then $\theta_c < \frac{1}{3}$ unless $H = 3$ and $m = 2$.*

7.5 SPECTRAL LOCALIZATION TO FREQUENCY BOXES WHERE $|\xi_i| \sim \lambda_i$: THE CASE WHERE NOT ALL λ_iS ARE COMPARABLE

Observe next that by (7.8) the Fourier transform $\widehat{\nu_\delta}$ of ν_δ can be written as

$$\widehat{\nu_\delta}(\xi) = \int e^{-i\Phi(x,\delta,\xi)}\,\eta(x)\,dx,$$

where the complete phase Φ corresponding to ϕ_δ is given by

$$\Phi(x,\delta,\xi) := \xi_1 x_1 + \xi_2(\delta_0 x_2 + x_1^m \omega(\delta_1 x_1)) + \xi_3 \phi_\delta(x_1, x_2),$$

with

$$\phi_\delta(x) = x_2^B\, b(x_1, x_2, \delta) + x_1^n \alpha(\delta_1 x_1) + r(x_1, x_2, \delta),$$

so that

$$\begin{aligned}\Phi(x,\delta,\xi) = {} & \xi_1 x_1 + \xi_2 x_1^m \omega(\delta_1 x_1) + \xi_3 x_1^n \alpha(\delta_1 x_1) \\ & + \xi_2 \delta_0 x_2 + \xi_3 \left(x_2^B\, b(x_1, x_2, \delta) + r(x_1, x_2, \delta)\right).\end{aligned}$$

By T_δ we shall denote the operator of convolution with $\widehat{\nu_\delta}$, which we need to estimate as an operator from L^p to $L^{p'}$. In order to estimate the operator T_δ, we follow the approach in Section 4.1 (compare with (4.10), where 2^{-j} plays the same role that δ_0 does here) and decompose

$$\nu_\delta = \nu_k = \sum_\lambda \nu_\delta^\lambda,$$

where the sum is taken over all triples $\lambda = (\lambda_1, \lambda_2, \lambda_3)$ of dyadic numbers $\lambda_i \geq 1$ and where ν_k^λ is localized to frequencies ξ such that $|\xi_i| \sim \lambda_i$ if $\lambda_i > 1$ and $|\xi_i| \lesssim 1$ if $\lambda_i = 1$. The cases where $\lambda_i = 1$ for at least one λ_i can be dealt with in the same way as the corresponding cases where $\lambda_i = 2$, and therefore we shall always assume in the sequel that

$$\lambda_i > 1, \qquad i = 1, 2, 3.$$

The spectrally localized measure $\nu_\delta^\lambda(x)$ is then given by

$$\widehat{\nu_\delta^\lambda}(\xi) := \chi_1\left(\frac{\xi_1}{\lambda_1}\right) \chi_1\left(\frac{\xi_2}{\lambda_2}\right) \chi_1\left(\frac{\xi_3}{\lambda_3}\right) \widehat{\nu_\delta}(\xi),$$

that is,

$$\begin{aligned}\nu_\delta^\lambda(x) = {} & \lambda_1\lambda_2\lambda_3 \int \check{\chi}_1(\lambda_1(x_1 - y_1))\, \check{\chi}_1\left(\lambda_2(x_2 - \delta_0 y_2 - y_1^m \omega(\delta_1 y_1)\right) \\ & \times \check{\chi}_1(\lambda_3(x_3 - \phi_\delta(y))\, \eta(y)\, dy, \end{aligned} \tag{7.22}$$

where $\check{\chi}_1$ again denotes the inverse Fourier transform of χ_1. Recall also that

$$\operatorname{supp} \eta \subset \{y_1 \sim 1,\ |y_2| < \varepsilon\}, \qquad (\varepsilon \ll 1). \tag{7.23}$$

Arguing as in the proof of estimate (4.37) in Section 4.2, by making use of the localizations given by the first and the third factor of the integrand in (7.22) for the integrations in y_1 and y_2 by means of the van der Corput–type estimates of Lemma 2.1, we obtain

$$\|\nu_\delta^\lambda\|_\infty \lesssim \lambda_1\lambda_2\lambda_3\lambda_1^{-1}\lambda_3^{-1/B}.$$

Similarly, the localizations given by the first and second factor imply

$$\|\nu_\delta^\lambda\|_\infty \lesssim \lambda_1\lambda_2\lambda_3\lambda_1^{-1}(\lambda_2\delta_0)^{-1},$$

and consequently

$$\|\nu_\delta^\lambda\|_\infty \lesssim \min\{\lambda_2\lambda_3^{(B-1)/B},\ \lambda_3\delta_0^{-1}\}. \tag{7.24}$$

We have to distinguish various cases. Notice first that it is easy to see that the phase function Φ has no critical point with respect to x_1 if one of the components λ_i of λ is much bigger than the two others, so that integrations by parts yield that

$$\|\widehat{\nu_\delta^\lambda}\|_\infty \lesssim |\lambda|^{-N} \quad \text{for every } N \in \mathbb{N},$$

which easily implies estimates for the operator $T_\delta^\lambda : \varphi \mapsto \varphi * \widehat{\nu_\delta^\lambda}$ that are better than needed. We may therefore concentrate on the following, remaining cases.

Recall the interpolation parameter $\theta_c = 2/p_c' \le \frac{1}{3}$.

Case 1. $\lambda_1 \sim \lambda_3$, $\lambda_2 \ll \lambda_1$. First applying the method of stationary phase in x_1 and then van der Corput's lemma in x_2, we find that

$$\|\widehat{\nu_\delta^\lambda}\|_\infty \lesssim \lambda_1^{-1/2-1/B}.$$

By interpolation, using this estimate and the first one in (7.24), we obtain

$$\|T_\delta^\lambda\|_{p\to p'} \lesssim \lambda_1^{-1/B-1/2+3\theta/2}\lambda_2^\theta,$$

where $\theta = 2/p'$. Summation over all dyadic λ_2 with $\lambda_2 \ll \lambda_1$ yields

$$\sum_{\lambda_2\ll\lambda_1} \|T_\delta^\lambda\|_{p\to p'} \lesssim \lambda_1^{-1/B-1/2+5\theta/2}.$$

Notice that for $\theta := \theta_B$ we have

$$-\frac{1}{B} - \frac{1}{2} + \frac{5}{2}\theta = \frac{1}{4}\left(\frac{1}{B} - \frac{1}{3}\right) \le 0, \qquad \text{if } B \ge 3,$$

and that strict inequality holds when $\theta < \theta_B$. But, Corollary 7.5(a) shows that if $H \ge 3$, then indeed $\theta_c < \theta_B$, unless $m = 2$ and $H = B = 3$. Consequently, for $\theta := \theta_c$ we can sum over all dyadic λ_1 (unless $H = B = 3$ and $m = 2$) and obtain

$$\|T_\delta^I\|_{p_c\to p_c'} \lesssim \sum_{\lambda_1\sim\lambda_3,\ \lambda_2\ll\lambda_1} \|T_\delta^\lambda\|_{p_c\to p_c'} \lesssim 1, \tag{7.25}$$

where $T_\delta^I := \sum_{\lambda_1\sim\lambda_3,\ \lambda_2\ll\lambda_1} T_\delta^\lambda$ denotes the contribution by the operators T_δ^λ that arise in this case. The constant in this estimate does not depend on δ.

If $H = B = 3$ and $m = 2$, then we get only a uniform estimate,

$$\sum_{\lambda_2 \ll \lambda_1} \|T_\delta^\lambda\|_{p_c \to p_c'} \lesssim 1.$$

Finally, assume that $B = 2$. Since we assume that $\theta_c \le \frac{1}{3}$, we then again find that

$$-\frac{1}{B} - \frac{1}{2} + \frac{5}{2}\theta_c \le -1 + \frac{5}{2} \cdot \frac{1}{3} < 0,$$

so that (7.25) remains valid.

Let us return to the case where $H = B = 3$ and $m = 2$; hence, $\theta_c = \frac{1}{3}$, which will require more refined methods.

In a first step, we shall take the sum of the ν_δ^λ over all dyadic $\lambda_2 \ll \lambda_1$. Moreover, since $\lambda_1 \sim \lambda_3$, we may reduce this to the case where $\lambda_3 = 2^M \lambda_1$, where $M \in \mathbb{N}$ is fixed and not too large. For the sake of simplicity of notation, we then assume that $M = 0$. All this then amounts to considering the functions $\sigma_\delta^{\lambda_1}$ given by

$$\widehat{\sigma_\delta^{\lambda_1}}(\xi) = \chi_1\left(\frac{\xi_1}{\lambda_1}\right)\chi_0\left(\frac{\xi_2}{\lambda_1}\right)\chi_1\left(\frac{\xi_3}{\lambda_1}\right)\widehat{\nu_\delta}(\xi),$$

where now χ_0 is smooth, compactly supported in an interval $[-\varepsilon, \varepsilon]$, where $\varepsilon > 0$ is sufficiently small, and $\chi_0 \equiv 1$ in the interval $[-\varepsilon/2, \varepsilon/2]$. In particular, $\sigma_\delta^{\lambda_1}(x)$ is given again by expression (7.22), only with the second factor $\check{\chi}_1(\lambda_2(x_2 - \delta_0 y_2 - y_1^m \omega(\delta_1 y_1))$ in the integrand replaced by $\check{\chi}_0\big(\lambda_1(x_2 - \delta_0 y_2 - y_1^m \omega(\delta_1 y_1))\big)$ and λ_2 replaced by λ_1. Thus we obtain the same type of estimates as in (7.24), that is,

$$\|\widehat{\sigma_\delta^{\lambda_1}}\|_\infty \lesssim \lambda_1^{-5/6}, \qquad \|\sigma_\delta^{\lambda_1}\|_\infty \lesssim \lambda_1^{2/3}\min\{\lambda_1, \delta_0^{-1}\lambda_1^{1/3}\} = \lambda_1 \min\{\lambda_1^{2/3}, \delta_0^{-1}\}. \tag{7.26}$$

By $T_\delta^{\lambda_1}$ we shall denote the operator of convolution with $\widehat{\sigma_\delta^{\lambda_1}}$.

In view of (7.26), we shall distinguish between two subcases.

1.1. The subcase where $\lambda_1 \le \delta_0^{-3/2}$. In this case, by (7.26) we have

$$\|\widehat{\sigma_\delta^{\lambda_1}}\|_\infty \lesssim \lambda_1^{-5/6}, \qquad \|\sigma_\delta^{\lambda_1}\|_\infty \lesssim \lambda_1^{5/3}, \tag{7.27}$$

so that

$$\|T_\delta^{\lambda_1}\|_{p_c \to p_c'} \lesssim \lambda_1^{-5/9}\lambda_1^{5/9} = 1,$$

and summing these estimates does not lead to the desired uniform estimate. Let us denote by

$$T_\delta^{I_1} := \sum_{\lambda_1 \le \delta_0^{-3/2}} T_\delta^{\lambda_1}$$

the contribution by the operators T_δ^λ that arise in this subcase. In order to prove the desired estimate

$$\|T_\delta^{I_1}\|_{p_c \to p_c'} \lesssim 1, \tag{7.28}$$

we shall therefore have to apply an interpolation argument (see Subsection 7.6.1).

1.2. The subcase where $\lambda_1 > \delta_0^{-3/2}$. In this case we have $\|\sigma_\delta^{\lambda_1}\|_\infty \lesssim \delta_0^{-1}\lambda_1$, and interpolation yields

$$\|T_\delta^{\lambda_1}\|_{p_c\to p_c'} \lesssim \delta_0^{-1/3}\lambda_1^{-2/9}.$$

If we denote the contribution by the operators T_δ^{λ} that arise in this subcase by $T_\delta^{I_2} := \sum_{\lambda_1>\delta_0^{-3/2}} T_\delta^{\lambda_1}$, we thus obtain

$$\|T_\delta^{I_2}\|_{p_c\to p_c'} \lesssim \sum_{\lambda_1>\delta_0^{-3/2}} \delta_0^{-1/3}\lambda_1^{-2/9} \lesssim 1. \tag{7.29}$$

Case 2. $\lambda_2 \sim \lambda_3$ and $\lambda_1 \ll \lambda_2$. Here, we can estimate $\widehat{\nu_\delta^\lambda}$ in the same way as in the previous case and obtain $\|\widehat{\nu_\delta^\lambda}\|_\infty \lesssim \lambda_2^{-1/2}\lambda_3^{-1/B} \sim \lambda_2^{-1/2-1/B}$. Moreover, by (7.24), we have $\|\nu_\delta^\lambda\|_\infty \lesssim \lambda_2 \min\{\lambda_2^{(B-1)/B}, \delta_0^{-1}\}$. Both these estimates are independent of λ_1. Assuming here without loss of generality that $\lambda_2 = \lambda_3$, we therefore consider the sum over all ν_δ^λ such that $\lambda_1 \ll \lambda_2$, by putting $\sigma_\delta^{\lambda_2} := \sum_{\lambda_1\ll\lambda_2} \nu_\delta^{(\lambda_1,\lambda_2,\lambda_2)}$. This means that

$$\widehat{\sigma_\delta^{\lambda_2}}(\xi) = \chi_0\left(\frac{\xi_1}{\lambda_2}\right)\chi_1\left(\frac{\xi_2}{\lambda_2}\right)\chi_1\left(\frac{\xi_3}{\lambda_2}\right)\widehat{\nu_\delta}(\xi),$$

where now χ_0 is smooth and compactly supported in an interval $[-\varepsilon, \varepsilon]$, with $\varepsilon > 0$ sufficiently small. In particular, $\sigma_\delta^{\lambda_2}(x)$ is given again by expression (7.22), only with the first factor $\check{\chi}_1(\lambda_1(x_1 - y_1))$ in the integrand replaced by $\check{\chi}_0(\lambda_2(x_1 - y_1))$ and λ_1 replaced by λ_2. Thus we obtain the same type of estimates

$$\|\widehat{\sigma_\delta^{\lambda_2}}\|_\infty \lesssim \lambda_2^{-1/2-1/B}, \qquad \|\sigma_\delta^{\lambda_2}\|_\infty \lesssim \lambda_2 \min\{\lambda_2^{(B-1)/B}, \delta_0^{-1}\}. \tag{7.30}$$

Denote by $T_\delta^{\lambda_2}$ the operator of convolution with $\widehat{\sigma_\delta^{\lambda_2}}$.

Interpolating between the first estimate in (7.30) and the the estimate $\|\sigma_\delta^{\lambda_2}\|_\infty \lesssim \lambda_2\lambda_2^{(B-1)/B}$, we get

$$\|T_\delta^{\lambda_2}\|_{p\to p'} \lesssim \lambda_2^{-1/B-1/2+5\theta/2}.$$

Arguing as in the first case, we see that this still suffices to sum over all dyadic λ_2 for $\theta = \theta_c = 2/p_c'$ to obtain the desired estimate

$$\|T_\delta^{II}\|_{p_c\to p_c'} \lesssim \sum_{\lambda_2} \|T_\delta^{\lambda_2}\|_{p_c\to p_c'} \lesssim 1, \tag{7.31}$$

unless $H = B = 3$ and $m = 2$. Here T_δ^{II} denotes the contribution by the operators T_δ^λ that arise in this case.

So, assume that $H = B = 3$ and $m = 2$, so that $\theta_c = \frac{1}{3}$. Then we distinguish two subcases.

2.1. The subcase where $\lambda_2 \leq \delta_0^{-3/2}$. In this case, (7.30) reads

$$\|\widehat{\sigma_\delta^{\lambda_2}}\|_\infty \lesssim \lambda_2^{-5/6}, \qquad \|\sigma_\delta^{\lambda_2}\|_\infty \lesssim \lambda_2^{5/3}, \tag{7.32}$$

which implies our previous estimate,

$$\|T_\delta^{\lambda_2}\|_{p_c\to p_c'} \lesssim 1.$$

Let us denote by

$$T_\delta^{II_1} := \sum_{\lambda_2 \le \delta_0^{-3/2}} T_\delta^{\lambda_2}$$

the contribution by the operators T_δ^λ which arise in this subcase. In order to prove the desired estimate

$$\|T_\delta^{II_1}\|_{p_c \to p_c'} \lesssim 1, \tag{7.33}$$

we shall thus have to apply an interpolation argument once more (see Subsection 7.6.1).

2.2. The subcase where $\lambda_2 > \delta_0^{-3/2}$. Then (7.30) implies that $\|\sigma_\delta^{\lambda_2}\|_\infty \lesssim \lambda_2 \delta_0^{-1}$, hence

$$\|T_\delta^{\lambda_2}\|_{p_c \to p_c'} \lesssim \delta_0^{-\theta_c} \lambda_2^{(3/2+1/3)\theta_c - 1/2 - 1/3} = \delta_0^{-1/3} \lambda_2^{-2/9}.$$

As in Subcase 1.2, this implies the desired estimate,

$$\|T_\delta^{II_2}\|_{p_c \to p_c'} \lesssim 1, \tag{7.34}$$

for the contributions $T_\delta^{II_2}$ of the operators T_δ^λ, with λ satisfying the assumptions of this subcase, to the operator to T_δ.

Case 3. $\lambda_1 \sim \lambda_2$ and $\lambda_3 \ll \lambda_1$. If $\lambda_3 \ll \lambda_2 \delta_0$, then the phase function has no critical point in x_2, and so integrations by parts in x_2 and the stationary phase method in x_1 yield

$$\|\widehat{\nu_\delta^\lambda}\|_\infty \lesssim \lambda_1^{-1/2} (\lambda_2 \delta_0)^{-N} \lesssim \lambda_2^{-1/2} \lambda_3^{-N}$$

for every $N \in \mathbb{N}$, and the second estimate in (7.24) implies that

$$\|\nu_\delta^\lambda\|_\infty \lesssim \lambda_3 \delta_0^{-1}.$$

Interpolating these estimates, we obtain

$$\|T_\delta^\lambda\|_{p \to p'} \lesssim \lambda_3^{-N'} \lambda_2^{-(1-\theta)/2} \delta_0^{-\theta},$$

where N' can be chosen arbitrarily large if $\theta < 1$. But, if $\theta = \theta_c$, then $\theta \le \frac{1}{3}$, and since $\lambda_2 \delta_0 \ge 1$ if $\lambda_3 \ll \lambda_2 \delta_0$, we see that

$$\sum_{\lambda_1 \sim \lambda_2,\ \lambda_3 \ll \lambda_2 \delta_0} \|T_\delta^\lambda\|_{p_c \to p_c'} \lesssim \delta_0^{(1-3\theta)/2} \lesssim 1.$$

Let us therefore assume from now on that in addition $\lambda_3 \gtrsim \lambda_2 \delta_0$. Then we can first apply the method of stationary phase to the integration in x_1 and, subsequently, van der Corput's estimate to the x_2-integration and obtain the estimate

$$\|\widehat{\nu_\delta^\lambda}\|_\infty \lesssim \lambda_2^{-1/2} \lambda_3^{-1/B}. \tag{7.35}$$

In view of (7.24), we distinguish two subcases.

3.1. The subcase where $\lambda_3^{1/B} > \lambda_2\delta_0$. Then interpolation of the first estimate in (7.24) with (7.35) yields

$$\|T_\delta^\lambda\|_{p\to p'} \lesssim \lambda_2^{3\theta-1/2}\lambda_3^{\theta-1/B}.$$

Since $\theta_c \le \frac{1}{3}$, we have $3\theta_c - 1 \le 0$ (even with strict inequality, unless $B=2$ or $H = B = 3$ and $m = 2$, because of Corollary 7.5(c)). If $3\theta_c - 1 < 0$, we can sum over all dyadic $\lambda_2 \gg \lambda_3$ and obtain

$$\sum_{\{\lambda_2:\lambda_3\ll\lambda_2<\delta_0^{-1}\lambda_3^{\frac{1}{B}}\}} \|T_\delta^\lambda\|_{p_c\to p_c'} \lesssim \lambda_3^{-1/B-1/2+5\theta/2}.$$

If

$$T_\delta^{III_1} := \sum_{\lambda_1\sim\lambda_2,\ \lambda_3\ll\delta_0^{-B/(B-1)},\ \lambda_3\ll\lambda_2<\delta_0^{-1}\lambda_3^{1/B}} T_\delta^\lambda$$

denotes the contribution by the operators T_δ^λ that arise in this subcase, this implies in a similar way as before that

$$\|T_\delta^{III_1}\|_{p_c\to p_c'} \lesssim 1. \tag{7.36}$$

If $H = B = 3$ and $m = 2$, then we get only a uniform estimate

$$\|T_\delta^\lambda\|_{p_c\to p_c'} \lesssim 1,$$

and in order to establish (7.36), we shall apply a complex interpolation argument in Subsection 7.6.2.

If $B = 2$ and $\theta_c = \frac{1}{3}$ (hence $m = 2$), then we can first sum over the dyadic λ_3 with $\lambda_3 > (\lambda_2\delta_0)^2$, provided $\lambda_2\delta_0 \gtrsim 1$, because $\theta_c - 1/B = -\frac{1}{6}$, and then we sum over the λ_2 for which $\lambda_2\delta_0 \gtrsim 1$. So, in order to prove (7.36) in this case, we are left with the estimation of the operator

$$T_\delta^{III_0} := \sum_{\lambda_1\sim\lambda_2,\ \lambda_3\ll\delta_0^{-1},\ \lambda_3\ll\lambda_2\ll\delta_0^{-1}} T_\delta^\lambda.$$

This will also be done by means of complex interpolation in Subsection 7.6.2.

3.2. The subcase where $\lambda_3^{1/B} \le \lambda_2\delta_0$. Assuming without loss of generality (in a similar way as before) that $\lambda_1 = \lambda_2$, then the second estimate in (7.24) implies that $\|\nu_\delta^\lambda\|_\infty \lesssim \lambda_3\delta_0^{-1}$, and in combination with (7.35) we obtain

$$\|T_\delta^\lambda\|_{p_c\to p_c'} \lesssim \lambda_2^{-(1-\theta)/2}\lambda_3^{[(B+1)\theta-1]/B}\delta_0^{-\theta}, \tag{7.37}$$

where $\theta := \theta_c$. Notice that in this subcase

$$\lambda_3 \le \min\{(\lambda_2\delta_0)^B, \lambda_2\}, \quad \text{and} \quad \lambda_2\delta_0 \lesssim \lambda_3.$$

In view of this, we shall distinguish two cases.

(a) $\lambda_2 \ge \delta_0^{-B/(B-1)}$. Assume first that $B \ge 3$. Then

$$(B+1)\theta_B - 1 = \frac{B^2 - 2B + 3}{6B} > 0$$

for $B \geq 2$, and since $\theta = \theta_c \leq \theta_B = 1/(2B) + \frac{1}{6}$ by Corollary 7.5, we see that we can sum the estimates in (7.37) over all dyadic $\lambda_3 \ll \lambda_2$ and get

$$\sum_{\lambda_3 \ll \lambda_2} \|T_\delta^\lambda\|_{p_c \to p_c'} \lesssim \lambda_2^{[(3B+2)\theta - B - 2]/2B} \delta_0^{-\theta}.$$

Again using that $\theta \leq \theta_B$, we find that for $B \geq 2$,

$$\begin{aligned}(3B+2)\theta - B - 2 &\leq (3B+2)\left(\frac{1}{6} + \frac{1}{2B}\right) - B - 2 \\ &= \frac{-3B^2 - B + 6}{6B} < 0,\end{aligned}$$

so we can also sum in $\lambda_2 \geq \delta_0^{-B/(B-1)}$ and find that $\|T_\delta^{III_{2a}}\|_{p_c \to p_c'} \lesssim \delta_0^{-[5B\theta - B - 2]/2(B-1)}$, where $T_\delta^{III_{2a}} := \sum_{\lambda_1 \sim \lambda_2 \geq \delta_0^{-B/(B-1)},\, \lambda_3 \ll \lambda_2} T_\delta^\lambda$. But,

$$5B\theta - B - 2 \leq 5B\theta_B - B - 2 = 5B\left(\frac{1}{2B} + \frac{1}{6}\right) - B - 2 = \frac{3-B}{6} \leq 0 \tag{7.38}$$

if $B \geq 3$, and thus for $B \geq 3$ we get

$$\|T_\delta^{III_{2a}}\|_{p_c \to p_c'} \lesssim 1. \tag{7.39}$$

There remains the case $B = 2$. Here, for $\theta = \theta_c$, we have $(B+1)\theta - 1 = 3\theta - 1 \leq 0$, and

$$\|T_\delta^\lambda\|_{p_c \to p_c'} \lesssim \lambda_2^{-(1-\theta)/2} \lambda_3^{(3\theta - 1)/2} \delta_0^{-\theta}.$$

Assume first that $\theta_c < \frac{1}{3}$. Then we can first sum in $\lambda_3 \geq \lambda_2\delta_0$ (notice that $\lambda_2\delta_0 > 1$) and obtain

$$\sum_{\lambda_3 \geq \lambda_2 \delta_0} \|T_\delta^\lambda\|_{p_c \to p_c'} \lesssim \lambda_2^{2\theta - 1} \delta_0^{(\theta - 1)/2}.$$

Then we sum over $\lambda_2 \geq \delta_0^{-1}$ and get an estimate by $C\delta_0^{(1-3\theta)/2} \lesssim 1$, so that (7.39) remains true also in this case.

Assume finally that $\theta_c = \frac{1}{3}$. Then $\|T_\delta^\lambda\|_{p_c \to p_c'} \lesssim \lambda_2^{-1/3} \delta_0^{-1/3}$. Summing first in $\lambda_3 \ll \lambda_2$, we get an estimate by $C(\log \lambda_2)\, \lambda_2^{-1/3} \delta_0^{-1/3}$, and summation over all $\lambda_2 \geq \delta_0^{-B/(B-1)} = \delta_0^{-2}$ leads to an estimate of order $\log(1/\delta_0)\delta_0^{1/3} \lesssim 1$. Thus, (7.39) again holds true.

(b) $\lambda_2 < \delta_0^{-B/(B-1)}$. Then we use $\lambda_3 \leq (\lambda_2\delta_0)^B$, that is, $\lambda_2 \geq \lambda_3^{1/B} \delta_0^{-1}$, and summation of the estimate (7.37) over these λ_2 yields

$$\sum_{\{\lambda_2 : \lambda_2 \geq \lambda_3^{1/B} \delta_0^{-1}\}} \|T_\delta^\lambda\|_{p_c \to p_c'} \lesssim \lambda_3^{[(3+2B)\theta - 3]/2B} \delta_0^{(1-3\theta)/2}.$$

If the exponent of λ_3 on the right-hand side of this estimate is strictly negative, then we see that

$$\|T_\delta^{III_{2b}}\|_{p_c \to p_c'} \lesssim 1, \tag{7.40}$$

where

$$T_\delta^{III_{2b}} := \sum_{\lambda_1 \sim \lambda_2 < \delta_0^{-B/(B-1)},\, \lambda_3 \ll (\lambda_2\delta_0)^B} T_\delta^\lambda.$$

So, assume that the exponent is nonnegative and notice that our assumptions in this case imply that $\lambda_3 \le \delta_0^{-B/(B-1)}$. Summation over all these dyadic λ_3 then leads to

$$\sum_{\lambda_2 \ge \lambda_3^{1/B}\delta_0^{-1},\, \lambda_3 \le \delta_0^{-B/(B-1)}} \|T_\delta^\lambda\|_{p_c \to p_c'} \lesssim \left(\log \frac{1}{\delta_0}\right) (\delta_0^{-B/(B-1)})^{[(3+2B)\theta-3]/2B} \delta_0^{(1-3\theta)/2}$$

$$= \left(\log \frac{1}{\delta_0}\right) \delta_0^{-(5B\theta - B - 2)/2(B-1)}.$$

However, if we assume without loss of generality that $\theta = \theta_B$, then we have seen in (7.38) that $5B\theta - B - 2 \le 0$ if $B \ge 3$. Moreover, by Corollary 7.5(a) we know that $\theta_c < \theta_B$, unless $B = H = 3$ and $m = 2$. Thus, using $\theta = \theta_c$ in place of θ_B, when $B \ge 3$ we have $5B\theta_c - B - 2 < 0$, and we again obtain (7.40), unless $B = H = 3$ and $m = 2$. Note that in the latter case $\theta_c = \frac{1}{3}$, so that

$$\sum_{\{\lambda_2 : \lambda_2 \ge \lambda_3^{1/3}\delta_0^{-1}\}} \|T_\delta^\lambda\|_{p_c \to p_c'} \lesssim 1.$$

The proof of (7.40) in this particular case, where we may write

$$T_\delta^{III_{2b}} = \sum_{\delta_0^{-1}\lambda_3^{1/3} \ll \lambda_1 \sim \lambda_2 < \delta_0^{-3/2}} T_\delta^\lambda,$$

will therefore again require the use of some interpolation argument in order to control the summation in λ_3. We remark that in this case, estimate (7.37) reads as $\|T_\delta^\lambda\|_{p_c \to p_c'} \lesssim \lambda_1^{-1/3}\lambda_3^{1/9}\delta_0^{-1/3}$, and

$$\sum_{1 \le \lambda_3 \lesssim \lambda_1\delta_0} \sum_{\lambda_1 < \delta_0^{-3/2}} \lambda_1^{-1/3}\lambda_3^{1/9}\delta_0^{-1/3} \lesssim 1$$

since $\lambda_1\delta_0 \ge 1$ in this double sum.

This shows that if $B = H = 3$ and $m = 2$ (hence $\theta_c = \frac{1}{3}$), then what remains to be estimated is the operator

$$T_\delta^{III_2} := \sum_{\delta_0^{-1} \ll \lambda_1 < \delta_0^{-3/2}} \sum_{\lambda_1\delta_0 \ll \lambda_3 \ll (\lambda_1\delta_0)^3} T_\delta^\lambda,$$

This will be done in Subsection 7.6.4.

Assume finally that $B = 2$. The case where $\theta_c < \frac{1}{3}$ can be treated as in the previous case (a), so assume that $\theta_c = \frac{1}{3}$. Then $\|T_\delta^\lambda\|_{p_c \to p_c'} \lesssim (\lambda_2\delta_0)^{-1/3}$. Summing first over all λ_2 such that $\lambda_2\delta_0 \ge \lambda_3^{1/2}$ leads to an estimate of order $\lambda_3^{-1/6}$, which then allows us also to sum in λ_3. Thus, (7.40) again holds true.

7.6 INTERPOLATION ARGUMENTS FOR THE OPEN CASES WHERE $m = 2$ AND $B = 2$ OR $B = 3$

Let us assume in this section that $m = 2$ and $B = 3$ or $B = 2$. Our goal will be to establish the estimates (7.28), (7.33), (7.36), and (7.40) for the operators $T_\delta^{I_1}$, $T_\delta^{II_1}$, $T_\delta^{III_1}$, and $T_\delta^{III_0}$, as well as $T_\delta^{III_{2b}}$, also at the endpoint p_c corresponding to $\theta_c = \frac{1}{3}$ (hence $p'_c = 2/\theta_c = 6$). These cases had been left open in the previous section.

The estimates for the operators $T_\delta^{III_1}$, $T_\delta^{III_0}$, and $T_\delta^{III_{2b}}$ will be established by means of complex interpolation, roughly in analogy with the proof of Proposition 4.2(a) and (b).

Also, the operators $T_\delta^{I_1}$ and $T_\delta^{II_1}$ could be handled by means of complex interpolation. We prefer to give a shorter proof, based on Proposition 2.6. Notice, however, that this approach requires us to assume the validity of the expected estimates for the operators $T_\delta^{I_1}$, $T_\delta^{II_1}$ as well as of several further operators.

7.6.1 Estimation of $T_\delta^{I_1}$ and $T_\delta^{II_1}$: Real interpolation

The estimations of the operators $T_\delta^{I_1}$ and $T_\delta^{II_1}$ will follow the same scheme, so let us consider only $T_\delta^{I_1}$, which is the operator of convolution with $\widehat{\sigma_\delta^{I_1}}$, where $\sigma_\delta^{I_1}$ denotes the measure

$$\sigma_\delta^{I_1} := \sum_{1 \leq 2^j \leq \delta_0^{-3/2}} \sigma_\delta^{2^j}.$$

Recall that if μ is any bounded, complex Borel measure on $\mathbb{R}^d$, then we denote by T_μ the convolution operator

$$T_\mu : \varphi \mapsto \varphi * \hat{\mu}.$$

As already mentioned in Chapter 2, we cannot apply the restriction estimate given by Theorem 1.1 in [BS11], whose proof is based on a real interpolation argument, directly to the measure $\mu = \sigma_\delta^{I_1}$, because an essential assumption in [BS11] is that μ is a bounded positive measure, whereas the measures $\sigma_\delta^{I_1}$ will not be positive. However, we can make use of Proposition 2.6.

Indeed, to this end let us return to our measure ν_δ. This is a positive measure, so we can choose $\mu := \nu_\delta$ in Proposition 2.6. The spectral decompositions of the measure ν_δ in Section 7.5 (as well as in the later Section 7.7) amounts to a decomposition of the measure ν_δ into a finite sum of complex measures

$$\nu_\delta = \sum_{j \in J} \nu^j + \sum_{i \in I} \mu^i \tag{7.41}$$

in such a way that the convolution operators T_{ν^j} corresponding to the measures ν^j $(j \in J)$ from the first class will be bounded from L^{p_c} to $L^{p'_c}$, whereas the measure μ^i $(i \in I)$ from the second class will satisfy the conditions required of the measures μ^i in Proposition 2.6. For instance, the operators $T_\delta^{I_2}$, $T_\delta^{II_2}$, and $T_\delta^{III_{2a}}$ from Section 7.5 belong to the first class (compare the estimates (7.29),

(7.34), (7.39)), but also $T_\delta^{III_1}$ and $T_\delta^{III_2}$ (the corresponding estimates (7.36) and (7.40) will be established by means of complex interpolation in the next subsection), whereas the operators $T_\delta^{I_1}$ and $T_\delta^{II_1}$ will belong to the second class.

We may then put $\mu^b := \sum_{j\in J} \nu^j$ in Proposition 2.6. Let us show, for instance, that the measure $\mu^i := \sigma_\delta^{I_1}$ corresponding to the operator $T_\delta^{I_1}$ satisfies the assumptions of that proposition.

Recall to this end that $\mu_j^i := \sigma_\delta^{2^j} = \nu_\delta * \phi_j$, where the Fourier transform of ϕ_j is given by $\widehat{\phi_j}(\xi) = \chi_1\left(\frac{\xi_1}{\lambda_1}\right)\chi_0\left(\frac{\xi_2}{\lambda_1}\right)\chi_1\left(\frac{\xi_3}{\lambda_1}\right)$. This implies a uniform estimate of the L^1-norms of the ϕ_j of the form (2.9) (possibly not with constant 1, but a fixed constant, which does not matter). Moreover, the estimates (2.10) and (2.11) are satisfied because of (7.27), with exponents $a_i := \frac{5}{3}$ and $b_i := \frac{5}{6}$, so that $p_0 = \frac{6}{5} = p_c$.

Similar arguments also apply to the measure $\mu^i := \sigma_\delta^{II_1}$ corresponding to the operator $T_\delta^{II_1}$, where the exponents a_i and b_i will be the same (compare (7.32)), as well as to the other measures of the first class that will appear later on.

7.6.2 Estimation of $T_\delta^{III_1}$: Complex interpolation

The discussion of this operator will somewhat resemble the one in Section 5.3 of the operator $T_{\delta,j}^V$ from Proposition 4.2, which arose from the same Subcase 3.1 of Section 4.2, with 2^{-j} playing the role of δ_0 here and where we had $B = 2$ in place of $B = 3$ here.

Recall that we have to consider the case where $m = 2$ and $H = B = 3$. The latter assumption implies that the left endpoint of the principal face $\pi(\tilde\phi^a)$ is given by $(0, 3) = (0, B)$. By Proposition 7.3, we conclude that a and n must be an integers, that $\tilde\phi^a$ is smooth, and that we may assume that our adapted coordinates are chosen so that the term of order $j = B - 1 = 2$ vanishes in (7.10), that is, that

$$\phi_\delta(x) := x_2^3\, b(x_1, x_2, \delta) + x_1^n \alpha(\delta_1 x_1) + \delta_3 x_2 x_1^{n_1} \alpha_1(\delta_1 x_1), \qquad (x_1 \sim 1, |x_2| < \varepsilon). \tag{7.42}$$

Since $H = B = 3$, we may also assume (cf. (8.4)) that

$$b(x_1, x_2, \delta) = b_3(\delta_1 x_1, \delta_2 x_2), \quad b_3(0,0) = 1. \tag{7.43}$$

Moreover, $\alpha(0) \neq 0$, and either $\alpha_1(0) \neq 0$ and then n_1 is fixed or $\alpha_1(0) = 0$ and then we may assume that n_1 is as large as we please (notice also that by (8.3), $\delta_3 = 2^{-k(n_1\tilde\kappa_1 + \tilde\kappa_2 - 1)}$ is coupled with n_1), so that in particular in the latter case $\delta_3 \ll \delta_0$. Observe also that $\delta_2 \ll \delta_0$.

Useful tools will also be Lemma 2.7 (in the sharper version of Remark 2.8) and Lemma 2.9 from Chapter 2 on oscillatory sums and double sums.

Coming back to the operator $T_\delta^{III_1}$, observe that $\lambda_1 \sim \lambda_2$ in the definition of $T_\delta^{III_1}$, so that we may and shall assume without loss of generality that $\lambda_1 = \lambda_2$. In order to verify estimate (7.36), we then have to prove the following.

Proposition 7.6. *Let $m = 2$ and $B = 3$, and consider the measure*

$$\nu_\delta^{III_1} := \sum_{2^M \le \lambda_3 \le 2^{-M}\delta_0^{-3/2}} \; \sum_{2^M \lambda_3 \le \lambda_1 \le \delta_0^{-1}\lambda_3^{1/3}} \nu_\delta^{(\lambda_1,\lambda_1,\lambda_3)},$$

where summation is taken over all sufficiently large dyadic $\lambda_i \geq 2^M$ in the given range. If we denote by $T_\delta^{III_1}$ the operator of convolution with $\widehat{\nu_\delta^{III_1}}$, then, if $M \in \mathbb{N}$ is sufficiently large (and ε sufficiently small),

$$\|T_\delta^{III_1}\|_{6/5\to 6} \leq C, \tag{7.44}$$

with a constant C not depending on δ, for δ sufficiently small.

Proof. Recall that, by (7.35), $\|\widehat{\nu_\delta^\lambda}\|_\infty \lesssim \lambda_1^{-1/2}\lambda_3^{-1/3}$. We therefore define here, for ζ in the strip $\Sigma = \{\zeta \in \mathbb{C} : 0 \leq \operatorname{Re}\zeta \leq 1\}$, an analytic family of measures by

$$\mu_\zeta(x) := \gamma(\zeta) \sum_{2^M \leq 2^{k_3} \leq 2^{-M}\delta_0^{-3/2}} \sum_{2^{M+k_3} \leq 2^{k_1} \leq \delta_0^{-1} 2^{k_3/3}} 2^{(1-3\zeta)k_1/2} 2^{(1-3\zeta)k_3/3} \nu_\delta^{(2^{k_1},2^{k_1},2^{k_3})},$$

where $\gamma(\zeta)$ is an entire function that will serve a similar role as the function $\gamma(z)$ in the proof of Proposition 4.2(a). We shall choose $\gamma(\zeta) = \gamma_1(\zeta)\gamma_2(\zeta)\gamma_3(\zeta)$ as the product of three factors $\gamma_j(\zeta)$, whose precise definition will be given in the course of the proof. It will be uniformly bounded on Σ, and such that $\gamma(\theta_c) = \gamma(\frac{1}{3}) = 1$.

By T_ζ we denote the operator of convolution with $\widehat{\mu_\zeta}$. Observe that for $\zeta = \theta_c = \frac{1}{3}$, we have $\mu_{\theta_c} = \nu_\delta^{III_1}$; hence, $T_{\theta_c} = T_\delta^{III_1}$, so that, again by Stein's interpolation theorem for analytic families of operators, (7.44) will follow if we can prove the following estimates on the boundaries of the strip Σ:

$$\|\widehat{\mu_{it}}\|_\infty \leq C \qquad \forall t \in \mathbb{R},$$

$$\|\mu_{1+it}\|_\infty \leq C \qquad \forall t \in \mathbb{R}.$$

The first estimate is an immediate consequence of our estimate for $\widehat{\nu_\delta^\lambda}$, since these functions have essentially disjoint supports, so let us concentrate on the second estimate, that is, assume that $\zeta = 1 + it$, with $t \in \mathbb{R}$. We then have to prove that there is constant C such that

$$|\mu_{1+it}(x)| \leq C, \tag{7.45}$$

where C is independent of t, x and δ.

Let us introduce the measures $\mu_{\lambda_1,\lambda_3}$ given by

$$\mu_{\lambda_1,\lambda_3}(x) := \lambda_1^{-1}\lambda_3^{-2/3}\nu_\delta^{(\lambda_1,\lambda_1,\lambda_3)}(x),$$

which allow us to rewrite

$$\mu_{1+it}(x) = \gamma(1+it) \sum_{2^M \leq \lambda_3 \leq 2^{-M}\delta_0^{-3/2}} \sum_{2^M\lambda_3 \leq \lambda_1 \leq \delta_0^{-1}\lambda_3^{1/3}} \lambda_1^{-3it/2}\lambda_3^{-it}\mu_{\lambda_1,\lambda_3}(x). \tag{7.46}$$

Recall also from (7.22) (in combination with (7.42)) that

$$\begin{aligned}\mu_{\lambda_1,\lambda_3}(x) = \lambda_1\lambda_3^{1/3} \int & \check{\chi}_1\left(\lambda_1(x_1 - y_1)\right) \check{\chi}_1\left(\lambda_1(x_2 - \delta_0 y_2 - y_1^2\omega(\delta_1 y_1))\right) \\ & \times \check{\chi}_1\left(\lambda_3\left(x_3 - y_2^3 b(y_1, y_2, \delta) - y_1^n\alpha(\delta_1 y_1) - \delta_3 y_2 y_1^{n_1}\alpha_1(\delta_1 y_1)\right)\right) \\ & \times \eta(y)\, dy,\end{aligned} \tag{7.47}$$

where η is supported where $y_1 \sim 1$ and $|y_2| < \varepsilon$. Assume first that $|x_1| \gg 1$ or $|x_1| \ll 1$. Since $\check{\chi}_1$ is rapidly decreasing and $\lambda_3 \ll \lambda_1$, we easily see that

$|\mu_{\lambda_1,\lambda_3}(x)| \le C_N \lambda_1^{-N} \lambda_3^{-N}$ for every $N \in \mathbb{N}$, which immediately implies (7.45). A similar argument applies if $|x_2| \gg 1$. However, if $|x_1| + |x_2| \lesssim 1$ and $|x_3| \gg 1$, we can conclude (after scaling by $1/\lambda_1$ in y_1) only that $|\mu_{\lambda_1,\lambda_3}(x)| \le C_N \lambda_3^{-N}$, which allows to sum in λ_3, but the summation in λ_1 remains a problem.

Let us thus assume from now on that $|x_1| \sim 1$ and $|x_2| \lesssim 1$.

By means of the change of variables $y_1 \mapsto x_1 - y_1/\lambda_1$, $y_2 \mapsto y_2/\lambda_3^{1/3}$ and Taylor expansion around x_1, we may rewrite

$$\mu_{\lambda_1,\lambda_3}(x) = \iint \check{\chi}_1(y_1) F_\delta(\lambda_1, \lambda_3, x, y_1, y_2)\, dy_1\, dy_2, \tag{7.48}$$

where

$$\begin{aligned} F_\delta(\lambda_1, \lambda_3, x, y_1, y_2) &:= \eta(x_1 - \lambda_1^{-1} y_1, \lambda_3^{-1/3} y_2)\, \check{\chi}_1 (D - E y_2 + r_1(y_1)) \\ &\times \check{\chi}_1 \Big(A - B y_2 - y_2^3 b\Big(x_1 - \lambda_3^{-1}(\lambda_3 \lambda_1^{-1} y_1), \lambda_3^{-1/3} y_2, \delta\Big) \\ &+ \lambda_3 \lambda_1^{-1} \Big(r_2(y_1) + (\lambda_3^{-1/3} y_2) \delta_3 r_3(y_1)\Big)\Big). \end{aligned}$$

Here, the quantities A to E are given by

$$\begin{aligned} A = A(x, \lambda_3, \delta) &:= \lambda_3 Q_A(x), & B := B(x, \lambda_3, \delta) &:= \lambda_3^{2/3} Q_B(x), \\ D = D(x, \lambda_1, \delta) &:= \lambda_1 Q_D(x), & E = E(\lambda_1, \lambda_3, \delta) &:= \lambda_1 \lambda_3^{-1/3} \delta_0 \le 1, \end{aligned} \tag{7.49}$$

with

$$\begin{aligned} Q_A(x) := x_3 - x_1^n \alpha(\delta_1 x_1), \quad Q_B(x) := \delta_3 x_1^{n_1} \alpha_1(\delta_1 x_1), \\ Q_D(x) := x_2 - x_1^2 \omega(\delta_1 x_1), \end{aligned}$$

and do not depend on y_1, y_2. Moreover, the functions $r_i(y_1) = r_i(y_1; \lambda_1^{-1}, x_1, \delta)$, $i = 1, 2, 3$, are smooth functions of y_1 (and λ_1^{-1} and x_1) satisfying estimates of the form

$$|r_i(y_1)| \le C|y_1|, \qquad \left|\left(\frac{\partial}{\partial(\lambda_1^{-1})}\right)^l r_i(y_1; \lambda_1^{-1}, x_1, \delta)\right| \le C_l |y_1|^{l+1} \quad \text{for every} \quad l \ge 1. \tag{7.50}$$

Notice that we may here assume that $|y_1| \lesssim \lambda_1$ because of our assumption $|x_1| \lesssim 1$ and the support properties of η. It will also be important to observe that $E = \delta_0 \lambda_1 \lambda_3^{-1/3} \le 1$ for the index set of λ_1, λ_3 over which we sum in (7.46). Notice also that $|\lambda_3^{-1/3} y_2| \le \varepsilon$.

Let us choose $c > 0$ so that $|r_i(y_1)| \le c(1 + |y_1|)$, $i = 1, 2, 3$.

In order to verify (7.45), given x, we shall split the sum in (7.46) into four parts, which will be treated subsequently in different ways.

1. The part where $\max\{|A|, |B|\} \ge 1$ **and** $|D| \ge 4c$**.** Denote by $\mu^1_{1+it}(x)$ the contribution to $\mu_{1+it}(x)$ by the terms for which $\max\{|A(x, \lambda_3, \delta)|, |B(x, \lambda_3, \delta)|\} \ge 1$ and $|D(x, \lambda_1, \delta)| \ge 4c$.

We claim that here

$$|\mu_{\lambda_1,\lambda_3}(x)| \lesssim |D|^{-1/4} \max\{|A|^{1/3}, |B|^{1/2}\}^{-1/4}. \tag{7.51}$$

In view of (7.49), this estimate will allow us to sum over all dyadic λ_1, λ_3 for which the corresponding quantities A, B, and D satisfy the conditions of this subcase, and we obtain the right kind of estimate $|\mu^1_{1+it}(x)| \leq C$, in agreement with (7.45).

In order to prove (7.51), let us first consider the contribution $\mu^1_{\lambda_1,\lambda_3}(x)$ to the integral defining $\mu_{\lambda_1,\lambda_3}(x)$ by the region where $|y_1| > |D|/4c$. Here we may estimate $|\check{\chi}_1(y_1)| \lesssim |D|^{-N}$ for every $N \in \mathbb{N}$. Moreover, if $|x_3| \gg 1$, then $|A| \gg \lambda_3 \gg |B|^{3/2}$, and A becomes the dominant term in the argument of the last factor of F_δ. Therefore, we may estimate

$$|\mu^1_{\lambda_1,\lambda_3}(x)| \leq C_N |D|^{-N} \lambda_3^{1/3} |A|^{-N}$$

for every $N \in \mathbb{N}$, which is stronger than (7.51).

And, if $|x_3| \lesssim 1$, then we may apply Lemma 2.4 (with $T := \lambda_3^{1/3}$, $\epsilon := 0$) and obtain the estimate

$$|\mu^1_{\lambda_1,\lambda_3}(x)| \lesssim |D|^{-N} | \max\{|A|^{1/3}, |B|^{1/2}\}^{-1/2},$$

which is still stronger than required in (7.51).

Denote next by $\mu^2_{\lambda_1,\lambda_3}(x)$ the contribution by the region where $|y_1| \leq |D|/4c$. Then $|r_i(y_1)| \leq |D|/2$, and thus if, in addition, $|Ey_2| \leq |D|/4$, or $|Ey_2| > 2|D|$, then $\left|D - Ey_2 + r_1(y_1)\right| \geq |D|/4$. We may then estimate the second factor of F_δ by $C_N|D|^{-N}$, which allows to argue as before. So, let us assume that $|r_1(y_1)| \leq |D|/2$ and $|D|/4 \leq |Ey_2| \leq 2|D|$. Then $|y_2| \sim |D|/|E| \geq |D| \gtrsim 1$. In case that $|D| \lesssim \max\{|A|^{1/3}, |B|^{1/2}\}$, we may apply Lemma 2.4 (assuming again that $|x_3| \lesssim 1$; the other case is again easier) and find that

$$|\mu^2_{\lambda_1,\lambda_3}(x))| \lesssim \max\{|A|^{1/3}, |B|^{1/2}\}^{-1/2} \leq |D|^{-1/4} \max\{|A|^{1/3}, |B|^{1/2}\}^{-1/4},$$

as desired. So assume that $|D| \gg \max\{|A|^{1/3}, |B|^{1/2}\}$. Then one easily sees that

$$\begin{aligned}\Big|A - By_2 - y_2^3 b\left(x_1 - \lambda_1^{-1}y_1, \lambda_3^{-1/3}y_2, \delta\right) \\ + \lambda_3\lambda_1^{-1}\left(r_2(y_1) + (\lambda_3^{-1/3}y_2)\delta_3 r_3(y_1)\right)\Big| \gtrsim |y_2|D^2,\end{aligned}$$

so that

$$\begin{aligned}|\mu^2_{\lambda_1,\lambda_3}(x))| &\lesssim \iint (1 + |y_1|)^{-N}(1 + |y_2|D^2)^{-N}\, dy_1\, dy_2 \\ &\lesssim D^{-2} \lesssim |D|^{-1} \max\{|A|^{1/3}, |B|^{1/2}\}^{-1},\end{aligned}$$

which is again stronger then required in (7.51).

2. The part where $\max\{|A|, |B|\} \geq 1$ **and** $|D| < 4c$**.** Denote by $\mu^2_{1+it}(x)$ the contribution to $\mu_{1+it}(x)$ by the terms for which $\max\{|A(x, \lambda_3, \delta)|, |B(x, \lambda_3, \delta)|\} \geq 1$ and $|D(x, \lambda_1, \delta)| < 4c$.

Let us fix λ_3 satisfying $2^M \leq \lambda_3 \leq 2^{-M}\delta_0^{-3/2}$ and $\max\{|A(x, \lambda_3, \delta)|, |B(x, \lambda_3, \delta)|\} \geq 1$ in the first sum in (7.46). In order to compute $\mu^2_{1+it}(x)$, we then have to study the following sum in $\lambda_1 = 2^{k_1}$:

$$\sigma^2(\lambda_3, t, x) := \sum_{\{\lambda_1 : 2^M\lambda_3 \leq \lambda_1 \leq \delta_0^{-1}\lambda_3^{1/3},\ \lambda_1|Q_D(x)| < 4c\}} \lambda_1^{-3it/2} \mu_{\lambda_1,\lambda_3}(x).$$

Indeed, we have $\mu^2_{1+it}(x) = \sum_{\lambda_3} \lambda_3^{-it} \sigma^2(\lambda_3, t, x)$, where summation is over all dyadic λ_3 in the range described before.

The oscillatory sum defining $\sigma^2(\lambda_3, t, x)$ can essentially be written in the form (2.15), with $\alpha := -\frac{3}{2}, l = k_1$ and

$$u_1 = 2^{\beta^1 l} a_1 := \lambda_1^{-1}\lambda_3, \; u_2 = 2^{\beta^2 l} a_2 := \lambda_1 Q_D(x), \; u_3 = 2^{\beta^3 l} a_3 := \lambda_1(\lambda_3^{-1/3}\delta_0),$$

and where the function $H = H_{\lambda_3, x, \delta}$ of $u := (u_1, u_2, u_3)$ is given by

$$\begin{aligned} H(u) := \iint & \check{\chi}_1(y_1)\eta(x_1 - u_1\lambda_3^{-1} y_1, \lambda_3^{-1/3} y_2)\, \check{\chi}_1\left(u_2 - u_3 y_2 + r_1(y_1; \lambda_3^{-1}u_1, x_1, \delta)\right) \\ & \times \check{\chi}_1\Big(A - B y_2 - y_2^3 b\left(x_1 - \lambda_3^{-1} u_1 y_1, \lambda_3^{-1/3} y_2, \delta\right) + u_1 r_2(y_1; \lambda_3^{-1}u_1, x_1, \delta) \\ & + (\lambda_3^{-1/3} y_2)\delta_3 r_3(y_1; \lambda_3^{-1}u_1, x_1, \delta)\Big)\, dy_1\, dy_2. \end{aligned}$$

Moreover, the cuboid Q in Lemma 2.7 is defined by the conditions

$$|u_1| \le 2^{-M}, \quad |u_2| < 4c, \quad |u_3| \le 1.$$

Let us estimate the C^1-norm of H on Q. If $|x_3| \gg 1$, then $|Q_A(x)| \gg 1$, whereas $|Q_B(x)| \lesssim 1$, so that $|A| \gg |B|$; hence, $\max\{|A|, |B|\} = |A|$. We even have that $|By_2| \ll \lambda_3 |Q_B(x)| \ll |A|$, as well as $|y_2^3 b(x_1 - \lambda_3^{-1}u_1 y_1, \lambda_3^{-1/3} y_2, \delta)| \ll |A|$. Making use also of the rapid decay of $\check{\chi}_1(y_1)$, this easily implies that

$$|H(u)| \lesssim |A|^{-N}$$

for every $N \in \mathbb{N}$, uniformly on Q.

On the other hand, if $|x_3| \lesssim 1$, then $|A| \lesssim \lambda_3$, so that the assumptions of Lemma 2.4 are satisfied (if we essentially put $T := \lambda_3^{1/3}$), and we may conclude that

$$|H(u)| \lesssim \max\{|A|^{1/3}, |B|^{1/2}\}^{-1/2}, \qquad \text{for all } u \in Q. \tag{7.52}$$

We have seen that this estimate holds true no matter which size $|x_3|$ may have.

We next consider partial derivatives of H. From our integral formula for $H(u)$, it is obvious that the partial derivative of H with respect to u_1 will essentially only produce additional factors of the form $\lambda_3^{-1} y_1$, $\lambda_3^{-1} y_1 y_2^3$, $\lambda_3^{-1/3} y_2$ and $\lambda_3^{-1} y_1$ under the double integral. However, powers of y_1 can be absorbed by the rapidly decaying factor $\check{\chi}_1(y_1)$, and $|\lambda_3^{-1} y_2^3| \le \varepsilon \ll 1$, so that $|\partial_{u_1} H(u)|$ will satisfy an estimate of the form (7.52) as well. It is also easy to see that $|\partial_{u_2} H(u)|$ satisfies such an estimate, too.

More of a problem is the partial derivative of H with respect to u_3. This will essentially produce an additional factor y_2 under the double integral. More precisely, let us put

$$g(y_1, y_2; u) := -\check{\chi}_1(y_1)\, y_2\, \check{\chi}_1'\left(u_2 - u_3 y_2 + r_1(y_1; \lambda_3^{-1}u_1, x_1, \delta)\right),$$

so that

$$\begin{aligned}\partial_{u_3} H(u) := \iint & \eta(x_1 - u_1\lambda_3^{-1} y_1, \lambda_3^{-1/3} y_2)\, g(y_1, y_2; u)\\ & \times \check{\chi}_1 \Big(A - By_2 - y_2^3 b\left(x_1 - \lambda_3^{-1} u_1 y_1, \lambda_3^{-1/3} y_2, \delta\right)\\ & + u_1 r_2(y_1; \lambda_3^{-1} u_1, x_1, \delta)\\ & + (\lambda_3^{-1/3} y_2)\delta_3 r_3(y_1; \lambda_3^{-1} u_1, x_1, \delta)\Big)\, dy_1\, dy_2.\end{aligned}$$

We claim that for every $\epsilon \in]0, 1]$, and $s \in [0, 1]$, we have

$$|g(y_1, y_2; su)| \leq C_N |u_3|^{\epsilon-1} |y_2|^{\epsilon}\, s^{\epsilon-1} (1 + |y_1|)^{-N} \qquad \text{for every } N \in \mathbb{N} \quad (7.53)$$

in the integrand. Indeed, if $|su_3 y_2| \gg (1 + |y_1|)$, then the third factor in $g(y_1, y_2; u)$ can be estimated by $C|su_3 y_2|^{\epsilon-1}$ because of the rapid decay of $\check{\chi}_1'$, and (7.53) follows; if $|su_3 y_2| \lesssim (1 + |y_1|)$, then $|y_2| \lesssim |y_2|^{\epsilon}(|su_3|^{-1}(1 + |y_1|))^{1-\epsilon}$, and (7.53) follows again.

By means of (7.53), we may now estimate

$$\begin{aligned}|\partial_{u_3} H(su)| \lesssim |u_3|^{\epsilon-1}\, s^{\epsilon-1} & \iint (1 + |y_1|)^{-N}\\ & \times \Big(1 + |A - By_2 - y_2^3 b\left(x_1 - \lambda_3^{-1} s u_1 y_1, \lambda_3^{-1/3} y_2, \delta\right)\\ & + s u_1 r_2(y_1; \lambda_3^{-1} s u_1, x_1, \delta) + (\lambda_3^{-1/3} y_2)\delta_3 r_3(y_1; \lambda_3^{-1} s u_1, x_1, \delta)|\Big)^{-N}\\ & \times |y_2|^{\epsilon}\, dy_1\, dy_2,\end{aligned}$$

and arguing from here on as before (distinguishing between the cases where $|x_3| \gg 1$ and where $|x_3| \lesssim 1$), we obtain by means of Lemma 2.4 that

$$|\partial_{u_3} H(su)| \lesssim |u_3|^{\epsilon-1}\, s^{\epsilon-1} \max\{|A|^{1/3}, |B|^{1/2}\}^{\epsilon-1/2}, \qquad \text{for all } u \in Q.$$

This implies for every sufficiently small $\epsilon > 0$ that for all $u \in Q$,

$$\int_0^1 \left|\frac{\partial H}{\partial u_k}(su)\right| ds \leq C|u_k|^{\epsilon-1} \max\{|A|^{1/3}, |B|^{1/2}\}^{\epsilon-1/2}.$$

By means of Lemma 2.7, we may thus conclude that

$$|\sigma^2(\lambda_3, t, x)| \lesssim \frac{1}{|2^{-3it/2t} - 1|} \max\left\{(\lambda_3 |Q_A(x)|)^{1/3}, (\lambda_3^{2/3} |Q_B(x)|)^{1/2}\right\}^{\epsilon-1/2}.$$

Finally, this estimate allows us also to sum in λ_3, and we conclude that $|\mu^2_{1+it}(x)| \leq C$, provided we choose the second factor in the definition of $\gamma(\zeta)$ as $\gamma_2(\zeta) := 2^{3(1-\zeta)/2} - 1$.

3. The part where $\max\{|A|, |B|\} < 1$ **and** $|D| \geq 4c$. Denote by $\mu^3_{1+it}(x)$ the contribution to $\mu_{1+it}(x)$ by the terms for which $\max\{|A(x, \lambda_3, \delta)|, |B(x, \lambda_3, \delta)|\} < 1$ and $|D(x, \lambda_1, \delta)| \geq 4c$.

This case can be treated again by means of Lemma 2.7, only with the roles of λ_1 and λ_3 interchanged. So, let us here fix λ_1 satisfying $2^{2M} \leq \lambda_1 \leq 2^{-M/3}\delta_0^{-3/2}$ and

$\lambda_1|Q_D(x)| \geq 4c$ and consider the remaining sum in λ_3 in (7.46), that is,

$$\sigma^3(\lambda_1, t, x) := \sum_{\{\lambda_3 : \delta_0^3\lambda_1^3 \leq \lambda_3 \leq 2^{-M}\lambda_1,\, \lambda_3|Q_A(x)|<1,\, \lambda_3^{2/3}|Q_B(x)|<1\}} \lambda_3^{-it}\mu_{\lambda_1,\lambda_3}(x).$$

Notice that then $\mu^3_{1+it}(x) = \sum_{\lambda_1} \lambda_1^{-3it/2}\sigma^3(\lambda_1, t, x)$, where summation is over all dyadic λ_1 in the range described before.

Also $\sigma^3(\lambda_1, t, x)$ can essentially be written in the form (2.15), with $\alpha := -1$ and $l = k_3$ (if $\lambda_3 = 2^{k_3}$), and

$$u_1 = 2^{\beta^1 l}a_1 := \lambda_3\lambda_1^{-1},\quad u_2 = 2^{\beta^2 l}a_2 := \lambda_3^{-1/3},\quad u_3 = 2^{\beta^3 l}a_3 := \lambda_3^{-1/3}(\lambda_1\delta_0),$$
$$u_4 = 2^{\beta^4 l}a_4 := \lambda_3 Q_A(x),\quad u_5 = 2^{\beta^5 l}a_5 := \lambda_3^{2/3} Q_B(x),$$

where the function $H = H_{\lambda_1,x,\delta}$ of $u := (u_1, \ldots, u_5)$ is given by

$$\begin{aligned} H(u) := \iint & \check{\chi}_1(y_1)\eta(x_1 - \lambda_1^{-1}y_1,\, u_2y_2)\,\check{\chi}_1\left(D - u_3y_2 + r_1(y_1; \lambda_1^{-1}, x_1, \delta)\right) \\ & \times \check{\chi}_1\left(u_4 - u_5y_2 - y_2^3 b\left(x_1 - \lambda_1^{-1}y_1, u_2y_2, \delta\right) + u_1r_2(y_1; \lambda_1^{-1}, x_1, \delta)\right. \\ & \left. +\ (u_2y_2)\delta_3 r_3(y_1; \lambda_1^{-1}, x_1, \delta)\right) dy_1\, dy_2. \end{aligned}$$

Moreover, the cuboid Q in Lemma 2.7 is here defined by the conditions

$$|u_1| \leq 2^{-M},\quad |u_2| \leq 2^{-M/3},\quad |u_3| \leq 1,\quad |u_4| \leq 1,\quad |u_5| \leq 1.$$

In order to estimate the C^1-norm of H on Q, observe first that here we may estimate

$$\begin{aligned} &\left|\check{\chi}_1(y_1)\,\check{\chi}_1\left(u_4 - u_5y_2 - y_2^3 b\left(x_1 - \lambda_1^{-1}y_1, u_2y_2, \delta\right) + u_1r_2(y_1; \lambda_1^{-1}, x_1, \delta)\right.\right. \\ &\quad \left.\left. +\ (u_2y_2)\delta_3r_3(y_1; \lambda_1^{-1}, x_1, \delta)\right)\right| \\ &\leq C_N(1 + |y_1|)^{-N}(1 + |y_2|)^{-N}, \end{aligned} \tag{7.54}$$

for every $N \in \mathbb{N}$ (just distinguish the cases where $|y_1| \ll |y_2|^3$, and $|y_1| \gtrsim |y_2|^3$). Notice that this estimate allows us, in particular, to absorb any powers of y_1 or y_2 in the upcoming estimations. Moreover, we find that

$$\begin{aligned} |H(u)| \leq C_N \iint & (1 + |y_1|)^{-N}(1 + |y_2|)^{-N} \\ & \times \left|\check{\chi}_1\left(D - u_3y_2 + r_1(y_1; \lambda_1^{-1}, x_1, \delta)\right)\right| dy_1\, dy_2. \end{aligned} \tag{7.55}$$

It is easy to see that this allows to estimate

$$|H(u)| \leq C_N|D|^{-N/2}, \qquad u \in Q.$$

Indeed, when $1 + |y_1| \gtrsim |D|$, then we can gain a factor $|D|^{-N/2}$ from the first factor in the integral in (7.55), and when $1 + |y_1| \ll |D|$ and $|u_3y_2| \leq |D|/2$, then the last factor in the integral can be estimated by $C'_N|D|^{-N}$. Finally, when $1 + |y_1| \ll |D|$ and $|u_3y_2| \geq |D|/2$, then $|y_2| \geq |D|/2$ because $|u_3| \leq 1$, and we can gain a factor $|D|^{-N/2}$ from the second factor in the integral in (7.55).

Similar estimates hold true also for partial derivatives of H, since these essentially produce only further factors of the order $|y_2|$, $|r_2(y_1)| \lesssim (1 + |y_1|)$ and

$|y_2 r_3(y_1)| \lesssim |y_2|(1+|y_1|)$ under the integral defining $H(u)$, and as we have observed before, such factors can easily be absorbed.

We thus find that $\|H\|_{C^1(Q)} \lesssim |D|^{-1}$, so that, by Lemma 2.7,

$$|\sigma^3(\lambda_1, t, x)| \lesssim \frac{1}{|2^{-it}-1|} (\lambda_1 |Q_D(x)|)^{-1}.$$

This estimate allows to sum in λ_1, since we are assuming that $\lambda_1 |Q_D(x)| \geq 4c$ in the definition of $\mu^3_{1+it}(x)$, and we conclude also that $|\mu^3_{1+it}(x)| \leq C$, provided we choose the second factor in the definition of $\gamma(\zeta)$ as $\gamma_2(\zeta) := (2^{1-\zeta}-1)/(2^{2/3}-1)$.

4. The part where $\max\{|A|, |B|\} < 1$ **and** $|D| < 4c$**.** Denote by $\mu^4_{1+it}(x)$ the contribution to $\mu_{1+it}(x)$ by the terms for which $\max\{|A(x, \lambda_3, \delta)|, |B(x, \lambda_3, \delta)|\} < 1$ and $|D(x, \lambda_1, \delta)| < 4c$.

Under the assumptions of this case, it is easily seen from formula (7.48), in combination with an estimate analogous to (7.55), that

$$\mu_{\lambda_1, \lambda_3}(x) = J(A, B, D, E, \lambda_1^{-1}, \lambda_3^{-1/3}, \lambda_3 \lambda_1^{-1}),$$

where J is a smooth function of all its (bounded) variables. We may thus invoke Lemma 2.9 on oscillatory double sums in order to also conclude that $|\mu^4_{1+it}(x)| \leq C$, provided we choose the third factor $\gamma_3(\zeta)$ of $\gamma(\zeta)$ according to Remark 2.10.

Since the details are very similar to the discussion of the corresponding case in the last part of the proof of Proposition 4.2, we shall skip the details.

Estimate (7.45) is a consequence of our estimates on the $\mu^j_{1+it}(x)$, $j = 1, \ldots, 4$, which completes the proof of Proposition 7.6. Q.E.D.

7.6.3 Estimation of $T_\delta^{III_0}$: Complex interpolation

The discussion of this operator is easier than the one in the preceding subsection. Observe that in place of (7.42), we here have

$$\phi_\delta(x) := x_2^2\, b(x_1, x_2, \delta) + x_1^n \alpha(\delta_1 x_1) + \delta_3 x_2 x_1^{n_1} \alpha_1(\delta_1 x_1), \qquad (x_1 \sim 1, |x_2| < \varepsilon),$$

since $B = 2$.

Assuming again without loss of generality that $\lambda_1 = \lambda_2$, we see that we have to prove the following.

Proposition 7.7. *Let $m = 2$ and $B = 2$, and consider the measure*

$$\nu_\delta^{III_0} := \sum_{2^M \leq \lambda_3 \leq 2^{-M}\delta_0^{-1}} \sum_{2^M \lambda_3 \leq \lambda_1 \leq 2^{-M}\delta_0^{-1}} \nu_\delta^{(\lambda_1, \lambda_1, \lambda_3)},$$

where summation is taken over all sufficiently large dyadic λ_i in the given range. If we denote by $T_\delta^{III_0}$ the operator of convolution with $\widehat{\nu_\delta^{III_0}}$, then, if $M \in \mathbb{N}$ is sufficiently large (and ε sufficiently small),

$$\|T_\delta^{III_0}\|_{6/5 \to 6} \leq C, \tag{7.56}$$

with a constant C not depending on δ, for δ sufficiently small.

Proof. For fixed λ_3 satisfying $2^M \le \lambda_3 \le 2^{-M}\delta_0^{-1}$, we let

$$\sigma^{\lambda_3} := \sum_{\{\lambda_1 : 2^M\lambda_3 \le \lambda_1 \le 2^{-M}\delta_0^{-1}\}} \nu_\delta^{(\lambda_1,\lambda_1,\lambda_3)},$$

so that

$$\nu_\delta^{III_0} = \sum_{2^M \le \lambda_3 \le 2^{-M}\delta_0^{-1}} \sigma^{\lambda_3}.$$

We embed σ^{λ_3} into an analytic family of measures

$$\sigma_\zeta^{\lambda_3}(x) := \gamma(\zeta) \sum_{\{\lambda_1 : 2^M\lambda_3 \le \lambda_1 \le 2^{-M}\delta_0^{-1}\}} \lambda_1^{(1-3\zeta)/2} \nu_\delta^{(\lambda_1,\lambda_1,\lambda_3)}, \qquad \zeta \in \Sigma,$$

where $\gamma(\zeta) := 2^{3(1-\zeta)/2} - 1$, so that $\sigma_{1/3}^{\lambda_3} = \sigma^{\lambda_3}$. From (7.35) we obtain that

$$\|\widehat{\sigma_{it}^{\lambda_3}}\|_\infty \le C\lambda_3^{-1/2} \qquad \forall t \in \mathbb{R}.$$

We shall also prove that for every sufficiently small $\epsilon > 0$, there is a constant C_ϵ such that

$$\|\sigma_{1+it}^{\lambda_3}\|_\infty \le C_\epsilon \lambda_3^{(1+\epsilon)/2} \qquad \forall t \in \mathbb{R}.$$

By Stein's interpolation theorem for analytic families of operators, these estimates easily imply that

$$\|T^{\lambda_3}\|_{p_c \to p_c'} \lesssim (\lambda_3^{-1/2})^{2/3} (\lambda_3^{(1+\epsilon)/2})^{1/3} = \lambda_3^{(\epsilon-1)/6},$$

where T^{λ_3} denotes the operator of convolution with $\widehat{\sigma^{\lambda_3}}$. Thus, if we choose ϵ sufficiently small, we can also sum in λ_3 and obtain (7.56).

Our goal will thus be to show that for ϵ sufficiently small, we have

$$|\sigma_{1+it}^{\lambda_3}(x)| \le C_\epsilon \lambda_3^{(1+\epsilon)/2}, \tag{7.57}$$

where C_ϵ is independent of t, x, δ and λ_3.

To this end, observe that

$$\sigma_{1+it}^{\lambda_3}(x) := \gamma(1+it) \sum_{\{\lambda_1 : 2^M\lambda_3 \le \lambda_1 \le 2^{-M}\delta_0^{-1}\}} \lambda_1^{-3it/2} \lambda_1^{-1} \nu_\delta^{(\lambda_1,\lambda_1,\lambda_3)}(x),$$

and, by (7.22), (7.23),

$$\begin{aligned}\lambda_1^{-1}\nu_\delta^{(\lambda_1,\lambda_1,\lambda_3)}(x) = \lambda_1\lambda_3 \int & \check{\chi}_1(\lambda_1(x_1 - y_1))\, \check{\chi}_1\big(\lambda_1(x_2 - \delta_0 y_2 - y_1^2\omega(\delta_1 y_1))\big) \\ & \times \check{\chi}_1\big(\lambda_3\big(x_3 - y_2^2 b(y_1, y_2, \delta) - y_1^n\alpha(\delta_1 y_1) - \delta_3 y_2 y_1^{n_1}\alpha_1(\delta_1 y_1)\big)\big) \\ & \times \eta(y)\, dy,\end{aligned} \tag{7.58}$$

where η is supported where $y_1 \sim 1$ and $|y_2| < \varepsilon$. Now, if $|x_1| \gg 1$, or $|x_2| \gg 1$, then similar arguments to those in the preceding subsection show that $|\lambda_1^{-1}\nu_\delta^{(\lambda_1,\lambda_1,\lambda_3)}(x)| \lesssim \lambda_1^{-N}\lambda_3 \le \lambda_3^{-1}\lambda_1^{2-N}$ for every $N \ge 2$, which implies (7.57).

We shall therefore assume from now on that $|x_1| + |x_2| \lesssim 1$. The change of variables $y_1 \mapsto x_1 - y_1/\lambda_1$, $y_2 \mapsto y_2/\lambda_3^{1/2}$ then leads in a similar way as in the previous subsection to

$$\lambda_1^{-1} \nu_\delta^{(\lambda_1,\lambda_1,\lambda_3)}(x) = \lambda_3^{1/2} \mu_{\lambda_1,\lambda_3}(x),$$

where

$$\mu_{\lambda_1,\lambda_3}(x) := \iint \check{\chi}_1(y_1) F_\delta(\lambda_1, \lambda_3, x, y_1, y_2)\, dy_1\, dy_2,$$

with

$$\begin{aligned} F_\delta(\lambda_1, \lambda_3, x, y_1, y_2) := &\, \eta(x_1 - \lambda_1^{-1} y_1, \lambda_3^{-1/2} y_2)\, \check{\chi}_1(D - Ey_2 + r_1(y_1)) \\ &\times \check{\chi}_1(A - By_2 - y_2^2 b(x_1 - \lambda_3^{-1}(\lambda_3\lambda_1^{-1} y_1), \lambda_3^{-1/2} y_2, \delta) \\ &+ \lambda_3\lambda_1^{-1}(r_2(y_1) + (\lambda_3^{-1/2} y_2)\delta_3 r_3(y_1))). \end{aligned}$$

Here, the quantities A to E are given by

$$\begin{aligned} A &:= \lambda_3 Q_A(x), & B &:= \lambda_3^{1/2} Q_B(x), \\ D &:= \lambda_1 Q_D(x), & E &:= (\delta_0 \lambda_1)\lambda_3^{-1/2}, \end{aligned}$$

with $Q_A(x)$, $Q_B(x)$ and $Q_D(x)$ as in (7.49). The functions $r_i(y_1)$ have properties as before, and we again choose $c > 0$ so that $|r_i(y_1)| \leq c(1 + |y_1|)$, $i = 1, 2, 3$. Observe also that in this integral, $|y_1| \lesssim \lambda_1$ and $|y_2| \ll \lambda_3^{1/2}$ and that only D and E depend on the summation variable λ_1.

Observe also that by the van der Corput estimate in Lemma 2.1(b) we may estimate

$$\int \left|\chi_1(A - By_2 - y_2^2 b(x_1, \dots, \delta) + \lambda_3\lambda_1^{-1}(r_2(y_1) + (\lambda_3^{-1/2} y_2)\delta_3 r_3(y_1)))\right| dy_2 \leq C, \tag{7.59}$$

with a constant C that does not depend on A, B, x, y_1, the λ_j and δ.

1. The part where $|D| \geq 4c$. Denote by $\sigma_{1+it,1}^{\lambda_3}(x)$ the contribution to $\sigma_{1+it}^{\lambda_3}(x)$ by the terms for which $|D(x, \lambda_1, \delta)| \geq 4c$. We claim that, for every $N \in \mathbb{N}$,

$$|\mu_{\lambda_1,\lambda_3}(x)| \lesssim |D|^{-N}. \tag{7.60}$$

Clearly, this estimate will allow us to sum in λ_1 and obtain the right kind of estimate $|\sigma_{1+it,1}^{\lambda_3}(x)| \leq C\lambda_3^{1/2}$, in agreement with (7.57).

In order to prove (7.60), let us first consider the contribution $\mu_{\lambda_1,\lambda_3}^1(x)$ to the integral defining $\mu_{\lambda_1,\lambda_3}(x)$ by the region where $|y_1| > |D|/4c$. Here we may estimate $|\check{\chi}_1(y_1)| \lesssim |D|^{-N}(1 + |y_1|)^{-N}$ for every $N \in \mathbb{N}$, and combining this with (7.59) clearly yields (7.60).

Denote next by $\mu_{\lambda_1,\lambda_3}^2(x)$ the contribution by the region where $|y_1| \leq |D|/4c$. Then $|r_i(y_1)| \leq |D|/2$. But notice also that

$$|Ey_2| \ll |E|\lambda_3^{1/2} \leq \lambda_1\delta_0 \ll 1,$$

which shows that $|D - Ey_2 + r_1(y_1)| \geq |D|/4$. Thus, the second factor in F_δ can be estimated by $C_N|D|^{-N}$, and we again arrive at (7.60).

2. The part where $|D| < 4c$**.** Denote by $\sigma^{\lambda_3}_{1+it,2}(x)$ the contribution to $\sigma^{\lambda_3}_{1+it}(x)$ by the terms for which $|D(x, \lambda_1, \delta)| < 4c$.

As in the discussion in the previous subsection (part 2.) we see that the oscillatory sum defining $\lambda_3^{-1/2}\sigma^{\lambda_3}_{1+it,2}(x)$ can essentially be written in the form (2.15), with $\alpha := -\frac{3}{2}, l = k_1$ and

$$u_1 = 2^{\beta^1 l}a_1 := \lambda_1^{-1}\lambda_3,\ u_2 = 2^{\beta^2 l}a_2 := \lambda_1 Q_D(x),\ u_3 = 2^{\beta^3 l}a_3 := \lambda_1(\lambda_3^{-1/2}\delta_0),$$

and where the function $H = H_{\lambda_3,x,\delta}$ of $u := (u_1, u_2, u_3)$ is now given by

$$\begin{aligned} H(u) := \iint & \check{\chi}_1(y_1)\eta(x_1 - u_1\lambda_3^{-1}y_1,\ \lambda_3^{-1/2}y_2)\,\check{\chi}_1(u_2 - u_3 y_2 + r_1(y_1; \lambda_3^{-1}u_1, x_1, \delta)) \\ & \times \check{\chi}_1(A - By_2 - y_2^2 b(x_1 - \lambda_3^{-1}u_1 y_1, \lambda_3^{-1/2}y_2, \delta) + u_1 r_2(y_1; \lambda_3^{-1}u_1, x_1, \delta) \\ & + (\lambda_3^{-1/2}y_2)\delta_3 r_3(y_1; \lambda_3^{-1}u_1, x_1, \delta))\, dy_1\, dy_2. \end{aligned}$$

Moreover, the cuboid Q in Lemma 2.7 is here defined by the conditions

$$|u_1| \leq 2^{-M}, \quad |u_2| < 4c, \quad |u_3| \leq 2^{-3M/2}.$$

Let us estimate the C^1-norm of H on Q. Because of (7.59), we clearly have $\|H\|_{C(Q)} \lesssim 1$. We next consider partial derivatives of H. From our integral formula for $H(u)$, it is obvious that the partial derivative of H with respect to u_1 will essentially produce only additional factors of the form $\lambda_3^{-1}y_1, \lambda_3^{-1}y_1y_2^2, \lambda_3^{-1/2}y_2$, and $\lambda_3^{-1}y_1$ under the double integral. However, powers of y_1 can be absorbed by the rapidly decaying factor $\check{\chi}_1(y_1)$, and $|\lambda_3^{-1}y_2^2| \leq \varepsilon \ll 1$, so that $|\partial_{u_1}H(u)| \lesssim 1$ too, and the same applies to $|\partial_{u_2}H(u)|$. The main problem is again caused by the partial derivative with respect to u_3, which produces an additional factor y_2.

However, arguing as in the preceding subsection, we find that for $\epsilon \in]0, 1]$ and $s \in [0, 1]$,

$$\begin{aligned} |\partial_{u_3}H(su)| \lesssim |u_3|^{\epsilon-1}s^{\epsilon-1} & \iint_{|y_2|\leq\lambda_3^{1/2}} (1 + |y_1|)^{-N} \\ & \times \Big|\check{\chi}_1(A - By_2 - y_2^2 b(x_1 - \lambda_3^{-1}su_1y_1, \lambda_3^{-1/2}y_2, \delta) \\ & + su_1r_2(y_1; \lambda_3^{-1}su_1, x_1, \delta) + (\lambda_3^{-1/2}y_2)\delta_3r_3(y_1; \lambda_3^{-1}su_1, x_1, \delta)\Big| \\ & \times |y_2|^\epsilon\, dy_1\, dy_2. \end{aligned}$$

Estimating $|y_2|^\epsilon$ in a trivial way by $|y_2|^\epsilon \leq \lambda_3^{\epsilon/2}$, we see by means of (7.59) that

$$|\partial_{u_3}H(su)| \lesssim |u_3|^{\epsilon-1}s^{\epsilon-1}\lambda_3^{\epsilon/2}, \qquad \text{for all } u \in Q.$$

By means of Lemma 2.7 (and our choice of $\gamma(\zeta)$), this implies that

$$|\lambda_3^{-1/2}\sigma^{\lambda_3}_{1+it,2}(x)| \lesssim \lambda_3^{\epsilon/2},$$

which completes the proof of (7.57) and hence also of Proposition 7.7. Q.E.D.

7.6.4 Estimation of $T_\delta^{III_2}$: Complex interpolation

The discussion of this operator will somewhat resemble the one of the operator $T_{\delta,j}^{VI}$ in Section 5.3, which arose from the same Subcase 3.2(b) (in the latter case in Section 4.2), with δ_0 playing here the role that 2^{-j} played there, and where we did have $B = 2$ in place of $B = 3$ here.

Assuming again without loss of generality that $\lambda_1 = \lambda_2$, we see that here we have to prove the following.

PROPOSITION 7.8. *Let $m = 2$ and $B = 3$, and consider the measure*

$$\nu_\delta^{III_2} := \sum_{2^M\delta_0^{-1} \leq \lambda_1 < \delta_0^{-3/2}} \ \sum_{2^M \lambda_1\delta_0 \leq \lambda_3 \leq 2^{-M}(\lambda_1\delta_0)^3} \nu_\delta^{(\lambda_1,\lambda_1,\lambda_3)},$$

where summation is taken over all sufficiently large dyadic λ_i in the given range. If we denote by $T_\delta^{III_2}$ the operator of convolution with $\widehat{\nu_\delta^{III_2}}$, then, if $M \in \mathbb{N}$ is sufficiently large (and ε sufficiently small),

$$\|T_\delta^{III_2}\|_{6/5\to 6} \leq C, \tag{7.61}$$

with a constant C not depending on δ, for δ sufficiently small.

Proof. Recall that, by (7.35), $\|\widehat{\nu_\delta^\lambda}\|_\infty \lesssim \lambda_1^{-1/2}\lambda_3^{-1/3}$. In analogy to the proof of Proposition 7.6, we therefore define here, for ζ in the strip $\Sigma = \{\zeta \in \mathbb{C} : 0 \leq \mathrm{Re}\,\zeta \leq 1\}$, an analytic family of measures by

$$\mu_\zeta(x) := \gamma(\zeta) \sum_{2^M\delta_0^{-1} \leq 2^{k_1} < \delta_0^{-3/2}} \ \sum_{2^{M+k_1}\delta_0 \leq 2^{k_3} \leq 2^{-M}2^{3k_1}\delta_0^3} 2^{(1-3\zeta)k_1/2} 2^{(1-3\zeta)k_3/3} \nu_\delta^{(2^{k_1},2^{k_1},2^{k_3})},$$

where we put

$$\gamma(\zeta) := \frac{1 - 2^{9(1-z)/2}}{1 - 2^3}.$$

By T_ζ we denote again the operator of convolution with $\widehat{\mu_\zeta}$. Observe that for $\zeta = \theta_c = \frac{1}{3}$, we have $\mu_{\theta_c} = \nu_\delta^{III_2}$, hence, $T_{\theta_c} = T_\delta^{III_2}$, so that, arguing exactly as in the preceding subsection, by means of Stein's interpolation theorem, (7.61) will follow if we can prove that there is a constant C such that

$$|\mu_{1+it}(x)| \leq C, \tag{7.62}$$

where C is independent of t, x and δ.

Setting

$$\mu_{\lambda_1,\lambda_3}(x) := \lambda_1^{-1}\lambda_3^{-2/3}\nu_\delta^{(\lambda_1,\lambda_1,\lambda_3)}(x),$$

we may rewrite

$$\mu_{1+it}(x) = \gamma(1+it) \sum_{2^M\delta_0^{-1} \leq \lambda_1 < \delta_0^{-3/2}} \ \sum_{2^M \lambda_1\delta_0 \leq \lambda_3 \leq 2^{-M}(\lambda_1\delta_0)^3} \lambda_1^{-3it/2}\lambda_3^{-it}\mu_{\lambda_1,\lambda_3}(x). \tag{7.63}$$

Arguing as in the preceding subsection, by means of the identity (7.47) we see again that we may assume in the sequel that $|x_1| \sim 1$ and $|x_2| \lesssim 1$ (notice also that here we have $\lambda_3 \ll \lambda_1$).

By means of the change of variables $y_1 \mapsto x_1 - y_1/\lambda_1$, $y_2 \mapsto y_2/(\lambda_1\delta_0)$ and Taylor expansion around x_1 we may rewrite

$$\mu_{\lambda_1,\lambda_3}(x) = \frac{\lambda_3^{1/3}}{(\lambda_1\delta_0)}\tilde{\mu}_{\lambda_1,\lambda_3}(x),$$

where

$$\tilde{\mu}_{\lambda_1,\lambda_3}(x) = \iint \check{\chi}_1(y_1)\tilde{F}_\delta(\lambda_1,\lambda_3,x,y_1,y_2)\,dy_1\,dy_2, \tag{7.64}$$

with

$$\begin{aligned}\tilde{F}_\delta(\lambda_1,\lambda_3,x,y_1,y_2) := {}&\eta(x_1-\lambda_1^{-1}y_1,\ \delta_0^{-1}\lambda_1^{-1}y_2)\,\check{\chi}_1(D-y_2+r_1(y_1))\\ &\times \check{\chi}_1\big(A - By_2 - E\,y_2^3 b\,(x_1-\lambda_1^{-1}y_1)\delta_0^{-1}\lambda_1^{-1}y_2,\delta\big)\\ &+\lambda_3\lambda_1^{-1}\big(r_2(y_1)+(\delta_0^{-1}\lambda_1^{-1}y_2)\delta_3 r_3(y_1)\big)\big).\end{aligned}$$

Here, the quantities A to E are given by

$$\begin{aligned} &A = A(x,\lambda_3,\delta) := \lambda_3 Q_A(x), && B := B(x,\lambda_1,\lambda_3,\delta) := \frac{\lambda_3}{\lambda_1}Q_B(x),\\ &D = D(x,\lambda_1,\delta) := \lambda_1 Q_D(x), && E := \frac{\lambda_3}{(\delta_0\lambda_1)^3} \leq 2^{-M}, \end{aligned} \tag{7.65}$$

with

$$Q_A(x) := x_3 - x_1^n\alpha(\delta_1x_1),\quad Q_B(x) := \frac{\delta_3}{\delta_0}x_1^{n_1}\alpha_1(\delta_1x_1),\quad Q_D(x) := x_2 - x_1^2\omega(\delta_1x_1).$$

Again, the functions $r_i(y_1) = r_i(y_1;\lambda_1^{-1},x_1,\delta)$, $i = 1,2,3$, are smooth functions of y_1 (and λ_1^{-1} and x_1) satisfying estimates of the form (7.50). Moreover, we may assume that $|y_1| \lesssim \lambda_1$ and $|\delta_0^{-1}\lambda_1^{-1}y_2| \leq \varepsilon$, because of our assumption $|x_1| \sim 1$ and the support properties of η.

The factor $\lambda_3^{1/3}/(\lambda_1\delta_0)$ by which $\mu_{\lambda_1,\lambda_3}(x)$ and $\tilde{\mu}_{\lambda_1,\lambda_3}(x)$ differ suggests we decompose the summation over k_3 into three arithmetic progressions, $k_3 = i + 3k_4$, $i = 0,1,2$ (for a similar discussion, compare Section 5.3). Restricting ourselves to any one of them, let us assume for simplicity that $i = 0$, so that $k_3 = 3k_4$, with $k_4 \in \mathbb{N}$. Let us also assume that δ_0 is a dyadic number (otherwise, replace δ_0 by the biggest dyadic number smaller or equal to δ_0). It is then convenient to introduce new summation variables (k_0,k_4) in place of (k_1,k_3) by requiring that $k_1 = k_0 + k_3/3 - \log_2(\delta_0) = k_0 + k_4 - \log_2(\delta_0)$. In terms of their exponentials $\lambda_0 := 2^{k_0}$ and $\lambda_4 := 2^{k_4}$, this means that

$$\lambda_1 = \frac{\lambda_0\lambda_4}{\delta_0},\quad \lambda_3 = \lambda_4^3,$$

and we can rewrite the conditions on the index sets for λ_1 and λ_3 over which we sum in (7.63) as

$$\lambda_0 \geq 2^{M/3},\quad 2^{M/2}\lambda_0^{1/2} \leq \lambda_4 \leq \delta_0^{-1/2}\lambda_0^{-1},$$

and correspondingly we shall rewrite (7.63) as

$$\mu_{1+it}(x) = \gamma(1+it)\,\delta_0^{3it/2} \sum_{\lambda_0 \geq 2^{M/3}} \sum_{2^{M/2}\lambda_0^{1/2} \leq \lambda_4 \leq \delta_0^{-1/2}\lambda_0^{-1}} \lambda_0^{-(1+3it/2)} \lambda_4^{-9it/2} \tilde{\mu}_{\lambda_0\lambda_4/\delta_0, \lambda_4^3}(x).$$

For λ_0 and x fixed, let us put

$$f_{\lambda_0,x}(\lambda_4) := \tilde{\mu}_{\frac{\lambda_0\lambda_4}{\delta_0}, \lambda_4^3}(x),$$

$$\rho_{t,\lambda_0}(x) := \gamma(1+it) \sum_{\{\lambda_4:\, 2^{M/2}\lambda_0^{1/2} \leq \lambda_4 \leq \delta_0^{-1/2}\lambda_0^{-1}\}} \lambda_4^{-9it/2} f_{\lambda_0,x}(\lambda_4).$$

The previous formula for $\mu_{1+it}(x)$ shows that in order to verify (7.62), it will suffice to prove the following uniform estimate: there exist constants $C > 0$ and $\epsilon \geq 0$ with $\epsilon < 1$, so that for all x such that $|x_1| + |x_2| \lesssim 1$ and δ sufficiently small, we have

$$|\rho_{t,\lambda_0}(x)| \leq C\lambda_0^{\epsilon} \qquad \text{for} \quad \lambda_0 \geq 2^{M/3}. \tag{7.66}$$

In order to prove this, observe that by (7.64)

$$f_{\lambda_0,x}(\lambda_4) = \iint \check{\chi}_1(y_1) F_\delta(\lambda_0, \lambda_4, x, y_1, y_2)\, dy_1\, dy_2, \tag{7.67}$$

where

$$\begin{aligned} F_\delta(\lambda_0, \lambda_4, x, y_1, y_2) := \eta\,&\big(x_1 - \delta_0(\lambda_0\lambda_4)^{-1}y_1,\ (\lambda_0\lambda_4)^{-1}y_2, \delta\big) \\ &\times \check{\chi}_1(D - y_2 + r_1(y_1))\, \check{\chi}_1\,(A - By_2 \\ &- E\, y_2^3 b\,\big(x_1 - \delta_0(\lambda_0\lambda_4)^{-1}y_1, (\lambda_0\lambda_4)^{-1}y_2, \delta\big) \\ &+ \delta_3\delta_0\lambda_4\lambda_0^{-2}\, y_2\, r_3(y_1) + \ \delta_0\lambda_4^2\lambda_0^{-1} r_2(y_1)) \end{aligned}$$

and

$$A = A(x, \lambda_4, \delta) := \lambda_4^3 Q_A(x), \qquad B := B(x, \lambda_0, \lambda_4, \delta) := \frac{\lambda_4^2}{\lambda_0} Q_B(x),$$

$$D = D(x, \lambda_0, \lambda_4, \delta) := \frac{\lambda_0\lambda_4}{\delta_0} Q_D(x), \qquad E := \lambda_0^{-3} \leq 2^{-M}, \tag{7.68}$$

with

$$Q_A(x) := x_3 - x_1^n \alpha(\delta_1 x_1), \quad Q_B(x) := \delta_3 x_1^{n_1} \alpha_1(\delta_1 x_1),$$
$$Q_D(x) := x_2 - x_1^2 \omega(\delta_1 x_1).$$

The functions $r_i(y_1) = r_i(y_1; \lambda_0^{-1}, \lambda_4^{-1}, x_1, \delta)$, $i = 1, 2, 3$, are smooth functions of y_1 (and $\lambda_1^{-1}, \lambda_4^{-1}$ and x_1), satisfying estimates of the form

$$|r_i(y_1)| \leq C|y_1|, \qquad \left| \left(\frac{\partial}{\partial(\lambda_4^{-1})} \right) r_i(y_1; \lambda_0^{-1}, \lambda_4^{-1}, x_1, \delta) \right| \leq C_l |y_1|^2 \tag{7.69}$$

(compare (7.50)).

Given x and λ_0, we shall split the summation in λ_4 into subintervals according to the (relative) sizes of the quantities A, B, and D, which are considered as functions of λ_4.

1. The part where $|D| \gg 1$**.** Denote by $\rho^1_{t,\lambda_0}(x)$ the contribution to $\rho_{t,\lambda_0}(x)$ by the terms for which $|D| \gg 1$.

We first consider the contribution to $f_{\lambda_0,x}(\lambda_4)$ given by integrating in (7.67) over the region where $|y_1| \gtrsim |D|^\varepsilon$ (where $\varepsilon > 0$ is assumed to be sufficiently small). Here, the rapidly decaying first factor $\check{\chi}_1(y_1)$ leads to an improved estimate of this contribution of the order $|D|^{-N}$ for every $N \in \mathbb{N}$, which allows to sum over the dyadic λ_4 for which $|D| \gg 1$, and the contribution to $\rho^1_{t,\lambda_0}(x)$ is of order $O(1)$, which is stronger than what is needed in (7.66).

We may therefore restrict ourselves in the sequel to the region where $|y_1| \ll |D|^\varepsilon$. Observe that, because of (7.69), this implies in particular that $|r_i(y_1)| \ll |D|^\varepsilon$, $i = 1, 2, 3$. By looking at the second factor in F_δ, we again see that the contribution by the regions where, in addition, $|y_2| < |D|/2$, or $|y_2| > 3|D|/2$ is again of the order $|D|^{-N}$ for every $N \in \mathbb{N}$, and their contributions to $\rho^1_{t,\lambda_0}(x)$ are again admissible.

What remains is the region where $|y_1| \ll |D|^\varepsilon$ and $|D|/2 \le |y_2| \le 3|D|/2$. In addition, we may assume that y_2 and D have the same sign, since otherwise we can estimate as before. Let us therefore assume, for example, that $D > 0$ and that $D/2 \le y_2 \le 3D/2$.

The change of variables $y_2 \mapsto Dy_2$ then allows us to rewrite the corresponding contribution to $f_{\lambda_0,x}(\lambda_4)$ as

$$\tilde{f}_{\lambda_0,x}(\lambda_4) := D \int_{|y_1| \ll |D|^\varepsilon} \int_{1/2 \le y_2 \le 3/2} \check{\chi}_1(y_1) \tilde{F}_\delta(\lambda_0, \lambda_4, x, y_1, y_2)\, dy_2\, dy_1,$$

where here

$$\begin{aligned} \tilde{F}_\delta(\lambda_0, \lambda_4, x, y_1, y_2) := &\, \eta\left(x_1 - \delta_0(\lambda_0\lambda_4)^{-1} y_1,\ (\lambda_0\lambda_4)^{-1} D y_2, \delta\right) \\ &\times \check{\chi}_1(D - Dy_2 + r_1(y_1))\, \chi_1(y_2) \\ &\times \check{\chi}_1\Big(A - BDy_2 - ED^3\, y_2^3 b\,(x_1 - \delta_0 \\ &\times (\lambda_0\lambda_4)^{-1} y_1, (\lambda_0\lambda_4)^{-1} D y_2, \delta) \\ &+\ \delta_3\delta_0\lambda_4\lambda_0^{-2} D\, r_3(y_1)\, y_2 + \delta_0\lambda_4^2\lambda_0^{-1} r_2(y_1)\Big) \end{aligned}$$

and where χ_1 is supported where $y_2 \sim 1$. In the subsequent discussion, we may and shall assume that $f_{\lambda_0,x}(\lambda_4)$ is replaced by $\tilde{f}_{\lambda_0,x}(\lambda_4)$.

Recall also from (7.43) that $b(x_1, x_2, \delta) = b_3(\delta_1 x_1, \delta_2 x_2)$, and that $\delta_2 \ll \delta_0$. The last estimate implies that

$$|\delta_2(\lambda_0\lambda_4)^{-1} D| = \frac{\delta_2}{\delta_0} |Q_D(x)| \ll 1,$$

which shows that the second derivative of the argument of the last factor of our function $\tilde{F}_\delta(\lambda_0, \lambda_4, x, y_1, y_2)$ with respect to y_2 is comparable to $|ED^3|$. We may therefore apply van der Corput's estimate to the integration in y_2 (case (i) in Lemma 2.1(b)) and obtain

$$|\tilde{f}_{\lambda_0,x}(\lambda_4)| \lesssim |D||ED^3|^{-1/2} = \lambda_0^{3/2}|D|^{-1/2}.$$

Interpolation with the trivial estimate $|\tilde{f}_{\lambda_0,x}(\lambda_4)| \lesssim 1$ then leads to $|\tilde{f}_{\lambda_0,x}(\lambda_4)| \lesssim \lambda_0^{1/2}|D|^{-1/6}$. The second factor allows us to sum in λ_4, since we are assuming that $|D| \gg 1$, and we obtain $|\rho^1_{t,\lambda_0}(x)| \leq C\lambda_0^{1/2}$, in agreement with (7.66).

We may thus in the sequel assume that $|D| \lesssim 1$**.** Here we go back to (7.67) and observe that $\check{\chi}_1(y_1)\check{\chi}_1(D - y_2 + r_1(y_1))$ can be estimated by $C_N(1+|y_1|)^{-N}(1+|y_2|)^{-N}$. This shows, in particular, that any power of y_1 or y_2 can be "absorbed" by these two factors.

We shall still have to distinguish between the cases where $|B| \geq 1$ and where $|B| < 1$.

2. The part where $|D| \lesssim 1$ **and** $|B| \geq 1$**.** Denote by $\rho^2_{t,\lambda_0}(x)$ the contribution to $\rho_{t,\lambda_0}(x)$ by the terms for which $|D| \lesssim 1$ and $|B| \geq 1$.

If $|y_2| \gtrsim (|B|/|E|)^{1/2}$, then we see that we can estimate the contribution to $f_{\lambda_0,x}(\lambda_4)$ by a constant times $(|B|/|E|)^{-1/2} = \lambda_0^{-3/2}|B|^{-1/2}$. Summing over all λ_4 such that $|B| \geq 1$ then leads to a uniform estimate for the contributions of these regions to $\rho^2_{t,\lambda_0}(x)$.

So, assume that $|y_2| \ll (|B|/|E|)^{1/2} =: H$. Applying the change of variables $y_2 \mapsto Hy_2$, we then see that we may replace $f_{\lambda_0,x}(\lambda_4)$ by

$$\tilde{f}_{\lambda_0,x}(\lambda_4) = H \iint \check{\chi}_1(y_1)\tilde{F}_\delta(\lambda_0, \lambda_4, x, y_1, y_2)\, dy_1\, dy_2,$$

where

$$\begin{aligned}\tilde{F}_\delta(\lambda_0, \lambda_4, x, y_1, y_2) :=\, & \eta\left(x_1 - \delta_0(\lambda_0\lambda_4)^{-1}y_1, (\lambda_0\lambda_4)^{-1}Hy_2, \delta\right) \\ & \times \check{\chi}_1(D - Hy_2 + r_1(y_1))\, \chi_0(y_2) \\ & \times \check{\chi}_1\big(A - HB\left(y_2 + y_2^3 \operatorname{sgn}(B)\right. \\ & \times\ b\left(x_1 - \delta_0(\lambda_0\lambda_4)^{-1}y_1, (\lambda_0\lambda_4)^{-1}Hy_2, \delta\right)\big) \\ & + \delta_3\delta_0\lambda_4\lambda_0^{-2}H\, y_2\, r_3(y_1) + \delta_0\lambda_4^2\lambda_0^{-1}r_2(y_1)\big).\end{aligned}$$

We claim that

$$|\tilde{f}_{\lambda_0,x}(\lambda_4)| \leq CH|HB|^{-1/2} = \lambda_0^{3/4}|B|^{-1/4}. \tag{7.70}$$

Since we are here assuming that $|B| \geq 1$, this estimate would imply the estimate $|\rho^2_{t,\lambda_0}(x)| \leq C\lambda_0^{3/4}$, again in agreement with (7.66).

In order to prove (7.70), observe first that the contribution to $\tilde{f}_{\lambda_0,x}(\lambda_4)$ by the region where $|y_1| > |HB|$ clearly can be estimated by the right-hand side of (7.70) because of the rapidly decaying factor $\check{\chi}_1(y_1)$ in the integrand. And, on the remaining region where $|y_1| \leq |HB|$, we have

$$|\delta_3\delta_0\lambda_4\lambda_0^{-2}H r_3(y_1)| \leq \delta_3\delta_0\lambda_4\lambda_0^{-2}|H|\,|HB| = \delta_3\frac{\delta_0\lambda_4^2}{\lambda_0}|Q_B(x)|^{1/2}|\,|HB|$$
$$\lesssim \lambda_0^{-3}\delta_3^{3/2}\,|HB| \ll |HB|.$$

Observe also that $(\lambda_0\lambda_4)^{-1}H = |Q_B(x)|^{1/2} \lesssim \delta_3^{1/2} \ll 1$. This shows that if γ denotes the argument of the last factor of $\tilde{F}_\delta(\lambda_0, \lambda_4, x, y_1, y_2)$, then there are constants $0 < C_1 < C_2$, such that

$$C_1|HB| \leq \left|\frac{\partial}{\partial y_2}\gamma(y_2)\right| + \left|\left(\frac{\partial}{\partial y_2}\right)^2 \gamma(y_2)\right| \leq C_2|HB|, \quad |y_2| \lesssim 1,$$

uniformly in x, y_1, and δ. We may thus apply a van der Corput–type estimate (see case (ii) in Lemma 2.1(b)) to the integration in y_2 and again arrive at an estimate by the right-hand side of (7.70) and also for the contribution by the region where $|y_1| \leq |HB|$.

3. The part where $|D| \lesssim 1, |B| < 1,$ **and** $|A| \gg 1$. Denote by $\rho^3_{t,\lambda_0}(x)$ the contribution to $\rho_{t,\lambda_0}(x)$ by the terms for which these conditions are satisfied. We claim that here we get

$$|f_{\lambda_0,x}(\lambda_4)| \leq C|A|^{-N}, \tag{7.71}$$

for every $N \in \mathbb{N}$. This estimate will imply the estimate $|\rho^3_{t,\lambda_0}(x)| \leq C\lambda_0^{3/4}$, again in agreement with (7.66).

In order to prove (7.71), observe that the contributions to $f_{\lambda_0,x}(\lambda_4)$ by the regions where $|y_1| \gtrsim |A|^{1/3}$ or $|y_2| \gtrsim |A|^{1/3}$ can be estimated by a constant times $(|A|^{1/3})^{-N}$, because of the rapid decay in y_1 and y_2 of F_δ. So, assume that $|y_1| + |y_2| \ll |A|^{1/3}$. Then we see that

$$|By_2| \ll |B||A|^{1/3} \ll |A| \quad \text{and} \quad |Ey_2^3| \ll |EA| \ll |A|,$$

as well as

$$|\delta_3\delta_0\lambda_4\lambda_0^{-2}\, y_2 r_3(y_1) + \delta_0\lambda_4^2\lambda_0^{-1} r_2(y_1)| \ll |A|^{2/3} \ll |A|,$$

and thus the last factor of F_δ is of order $|A|^{-N}$. Consequently, the contribution to $f_{\lambda_0,x}(\lambda_4)$ by the region where $|y_1| + |y_2| \ll |A|^{1/3}$ can also be estimated as in (7.71).

4. The part where $\max\{|A|, |B|, |D|\} \lesssim 1$. Denote by $\rho^4_{t,\lambda_0}(x)$ the contribution to $\rho_{t,\lambda_0}(x)$ by the terms for which $\max\{|A|, |B|, |D|\} \lesssim 1$. Then we can easily estimate $\rho^4_{t,\lambda_0}(x)$ by means of Lemma 2.7 in a very similar way as we did in the last part of Section 5.3 and obtain that $|\rho^4_{t,\lambda_0}(x)| \leq C$.

This completes the proof of estimate (7.66) (with $\epsilon := \frac{3}{4}$) and, hence, also the proof of Proposition 7.8.

Q.E.D.

7.7 THE CASE WHERE $\lambda_1 \sim \lambda_2 \sim \lambda_3$

We shall assume for the sake of simplicity that

$$\lambda_1 = \lambda_2 = \lambda_3 \gg 1.$$

The more general case where $\lambda_1 \sim \lambda_2 \sim \lambda_3 \gg 1$ can be treated in a very similar way. By changing notation slightly, we shall denote in this section by λ the common value of the λ_j.

We change coordinates from $\xi = (\xi_1, \xi_2, \xi_3)$ to s_1, s_2 and $s_3 := \xi_3/\lambda$, that is,

$$\xi_1 = s_1\xi_3 = \lambda s_1 s_3, \quad \xi_2 = \lambda s_2 \xi_3 = \lambda s_2 s_3, \quad \xi_3 = \lambda s_3,$$

and write in the sequel

$$s := (s_1, s_2, s_3) \quad s' := (s_1, s_2).$$

Then we may rewrite

$$\Phi(x, \delta, \xi) = \lambda s_3 \tilde{\Phi}(x, s'),$$

where

$$\begin{aligned} \tilde{\Phi}(x, s') := {} & s_1x_1 + s_2x_1^m\omega(\delta_1x_1) + x_1^n\alpha(\delta_1x_1) \\ & + s_2\delta_0x_2 + \left(x_2^B\, b(x_1, x_2, \delta) + r(x_1, x_2, \delta)\right), \end{aligned} \tag{7.72}$$

where $\omega(0) \neq 0$, $\alpha(0) \neq 0$, and $b(x_1, 0, \delta) \neq 0$, if $x_1 \sim 1$, and where δ and $r(x_1, x_2, \delta)$ are given by (7.11) and (7.10), respectively.

Now, the first part of $\tilde{\Phi}$ has at worst an Airy-type singularity with respect to x_1, and the derivative of order B with respect to x_2 does not vanish, so that we obtain

$$\|\widehat{\nu_\delta^\lambda}\|_\infty \lesssim \lambda^{-1/3-1/B} \tag{7.73}$$

(indeed, by localizing near a given point x^0 and looking at the corresponding Newton polyhedron of $\tilde{\Phi}$ at this point, this follows more precisely from the main result in [IM11b]). On the other hand, standard van der Corput–type arguments (compare Lemma 2.1) show that here

$$\|\nu_\delta^\lambda\|_\infty \lesssim \min\{\lambda^3\lambda^{-1}\lambda^{-1/B}, \lambda^3\lambda^{-1}(\lambda\delta_0)^{-1}\} = \lambda \min\{\lambda^{(B-1)/B}, \delta_0^{-1}\}. \tag{7.74}$$

We therefore distinguish the following cases:

Case A. $\lambda \leq \delta_0^{-B/(B-1)}$. Then $\|\nu_\delta^\lambda\|_\infty \lesssim \lambda^{2-1/B}$, and by interpolation we get

$$\|T_\delta^\lambda\|_{p_c \to p_c'} \lesssim \lambda^{-1/3-1/B+7\theta_c/3}, \tag{7.75}$$

again with $\theta_c = 2/p_c'$.

Case B: $\lambda > \delta_0^{-B/(B-1)}$. Then $\|\nu_\delta^\lambda\|_\infty \lesssim \lambda\delta_0^{-1}$, and by interpolation we get

$$\|T_\delta^\lambda\|_{p_c \to p_c'} \lesssim \lambda^{-1/3-1/B+(4/3+1/B)\theta_c}\,\delta_0^{-\theta_c}. \tag{7.76}$$

Observe that in both cases, the exponents of λ in these estimates become strictly smaller and the one of δ increases if we replace θ by a strictly smaller number.

7.7.1 The case where $h^r + 1 > B$

We observe that then $p_c' > 2B$, and thus

$$\theta_c < \frac{1}{B}.$$

This shows that

$$-\frac{1}{3} - \frac{1}{B} + \frac{7}{3}\theta_c < -\frac{1}{3} - \frac{1}{B} + \frac{7}{3} \cdot \frac{1}{B} = -\frac{B-4}{3B}.$$

Thus, if $B \geq 4$, then for $\theta = \theta_c$, the exponent of λ in (7.75) is strictly negative, so that in Case A we can sum the estimates for $p = p_c$ over all $\lambda \leq \delta_0^{-B/(B-1)}$ and obtain, for $B \geq 4$,

$$\sum_{\lambda \leq \delta_0^{-B/(B-1)}} \|T_\delta^\lambda\|_{p_c \to p_c'} \lesssim 1. \tag{7.77}$$

Similarly, since

$$-\frac{1}{3} - \frac{1}{B} + \left(\frac{4}{3} + \frac{1}{B}\right)\theta_c < -\frac{1}{3} - \frac{1}{B} + \left(\frac{4}{3} + \frac{1}{B}\right)\frac{1}{B} = -\frac{B^2 - B - 3}{3B^2},$$

where $B^2 - B - 3 > 0$ if $B \geq 3$, we see that (7.76) implies in Case B that

$$\sum_{\lambda > \delta_0^{-B/(B-1)}} \|T_\delta^\lambda\|_{p_c \to p_c'} \lesssim \delta_0^{(B+3-7B\theta_c)/3(B-1)} < \delta_0^{(B-4)/3(B-1)};$$

hence, for $B \geq 4$,

$$\sum_{\lambda > \delta_0^{-B/(B-1)}} \|T_\delta^\lambda\|_{p_c \to p_c'} \lesssim 1, \qquad (B \geq 4). \tag{7.78}$$

The case where $B = 3$ requires more refined estimates, whereas we shall see that the case $B = 2$ is rather easy to handle because of the estimate $\theta_c \leq \frac{1}{3}$ in (7.20).

Assume first that $B = 2$. Then, by (7.75), if $\lambda \leq \delta_0^{-2}$,

$$\|T_\delta^\lambda\|_{p_c \to p_c'} \lesssim \lambda^{-1/3-1/2+7\theta_c/3} = \lambda^{(14\theta_c - 5)/6},$$

and since $\theta_c \leq \frac{1}{3}$, the exponent of λ in this estimate is strictly negative, so that we can sum over λ and again obtain (7.77).

Similarly, by (7.76), if $\lambda > \delta_0^{-2}$,

$$\|T_\delta^\lambda\|_{p_c \to p_c'} \lesssim \lambda^{-1/3-1/2+(4/3+1/2)\theta_c} \delta_0^{-\theta_c} = \lambda^{(11\theta_c - 5)/6} \delta_0^{-\theta_c}.$$

But, $11\theta_c - 5 \leq \frac{11}{3} - 5 < 0$, and so we get

$$\sum_{\lambda > \delta_0^{-2}} \|T_\delta^\lambda\|_{p_c \to p_c'} \lesssim \delta_0^{(5-14\theta_c)/3} \leq 1,$$

so that (7.78) also holds true in this case.

Assume next that $B = 3$. Then in Case A, where $\lambda \leq \delta_0^{-3/2}$, we have by (7.75)

$$\|T_\delta^\lambda\|_{p_c \to p_c'} \lesssim \lambda^{(7\theta_c - 2)/3},$$

and thus, if $7\theta_c - 2 < 0$, then we can sum these estimates in $\lambda \le \delta_0^{-3/2}$ and obtain (7.78).

Let us therefore assume henceforth that $\theta_c \ge \frac{2}{7}$. Observe that by Lemma 7.4 we have $\theta_c < \tilde{\theta}_c$, unless $\tilde{h}^r = d$ and $h^r + 1 \ge H$, in which case we have $\theta_c = \tilde{\theta}_c$ and $\tilde{p}_c' = p_c'$. Thus

$$\|T_\delta^\lambda\|_{p_c \to p_c'} \lesssim \lambda^{(7\tilde{\theta}_c - 2)/3},$$

with $\tilde{\theta}_c > \frac{2}{7}$, unless $\theta_c = \tilde{\theta}_c = \frac{2}{7}$, $\tilde{h}^r = d$ and $h^r + 1 \ge H$. Note that in the latter case, $H = B = 3$, and since $\tilde{\theta}_c = \frac{2}{7}$, we find that $m = 5$ and $d = \frac{5}{2}$.

In this particular case, we get only a uniform estimate $\|T_\delta^\lambda\|_{p_c \to p_c'} \lesssim 1$. However, in this case have $\|\nu_\delta^\lambda\|_\infty \lesssim \lambda^{5/3}$, since we are in Case A, and $\|\widehat{\nu_\delta^\lambda}\|_\infty \lesssim \lambda^{-2/3}$, whereas $\theta_c = \frac{2}{7}$, and thus $-(1-\theta_c)2/3 + \theta_c 5/3 = 0$. Moreover, $\nu_\delta^\lambda = \nu_\delta * \phi_\lambda$, where the Fourier transform of ϕ_λ is given by $\widehat{\phi_\lambda}(\xi) = \chi_1\left(\frac{\xi_1}{\lambda}\right)\chi_1\left(\frac{\xi_2}{\lambda}\right)\chi_1\left(\frac{\xi_3}{\lambda}\right)$. This implies a uniform estimate of the L^1-norms of the ϕ_λ for all dyadic λ. We may thus estimate the operator T^{IV_1} of convolution with the Fourier transform of the complex measure

$$\nu_\delta^{IV_1} := \sum_{\lambda \le \delta_0^{-3/2}} \nu_\delta^\lambda$$

by means of the real interpolation Proposition 2.6, in the same way as we estimated the operators T^{I_1} and T^{II_1} in Section 7.6.1, by adding the measure $\nu_\delta^{IV_1}$ to the family of measures μ^i, $i \in I$ from the second class in (7.41).

So, assume that $\tilde{\theta}_c > \frac{2}{7}$. Then we find that

$$\sum_{\lambda \le \delta_0^{-3/2}} \|T_\delta^\lambda\|_{p_c \to p_c'} \lesssim \delta_0^{(2-7\tilde{\theta}_c)/2}.$$

Let us next turn to Case B, where $\lambda > \delta_0^{-3/2}$. Then, by (7.76),

$$\|T_\delta^\lambda\|_{p_c \to p_c'} \lesssim \lambda^{(5\theta_c - 2)/3}\, \delta_0^{-\theta_c}.$$

Since $\theta_c \le \frac{1}{3}$, we can sum in λ and obtain

$$\sum_{\lambda > \delta_0^{-3/2}} \|T_\delta^\lambda\|_{p_c \to p_c'} \lesssim \delta_0^{(2-7\theta_c)/2} \le \delta_0^{(2-7\tilde{\theta}_c)/2}.$$

Combining these estimates, we obtain

$$\sum_{\lambda \gg 1} \|T_\delta^\lambda\|_{p_c \to p_c'} \lesssim \delta_0^{(2-7\tilde{\theta}_c)/2}. \tag{7.79}$$

Observe next that

$$\frac{2 - 7\tilde{\theta}_c}{2} = 1 - \frac{7}{\tilde{p}_c'} = \frac{2\tilde{h}^r - 5}{\tilde{p}_c'},$$

and recall that $\delta_0 = 2^{-(\tilde{\kappa}_2 - m\tilde{\kappa}_1)k}$. In combination with the rescaling estimate (6.16), the estimate in (7.79) thus leads to

$$\begin{aligned}\Big(\int |\hat{f}|^2 d\mu_{1,k}\Big)^{1/2} &\lesssim 2^{-k\left(|\tilde{\kappa}|/2 - [\tilde{\kappa}_1(1+m)+1]/p_c' + (\tilde{\kappa}_2 - m\tilde{\kappa}_1)(2\tilde{h}^r - 5)/2\tilde{p}_c'\right)} \|f\|_{L^{p_c}} \\ &\le C 2^{-k\left(|[\tilde{\kappa}|/2 - \tilde{\kappa}_1(1+m)+1]/\tilde{p}_c' + (\tilde{\kappa}_2 - m\tilde{\kappa}_1)[2\tilde{h}^r - 5]/2\tilde{p}_c'\right)} \|f\|_{L^{p_c}}.\end{aligned} \tag{7.80}$$

where $\mu_{1,k}$ denotes the measure corresponding to the frequency domains that we are here considering; that is, $\mu_{1,k}$ corresponds to the rescaled measure

$$\nu_{1,\delta} := \sum_{\lambda \gg 1} \nu_\delta^{(\lambda,\lambda,\lambda)}.$$

But

$$\begin{aligned} E &:= 2\tilde{p}_c' \left(\frac{|\tilde{\kappa}|}{2} - \frac{\tilde{\kappa}_1(1+m)+1}{\tilde{p}_c'} + (\tilde{\kappa}_2 - m\tilde{\kappa}_1)\frac{2\tilde{h}^r - 5}{2\tilde{p}_c'} \right) \\ &= |\tilde{\kappa}|(2\tilde{h}^r + 2) - 2(\tilde{\kappa}_1(1+m)+1) + (\tilde{\kappa}_2 - m\tilde{\kappa}_1)(2\tilde{h}^r - 5) \\ &= \tilde{\kappa}_2(4\tilde{h}^r - 3) + \tilde{\kappa}_1(3m - 2\tilde{h}^r(m-1)) - 2, \end{aligned}$$

where

$$\begin{aligned} \tilde{\kappa}_2(4\tilde{h}^r - 3) &= \frac{4m}{m+1} - 3\tilde{\kappa}_2, \\ \tilde{\kappa}_1(3m - 2\tilde{h}^r(m-1)) &= m\frac{\tilde{\kappa}_1}{\tilde{\kappa}_2}\left(\frac{3}{H} - 2\frac{m-1}{m+1}\right). \end{aligned}$$

Since $H \ge B = 3$, we see that $3/H - 2(m-1)/(m+1) \le (3-m)/(m+1) \le 0$ if $m \ge 3$. Thus, if $m \ge 3$, then since $\tilde{\kappa}_1/\tilde{\kappa}_2 = 1/a < 1/m$, we see that

$$\tilde{\kappa}_1(3m - 2\tilde{h}^r(m-1)) \ge 3\tilde{\kappa}_2 - 2\frac{m-1}{m+1},$$

and altogether we find that $E \ge 0$ (even with strict inequality, if $H > 3$). We thus have proved

$$\Big(\int |\hat{f}|^2 d\mu_{1,k}\Big)^{1/2} \le C\|f\|_{L^{p_c}}, \tag{7.81}$$

with a constant C not depending on k, provided that $m \ge 3$.

Assume finally that $B = 3$ **and** $m = 2$**.** Recall also that we are still assuming that $h^r + 1 > B = 3$ and $\theta_c \ge \frac{2}{7}$, so that $3 < h^r + 1 \le \frac{7}{2}$.

We shall prove that the Newton polyhedron of $\tilde{\phi}^a$ (defined in Section 7.1), respectively, of ϕ, will have a particular structure. Indeed, if ϕ is analytic, then one can show that ϕ is of type Z, E, J, Q, and so on, in the sense of Arnol'd's classification of singularities (compare [AGV88]).

We shall, however, content ourselves with a little less precise information, which will nevertheless be sufficient for our purposes.

Recall from Chapter 1 the notion of *augmented Newton polyhedron* $\mathcal{N}^r(\tilde{\phi}^a)$ of $\tilde{\phi}^a$. If L denotes the principal line of $\mathcal{N}(\phi)$, then it is a supporting line to $\mathcal{N}(\tilde{\phi}^a)$ too, and if (A^+, B^+) is the right endpoint of the line segment $L \cap \mathcal{N}(\tilde{\phi}^a)$, then let us denote by L^+ the half line $L^+ \subset L$ contained in the principal line of $\mathcal{N}(\phi)$ whose right endpoint is given by (A^+, B^+). Then $\mathcal{N}^r(\tilde{\phi}^a)$ is the convex hull of the union of $\mathcal{N}(\tilde{\phi}^a)$ with the half line L^+. Recall also that $\mathcal{N}^r(\tilde{\phi}^a)$ and $\mathcal{N}^r(\phi^a)$ do agree within the closed half space bounded from below by the bisectrix Δ, so that $h^r + 1$ is the second coordinate of the point at which the line $\Delta^{(m)}$ intersects the boundary of $\mathcal{N}^r(\tilde{\phi}^a)$.

Proposition 7.9. *If $B = 3$, $m = 2$, and $3 < h^r + 1 \leq 3.5$, then $(A^+, B^+) = (1, 3)$, and $\mathcal{N}^r(\tilde{\phi}^a)$ has exactly two edges, L^+ and the line segment $[(1, 3), (0, n)]$, which is contained in the principal line L^a of $\mathcal{N}(\tilde{\phi}^a)$.*

In particular,

$$\kappa = \left(\frac{1}{7}, \frac{2}{7}\right), \quad h^r = d = \frac{7}{3}, \quad \text{and } \tilde{\kappa} = \left(\frac{1}{n}, \frac{n-1}{3n}\right), \tag{7.82}$$

where $n > 7$.

Proof. Denote by $(A', B') := (A'_{(0)}, B'_{(0)}) \in L^a$ the left endpoint of the principal face $\pi(\tilde{\phi}^a)$ of the Newton polyhedron of $\tilde{\phi}^a$. Then $B' \geq B = 3$. In a first step, we prove that $B' = 3$.

Assume, to the contrary, that we had $B' \geq 4$ (observe that B' is an integer). Since the line L^a has slope strictly less $1/m = \frac{1}{2}$, then it easily seen that the line L^a would intersect the line $\Delta^{(2)}$ at some point with second coordinate z_2 strictly bigger than 3.5, so that $h^r + 1 \geq z_2 > 3.5$, which would contradict our assumption (Figure 7.2).

Thus, $B' = B = 3$. In a second step, we show that $A' = 1$. To this end, let us here work with ϕ^a in place of $\tilde{\phi}^a$. Note the point (A', B') is also the left endpoint of the principal face of $\mathcal{N}(\phi^a)$, and that the principal faces of the Newton polyhedra of ϕ^a and $\tilde{\phi}^a$ both lie on the same line L^a, since the last step in the change to modified adapted coordinates (7.5) preserves the homogeneity $\tilde{\kappa}$. This shows that also $A' \in \mathbb{N}$. Moreover, $A' \geq 1$, for otherwise we had $A' = 0$ and thus $h^r + 1 = 3$.

Assume that $A' \geq 2$. We have to distinguish two cases.

(a) If the line L, which has slope $\frac{1}{2}$, contains the point (A', B'), then the assumption $A' \geq 2$ would imply that $h^r + 1 > 3.5$ (see Figure 7.3).

(b) If not, then $\pi(\phi^a)$ will have an edge $\gamma = [(A'', B''), (A', B')]$ with right endpoint (A', B'), and L must touch $\mathcal{N}(\phi^a)$ in a point contained in an edge strictly to the left of γ. But then the line L'' containing γ must have slope strictly less than the slope $\frac{1}{2}$ of L, and necessarily $B'' > B' = 3$; hence $B'' \geq 4$. It is then again easily seen that the line L'' would intersect the line $\Delta^{(2)}$ at some point with second coordinate z_2 strictly bigger than 3.5, so that again $h^r + 1 \geq z_2 > 3.5$, which would contradict our assumption (Figure 7.3).

We have thus found that $(A', B') = (1, 3)$. Assume finally that $\mathcal{N}(\phi)$ had a vertex (A'', B'') to the left of $(1, 3)$. Then, necessarily, $A'' = 0$ and $B'' \geq 4$, so that the line passing through the points (A'', B'') and $(1, 3)$ would have slope at least 1, a contradiction. We have seen that $\mathcal{N}(\phi)$ is contained in the half plane where

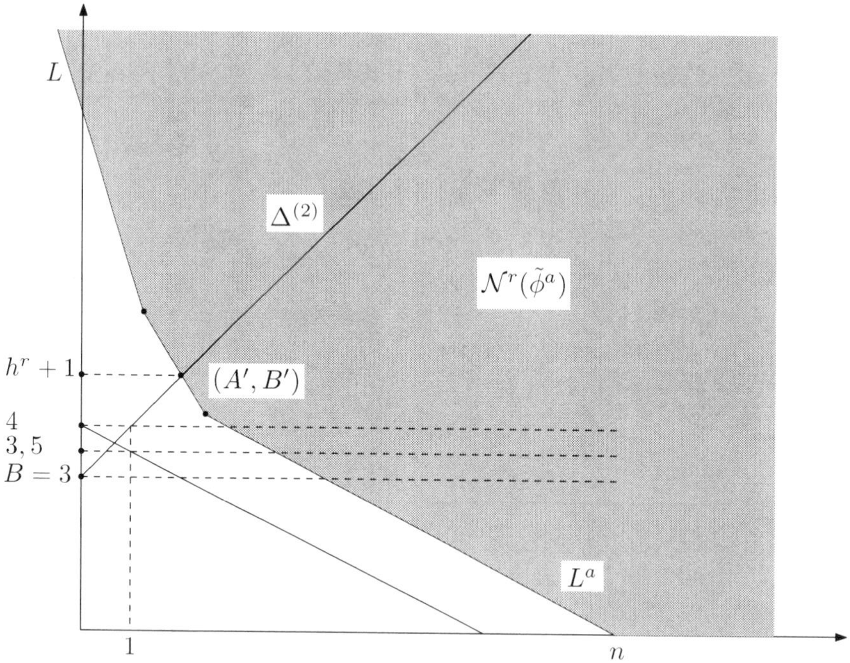

Figure 7.2

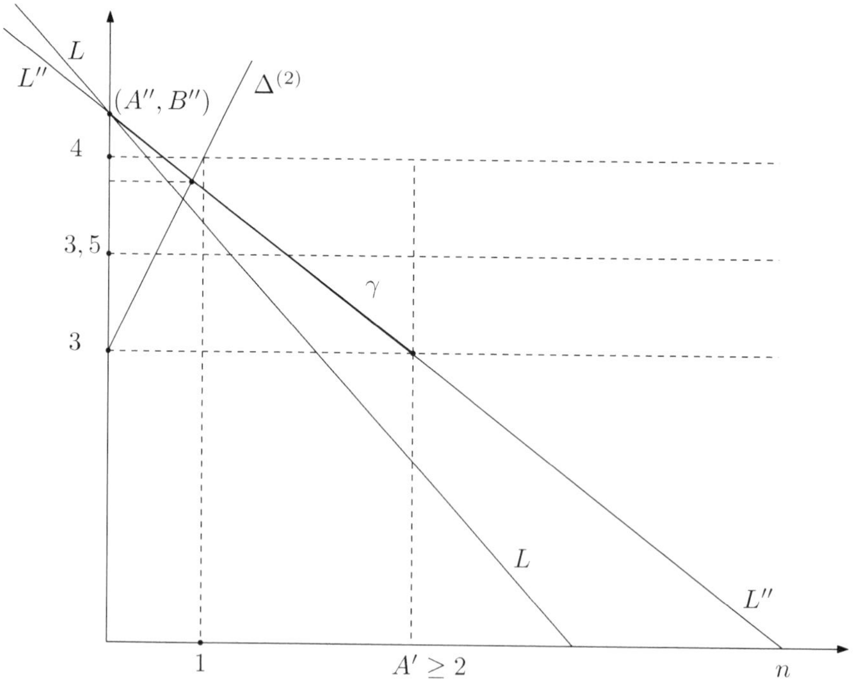

Figure 7.3

$t_1 \geq 1$, and thus the line L must pass through the point $(1, 3)$, and the claim on the structure of $\mathcal{N}^r(\tilde{\phi}^a)$ is now obvious.

But then clearly $\Delta^{(2)}$ will intersect the boundary of $\mathcal{N}^r(\tilde{\phi}^a)$ in a point of L^+, so that $h^r = d$. The remaining statements in (7.82) are now easily verified. Q.E.D.

With this structural result, we can now conclude the discussion of this case. Indeed, by Proposition 7.9 we have $\theta_c = \frac{3}{10} > \frac{2}{7}$, and, arguing as before, only with θ_c in place of $\tilde{\theta}_c$, we obtain

$$\sum_{\lambda \gg 1} \| T_\delta^\lambda \|_{p_c \to p_c'} \lesssim \delta_0^{(2-7\theta_c)/2}.$$

Following the rescaling arguments from the previous discussion, we see that this implies that here the right-hand side of the estimate (7.80) can be written as $2^{-kE}\| f \|_{L^{p_c}}$, with exponent E satisfying

$$\begin{aligned} 2p_c' E &:= 2p_c' \left(\frac{|\tilde{\kappa}|}{2} - \frac{\tilde{\kappa}_1(1+2)+1}{p_c'} + (\tilde{\kappa}_2 - 2\tilde{\kappa}_1)\frac{2h^r - 5}{2p_c'} \right) \\ &= \tilde{\kappa}_2(4h^r - 3) + \tilde{\kappa}_1(6 - 2h^r) - 2. \end{aligned}$$

By means of (7.82) one then computes that

$$2p_c' E = \frac{19}{9} \cdot \frac{n-1}{n} + \frac{4}{3} \cdot \frac{1}{n} - 2 = \frac{n-7}{9n} > 0,$$

so that the uniform estimate (7.81) remains valid also in this case.

7.7.2 The case where $h^r + 1 \leq B$

In this case, since $d < h \leq h^r + 1$, we have $d < B$, and since we are assuming that $d > 2$, we see that we may assume that $B \geq 3$.

Moreover, it is obvious from the structure of the Newton polyhedron of ϕ^a that necessarily $m + 1 \leq B$, and

$$h^r + 1 = h_{l_{\mathrm{pr}}} + 1 = \frac{1 + (m+1)\tilde{\kappa}_1}{|\tilde{\kappa}|}. \tag{7.83}$$

Indeed, this follows from the geometric interpretation of the notion of r-height given directly after Remarks 1.9, since the line $\Delta^{(m)}$ intersects the principal face $\pi(\phi^a)$ (cf. Figure 1.4).

Therefore, in passing from the measure ν_δ to μ_k, no further gain is possible in this situation in (7.14).

First consider Case A. Corollary 7.5(b) implies that $\theta_c \leq \tilde{\theta}_B$. Thus, by (7.75) we have

$$\| T_\delta^\lambda \|_{p_c \to p_c'} \lesssim \lambda^{-1/3 - 1/B + 7(m+1)/3(mB+m+1)} = \lambda^{-M(m,B)/3B(mB+m+1)},$$

where

$$M(m, B) := mB^2 - (3m+6)B + 3(m+1)$$

is increasing in B if $B \geq 3$ and $m \geq 2$. Since $B \geq m+1$, we thus have

$$M(m, B) \geq m^3 - m^2 - 5m - 3,$$

and the right-hand side of this inequality is increasing in m if $m \geq 2$ and assumes the value 0 if $m = 3$. Therefore, $M(m, B) \geq 0$ if $m \geq 3$ and even with strict inequality if $B > m+1$.

We thus see that our estimates for $\|T_\delta^\lambda\|_{p_c \to p_c'}$ do sum for $m \geq 3$ over all dyadic $\lambda \leq \delta_0^{-B/(B-1)}$ when $B > m+1$, and they are at least uniform, if $B = m+1$.

Finally, when $m = 2$, then $M(2, B) = 2B^2 - 12B + 9 = 2[(B-3)^2 - 9/2] > 0$, iff $B \geq 6$. We thus find that

$$\sum_{\lambda \leq \delta_0^{-B/(B-1)}} \|T_\delta^\lambda\|_{p_c \to p_c'} \lesssim 1,$$

except possibly when

$$m = 2,\ B = 3, 4, 5, \ \text{or}\ m = 3,\ B = 4. \tag{7.84}$$

Moreover, we have

$$\|T_\delta^\lambda\|_{p_c \to p_c'} \lesssim 1 \qquad \text{if}\ m = 3,\ B = 4.$$

But, recall that here $\theta_c < \tilde{\theta}_c \leq \tilde{\theta}_B$, unless $4 = B = H = h^r + 1 = d + 1$, so that $\theta_c = \frac{1}{4}$. Therefore, in those cases where $m = 3$ and $B = 4$ and in which we can obtain only a uniform estimate for the norms $\|T_\delta^\lambda\|_{p_c \to p_c'}$, we will have $\theta_c = \frac{1}{4}$. But, by (7.73) and (7.74) we have $\|\nu_\delta^\lambda\|_\infty \lesssim \lambda^{7/4}$, since we are in Case A, and $\|\widehat{\nu_\delta^\lambda}\|_\infty \lesssim \lambda^{-7/12}$, where $-(1-\theta_c)7/12 + \theta_c 7/4 = 0$. We may thus again use real interpolation in order to estimate the operator of convolution with the Fourier transform of the complex measure $\sum_{\lambda \leq \delta_0^{-4/3}} \nu_\delta^\lambda$, namely, by means of Proposition 2.6, in the same way as we did just before in the corresponding case where $m = 5$ and $B = 3$.

We are thus left with the cases where $m = 2$, $h^r + 1 \leq B$, and $B = 3, 4, 5$. So assume in the sequel that that $m = 2$ and $h^r + 1 \leq B$.

If $m = 2$ and $B = 3$, then $B = m + 1$, and since we are assuming $h^r + 1 \leq B$, a look at the Newton polyhedron of $\tilde{\phi}^a$ shows that necessarily $H = B = 3 = h^r + 1$.

Lemma 7.10. *Assume that $m = 2$ and $B = 4, 5$. Then we have*

$$-\frac{1}{3} - \frac{1}{B} + \frac{7}{3}\tilde{\theta}_c < 0,$$

provided

$$H > H(B) := \begin{cases} \frac{9}{2}, & \text{if } B = 4, \\ \frac{81}{16}, & \text{if } B = 5. \end{cases} \tag{7.85}$$

Proof. For $m = 2$ we have

$$-\frac{1}{3} - \frac{1}{B} + \frac{7}{3}\tilde{\theta}_c = -\frac{1}{3} - \frac{1}{B} + \frac{7}{2H+3} < 0$$

if and only if

$$H > \frac{21}{2}\frac{B}{B+3} - \frac{3}{2}, \tag{7.86}$$

and it is easily checked that for $B = 4, 5$, this holds true if and only if $H > H(B)$.
Q.E.D.

Since by Corollary 7.5 (b) $\theta_c \le \tilde{\theta}_c$, the previous lemma shows that for $m = 2$, (7.84) can be sharpened as follows:

$$\sum_{\lambda \le \delta_0^{-B/(B-1)}} \|T_\delta^\lambda\|_{p_c \to p_c'} \lesssim 1, \tag{7.87}$$

with the possible exceptions of the cases where $B = H = 3 = h^r + 1$, or $B = 4, 5, h^r + 1 \le B$ and $H \le H(B)$.

Finally, consider Case B. Then, since $\theta_c \le \tilde{\theta}_B$,

$$\begin{aligned}\|T_\delta^\lambda\|_{p_c \to p_c'} &\lesssim \lambda^{-1/3-1/B+(4/3+1/B)(m+1)/(mB+m+1)} \delta_0^{-\tilde{\theta}_B} \\ &= \lambda^{-B(mB-3)/3B(mB+m+1)} \delta_0^{-\tilde{\theta}_B}.\end{aligned}$$

The exponent of λ is negative, so we can sum these estimates in λ and obtain

$$\begin{aligned}\sum_{\lambda > \delta_0^{-B/(B-1)}} \|T_\delta^\lambda\|_{p_c \to p_c'} &\lesssim \delta_0^{[1/3+1/B-(4/3+1/B)\tilde{\theta}_B]B/(B-1)-\tilde{\theta}_B} \\ &= \delta_0^{(B+3-7B\tilde{\theta}_B)/3(B-1)} = \delta_0^{M(m,B)/3(B-1)(mB+m+1)}.\end{aligned}$$

But, our previous discussion of $M(m, B)$ shows that $M(m, B) \ge 0$, unless $m = 2$ and $B = 3, 4, 5$ (notice that the case $m = 3$, $B = 4$ still works here).

In the latter cases, we can improve our estimates again by using $\tilde{\theta}_c$ in place of $\tilde{\theta}_B$. Indeed, notice that the condition $B + 3 - 7B\tilde{\theta}_c > 0$ is equivalent to (7.86), which by Lemma 7.10 does hold true if $H \le H(B)$.

We thus find that

$$\sum_{\lambda > \delta_0^{-B/(B-1)}} \|T_\delta^\lambda\|_{p_c \to p_c'} \lesssim 1, \tag{7.88}$$

unless $m = 2$ and $B = H = 3 = h^r + 1$, or $B = 4, 5$ $h^r + 1 \le B$ and $H \le H(B)$.

Finally, we observe that the following consequence of Lemma 7.10.

Corollary 7.11. *Assume that $m = 2$ and that $B = H = 3$, or $B = 4, 5$, $h^r + 1 \le B$, and $H \le H(B)$. Then the left endpoint of the principal face of the Newton polyhedron of $\tilde{\phi}^a$ is of the form (A, B), where $A \in \{0, 1, \dots, B - 3\}$, and we have*

$$\tilde{\kappa} = \left(\frac{1}{n}, \frac{n - A}{Bn}\right) \quad \text{and } h^r + 1 = \frac{n + 3}{n + B - A} B,$$

where n must satisfy $n > 2B$ and $(Bn)/(n - A) = H \le H(B)$. Moreover, $\tilde{\phi}^a$ is smooth here, so that in particular a and n are integers.

Proof. Adopting the notation of the proof of Proposition 7.9, let (A, B') denote the left endpoint of $\pi(\tilde{\phi}^a)$, which, as we recall, is also the left endpoint of $\pi(\phi^a)$. If $B = 4, 5$, then $B \le B' \le H \le H(B) < B + 1$, so that $B' = B$. For $B = 3 = H$, the same conclusion applies. Next, since we assume that $h^r + 1 \leq B$, and since the line $\Delta^{(2)}$ intersects the line where $t_2 = B$ in the point $(B-3, B)$, we must have $0 \leq A \leq B-3$. This implies the first statement of the corollary, because A is an integer.

Moreover, Proposition 7.3 now shows that $\tilde{\phi}^a$ is smooth and that a and n are integers.

The remaining statements follow easily (for the identity for $h^r + 1$, recall (7.83)).

Q.E.D.

7.8 THE CASE WHERE $B = 5$

The case where $B = 5$ can now be treated quite easily. We begin by recalling that by (7.75) we have the estimate

$$\|T_\delta^\lambda\|_{p_c \to p_c'} \lesssim \lambda^{-1/3-1/5+7\theta_c/3}.$$

This estimate is valid in Case A as well as in Case B (in the latter case, an even stronger estimate holds true, but we won't need it). And, according to Corollary 7.11, if $B = 5$, then we have the precise formula,

$$\theta_c = \frac{n+5-A}{5(n+3)},$$

for θ_c, where $A \in \{0, 1, 2\}$ and $n > 2B = 10$. Since here n is an integer, we find that $n \geq 11$, and thus $\theta_c \leq (11+5)/70 = 8/35$, with strict inequality, unless $A = 0$ and $n = 11$. This shows that the exponent of λ in the estimate for $\|T_\delta^\lambda\|_{p_c \to p_c'}$ is strictly negative, so that we can sum these estimates over all dyadic $\lambda \geq 1$, unless $A = 0$ and $n = 11$, where we only get a uniform estimate

$$\|T_\delta^\lambda\|_{p_c \to p_c'} \lesssim 1.$$

However, if $A = 0$, $B = 5$, and $n = 11$, then $\theta_c = \frac{8}{35}$, and moreover, by (7.73) and (7.74), we have $\|\nu_\delta^\lambda\|_\infty \lesssim \lambda^{9/5}$ and $\|\widehat{\nu_\delta^\lambda}\|_\infty \lesssim \lambda^{-8/15}$, where $-(1-\theta_c)8/15 + \theta_c 9/5 = 0$. We may thus again estimate the operator of convolution with the Fourier transform of the complex measure $\sum_{\lambda \gg 1} \nu_\delta^\lambda$ by means of the real interpolation method of Proposition 2.6 by adding this measure to the list of measures μ^i, $i \in I$, of the second class in (7.41).

7.9 COLLECTING THE REMAINING CASES

Recall that the integer $B \geq 2$ is chosen so that (7.6) holds true, that is, that

$$\tilde{\phi}_{\tilde{\kappa}}^a(y_1, y_2) = y_2^B Q(y_1, y_2) + c_1 y_1^n, \quad c_1 \neq 0, \ Q(1, 0) \neq 0,$$

and that according to (7.21) $H = 1/\tilde{\kappa}_2$ is the second coordinate of the point of intersection of the principal line L^a with the t_2-axis. The quantity $H(B)$ has been defined in (7.85).

The estimates that remain to be established when $m = 2$ and $B = 3$ have essentially already been addressed in the previous sections. Their treatment will again require Airy-type analysis of a similar form as in Chapter 5, which had been devoted to the case where $m = 2$ and $B = 2$, but the discussion in the next chapter will be even more involved. Also in the case $m = 2$ and $B = 4$, we shall need some Airy-type analysis, but of a simpler form.

Combining all the previous estimates and applying (7.14), we see that we have proved the following.

Corollary 7.12. *Assume that m and H, B are not such that $m = 2$ and $B = H = 3 = h^r + 1$ or $m = 2$, $B = 4$, $h^r + 1 \leq 4$, and $H \leq H(4)$. Then the estimates in Proposition 7.2 hold true for $l = 1$.*

More precisely, in view of Corollary 7.11, what remains open are the following situations:

$$m = 2,\ B = 3, 4,\ h^r + 1 \leq B, \qquad \text{and } \lambda_1 \sim \lambda_2 \sim \lambda_3, \tag{7.89}$$

where the left endpoint of the principal face of the Newton polyhedron of $\tilde{\phi}^a$ is of the form (A, B), where $A \in \{0, \ldots, B-3\}$.

In these situations, we have

$$\tilde{\kappa} = \left(\frac{1}{n}, \frac{n-A}{Bn}\right) \quad \text{and } h^r + 1 = \frac{n+3}{n+B-A}B, \tag{7.90}$$

where n must be an integer satisfying $n > 2B$, and $(Bn)/(n-A) = H \leq H(B)$ if $B = 4$. Moreover, according to Corollary 7.11, in all these cases the exponent $a = \tilde{\kappa}_2/\tilde{\kappa}_1$ is an integer, and the modified adapted coordinates given by (7.5) are in fact smooth, so that we could use them as well as adapted coordinates for ϕ. In these adapted coordinates, ϕ is represented by the smooth function $\tilde{\phi}^a$.

The discussion of these cases will occupy the major part of remainder of this monograph. Before we come to this, let us first study also the contributions by the remaining domains $D'_{(l)}$, $l \geq 2$, which will turn out to be easier to handle.

7.10 RESTRICTION ESTIMATES FOR THE DOMAINS $D'_{(l)}, l \geq 2$

For the domans $D'_{(l)}$, $l \geq 2$, we can essentially argue as in the preceding sections by letting

$$\tilde{\phi}^a := \phi^{(l+1)}, \quad \tilde{\kappa} := \kappa^{(l)}, \quad D^a := D^a_{(l)}, \quad L^a := L_{(l)}, \quad \text{and so on.}$$

H and n are defined correspondingly.

We then have the following analogue of Lemma 7.4.

Lemma 7.13. (a) *If $l \geq 2$, then $p'_c > \tilde{p}'_c$.*

(b) *If $m \geq 3$ and $H \geq 2$ or $m = 2$ and $H \geq 3$, then*

$$\tilde{p}'_c \geq p'_H \geq p'_B.$$

Proof. (a) We can follow the proof of Lemma 7.4(a) and see again that $p'_c > \tilde{p}'_c$, unless the principal face $\pi(\tilde{\phi}^a)$ of $\tilde{\phi}^a$ is the edge $[(0, H), (n, 0)]$. However, for $l \geq 2$, we know from Section 7.1 that $\pi(\tilde{\phi}^a)$ is the edge $\gamma'_{(l)}$, which lies below the bisectrix Δ, so that we cannot have $p'_c = \tilde{p}'_c$.

(b) follows as before. Q.E.D.

This implies the following, stronger analogue of Corollary 7.5.

Corollary 7.14. *Assume that $l \geq 2$.*

(a) *If $m \geq 3$ and $H \geq 2$ or $m = 2$ and $H \geq 3$, then $\theta_c < \theta_B$.*
(b) *If $h^r + 1 \leq B$, then $\theta_c < \tilde{\theta}_c$, unless $B = H = h^r + 1 = d + 1$, where $\theta_c = \tilde{\theta}_c$.*
(c) *If $H \geq 3$, then $\theta_c < \frac{1}{3}$ unless $H = 3$ and $m = 2$.*

Finally, in place of Proposition 7.9, we have the following.

Proposition 7.15. *Assume that $l \geq 2$ and that $B = 3$, $m \geq 2$. Then $h^r + 1 > 3.5$.*

Proof. Assume we had $h^r + 1 \leq 3.5$, and denote by $(A', B') := (A'_{(l-1)}, B'_{(l-1)}) \in L^a$ the left endpoint of the principal face $\pi(\tilde{\phi}^a) = \gamma'_{(l)}$ of the Newton polyhedron of $\tilde{\phi}^a$. Then $B' \geq B = 3$. Arguing in the same way as in the first step of the proof of Proposition 7.5, we see that $B' = B = 3$.

Then the preceding edge $\gamma'_{(l-1)}$ will have a left endpoint (A'', B'') with $B'' \geq 4$. Moreover, since the line $L^{(1)}$ has slope $1/a < 1/m \leq \frac{1}{2}$, the line $L^{(l-1)}$ containing $\gamma'_{(l-1)}$ has slope strictly less than $\frac{1}{2}$. But then it will intersect the line $\Delta^{(2)}$ at some point with second coordinate z_2 strictly bigger than 3.5, so that we would again arrive at $h^r + 1 \geq z_2 > 3.5$, which contradicts our assumption. Q.E.D.

These results allow us to proceed exactly as in Sections 7.5 and 7.7, even with some simplifications. Indeed, a careful inspection of our arguments in these sections reveals that here all the series that appear do sum, and no further interpolation arguments are required.

This is because of the stronger estimate $p'_c > \tilde{p}'_c$ of Lemma 7.13 and the stronger statement of Proposition 7.15, which implies in particular that $\theta_c < \frac{1}{3}$ when $B = 3$.

Moreover, in Subsection 7.7.1, the more delicate case where $B = 3$ and $\theta_c \geq \frac{2}{7}$ does not appear anymore, since by Proposition 7.14 we have $\theta_c < \frac{2}{7}$.

Observe finally that if $m = 2$, $B = 3, 4, 5$, $h^r + 1 \leq B$, and $H \leq H(B)$, then Corollary 7.11, whose proof applies equally well when $l \geq 2$, shows that left endpoint of the principal face of the Newton polyhedron of $\tilde{\phi}^a$ is of the form (A, B), where $A \leq B - 3$. However, if $l \geq 2$, this endpoint must lie on or below the bisectrix, which leads to a contradiction. These cases therefore cannot arise when $l \geq 2$.

We therefore obtain the following.

Corollary 7.16. *The estimates in Proposition 7.2 hold true for every $l \geq 2$.*

Chapter Eight

The Remaining Cases Where $m = 2$ and $B = 3$ or $B = 4$

We finally turn to the discussion of the remaining cases for $l = 1$, which have not been covered yet by Corollary 7.12 and which are described by (7.89) and (7.90). In these cases, the exponent B in our expression (7.9) for the phase ϕ_δ is either 3 or 4.

Our basic approach will be similar to the one that we have developed for the case $h_{\text{lin}}(\phi) < 2$ in Chapter 5 (in which the exponent B would be given by $B = 2$). We shall again have to perform an additional dyadic frequency domain decomposition related to the distance to a certain Airy cone. This is needed in order to be able to control the integration with respect to the variable x_1 in the Fourier integral defining the Fourier transform of the complex measures ν_δ^λ. Recall that these are already localized to frequency domains on which $|\xi_i| \sim \lambda = 2^j$ for $i = 1, 2, 3$. We shall see by means of Proposition 7.3 that $\widehat{\nu_\delta^\lambda}(\xi)$ is an oscillatory integral with a phase function $-(\xi_3\phi_\delta(x) + \xi_1 x_1 + \xi_2 x_2)$, where the function ϕ_δ, which depends on a finite number of small parameters δ_i, can be assumed to be of the form

$$\phi_\delta(x) = x_2^B\, b(x_1, x_2, \delta) + x_1^n \alpha(\delta_1 x_1) + \sum_{j=1}^{B-2} x_2^j x_1^{n_j} \delta_{j+2} \alpha_j(\delta_1 x_1).$$

If $B = 2$, then the preceding formula shows that the integration with respect to the variable x_2 can easily be dealt with by means of the method of stationary phase; this is exactly what we did in Section 5.1. However, if $B = 3$ or $B = 4$, this is no longer possible, and a new complication arises, which will necessitate a considerable amount of additional and refined analysis in order to carry out the basic approach of Chapter 5.

Indeed, after applying a suitable translation in the x_1-coordinate in a similar way as in Section 5.1, basically by some critical point $x_1^c(\delta_1, s_2)$ of $\partial_{x_1}\phi_\delta$, in combination with a change of variables, it turns out that $\widehat{\nu_\delta^\lambda}(\xi)$ can be rewritten as an oscillatory integral with phase $\lambda\Phi^\sharp(x, \delta, s)$, where

$$\Phi^\sharp(x, \delta, s) = x_1^3\, B_3(s_2, \delta_1, x_1) - x_1\, B_1(s, \delta_1) + B_0(s, \delta_1) + \phi^\sharp(x, \delta, s_2),$$

with

$$\phi^\sharp(x, \delta, s_2) = x_2^B\, b(x, \delta, s_2) + \sum_{j=2}^{B-2} \delta_{j+2} x_2^j\, \tilde{\alpha}_j(x_1, \delta, s_2)$$
$$+ x_2\left(s_2\delta_0 + \delta_3\left(x_1^c(\delta_1, s_2) + x_1\right)^{n_1} \tilde{\alpha}_1(x_1, \delta_1, s_2)\right)$$

and with $\xi = \lambda(s_1 s_3, s_2 s_3, s_3)$. Here the s_i, $i = 1, 2, 3$, are integration variables of size $|s_i| \sim 1$. A major problem will then be caused by the term $\delta_3(x_1^c(\delta_1, s_2) + x_1)^{n_1} \times \tilde{\alpha}_1(x_1, \delta_1, s_2)$, which does not yet appear when $B = 2$. Indeed, without this term, the coefficient of the linear term in x_2 would be given by $s_2\delta_0$ and hence would be of a well-defined size δ_0. However, in the presence of the additional term depending on the small parameter δ_3, the situation becomes less clear. We therefore have to perform a more refined analysis of the phase $\Phi^\sharp$ in Section 8.2, which leads us to distinguish between a nondegenerate Case ND and a degenerate Case D in Lemma 8.3. In Case ND, we do have good control of the size of the coefficient of x_2, and the same is true of Case D, in which, however, an additional term, roughly of the form $\delta_0 x_1 x_2$, does appear. The new form of the phase arising in this way will depend on a slightly modified vector $\tilde{\delta}$ of perturbation parameters $\tilde{\delta}_i$ in place of δ.

Next, following the proof of Proposition 4.2 in Chapter 5, we decompose

$$\nu_\delta^\lambda = \nu_{\delta, Ai}^\lambda + \sum_{M_0 \le 2^l \le \lambda/M_1} \nu_{\delta, l}^\lambda,$$

where $\widehat{\nu_{\delta, Ai}^\lambda}(\xi) := \chi_0(\lambda^{2/3} B_1(s, \delta_1))\widehat{\nu_\delta^\lambda}(\xi)$ and $\widehat{\nu_{\delta, l}^\lambda}(\xi) := \chi_1((2^{-l}\lambda)^{2/3} B_1(s, \delta_1)) \times \widehat{\nu_\delta^\lambda}(\xi)$, and estimate the corresponding terms separately. This decomposition is motivated by the proof of Lemma 2.2 and it allows for a precise understanding of the integration with respect to the variable x_1 in the phase $\Phi^\sharp(x, \delta, s)$.

A major problem that remains is to also control the integration in x_2. We cannot simply view the phase $\Phi^\sharp$ as a perturbation of the phase given when $\tilde{\delta} = 0$ but need more precise estimates in terms of the (relative) sizes of λ and the components $\tilde{\delta}_i$ of $\tilde{\delta}$. Here we pick up some ideas going back to Duistermaat [Dui74] and define a kind of homogeneous gauge $\rho(\tilde{\delta})$ in such a way that under the natural dilations $\sigma_r(x_1, x_2) := (r^{1/3}x_1, r^{1/B}x_2)$, the rescaled phase

$$\Phi_r^\sharp(x_1, x_2, \delta, s) := r\Phi^\sharp\big(\sigma_{1/r}(x_1, x_2), \delta, s\big)$$

looks essentially like $\Phi^\sharp$, only with the coefficients $\tilde{\delta}_i$ replaced by coefficients $\tilde{\delta}_i^r$ in such a way that $\rho(\tilde{\delta}^r) = r\rho(\tilde{\delta})$. In particular, if we choose $r = 1/\rho(\tilde{\delta})$, then the rescaled coefficients will satisfy $\rho(\tilde{\delta}^r) = 1$, and thus at least one component of $\tilde{\delta}^r$ will be of size one. The effect is that the rescaled phase will have at worst a singularity of order $B - 1$ with respect to x_2, in contrast to the original phase, which had a singularity of order B, so that we can obtain a better decay of the corresponding oscillatory integral. Notice, however, that the price to be paid here is that we have to replace the parameter λ effectively by $\lambda/r = \lambda\rho(\tilde{\delta})$, and that moreover, we pick up the Jacobian of this change of coordinates.

This suggest that we distinguish between the cases where $\lambda\rho(\tilde{\delta}) \lesssim 1$ and where $\lambda\rho(\tilde{\delta}) \gg 1$. Moreover, matters are even more complicated, since we cannot really treat the variables x_1 and x_2 separately.

Anyway, in Section 8.3 we first treat the case where $\lambda\rho(\tilde{\delta}) \lesssim 1$. Here, the change of variables $x = \sigma_{1/\lambda}u = (\lambda^{-1/3}u_1, \lambda^{-1/B}u_2)$ turns out to be convenient for the treatment of the $\nu_{\delta, Ai}^\lambda$ (and related scalings will work for the treatment of the $\nu_{\delta, l}^\lambda$). In fact, this approach is analogous to the proof of part (a) in Lemma 2.2. Under the assumption $\lambda\rho(\tilde{\delta}) \lesssim 1$ one finds that the rescaled perturbation parameters $\tilde{\delta}^\lambda$ are still small, so that we do get a rather precise estimate for the oscillatory integrals arising in this situation. This will eventually lead to the desired restriction estimates, at least when $B = 4$ (compare Proposition 8.6), and when $B = 3$ we miss only the endpoint estimate for $p = p_c$. That case will therefore again require a complex interpolation argument in order to capture the endpoint as well.

Next, in Sections 8.4 and 8.5 we deal with the case where $\lambda\rho(\tilde{\delta}) \gg 1$ and $B = 4$. Here, we apply, in fact, the change of variables $x = \sigma_{\tilde{\rho}} u$, where $\tilde{\rho} := \rho(\tilde{\delta}) + |B_1(s, \delta_1)|^{3/2}$ captures not only the behavior of our phase with respect to the variable x_2, but also with respect to x_1 (compare Lemma 8.8). The basic underlying idea is, however, similar to the one explained before, only that we effectively use $\tilde{\rho}$ as a scaling parameter in place of $\rho(\tilde{\delta})$. It turns out that by means of these methods, we eventually obtain sufficiently strong estimates, which can even be summed absolutely over the dyadic λs for which $\lambda\rho(\tilde{\delta}) \gg 1$, and we arrive at the required restriction estimates.

Finally, we turn to the case where $B = 3$ in Section 8.6. Here, we can basically proceed in a similar way as for $B = 4$, but it turns out that by these methods we do miss the endpoint estimates for $p = p_c$ in a few cases. These remaining cases are listed in Proposition 8.12. The treatment of the remaining four cases in Sections 8.7 and 8.8 will again be based on complex interpolation arguments, similar in spirit to the ones that we had employed in the proof of Proposition 4.2 in Chapter 4. Regretfully, the technical problems to be overcome here will be substantially more involved than in Chapter 4.

8.1 PRELIMINARIES

Recall from Corollary 7.11 that in the cases under consideration in this chapter, the function $\tilde{\phi}^a$ is smooth, so that in particular $a = \tilde{\kappa}_2/\tilde{\kappa}_1$ and n are integers. Moreover, in the proof of Proposition 7.3, whose assumptions are satisfied here, we have seen (cf. (7.16)) that we may assume that

$$\tilde{\phi}^a_{\tilde{\kappa}}(y) = y_1^A y_2^B + c_1 y_1^n, \qquad c_1 \neq 0,$$

so that the function Q in (7.6) is given by $Q(y_1, y_2) = y_1^A$. Then, by (7.9) and (7.12), we may write

$$\phi_\delta(x) := x_2^B\, b(x_1, x_2, \delta) + x_1^n \alpha(\delta_1 x_1) + r(x_1, x_2, \delta), \quad (x_1 \sim 1, |x_2| < \varepsilon), \quad (8.1)$$

where $b(x_1, x_2, 0) = x_1^A \sim 1$ and $\alpha(0) \neq 0$, and where $r(x_1, x_2, \delta)$ is given by (7.10). Notice, however, that Proposition 7.3 shows that, by passing to alternative smooth adapted coordinates, we may assume without loss of generality that the term of order $j = B - 1$ in (7.10) vanishes. In the sequel, we shall therefore assume that $r(x_1, x_2, \delta)$ is of the form

$$r(x_1, x_2, \delta) = \sum_{j=1}^{B-2} x_2^j x_1^{n_j} \delta_{j+2} \alpha_j(\delta_1 x_1). \qquad (8.2)$$

Moreover, either $\alpha_j(0) \neq 0$, and then n_j is fixed (n_j is then the type of the finite-type function b_j), or $\alpha_j(0) = 0$, and then we may assume that n_j is as large as we please.

In particular, in view of (7.11), we may then assume that $\delta = (\delta_0, \delta_1, \delta_2, \delta_3, \dots, \delta_B)$ is given by

$$\delta := (2^{-k(\tilde{\kappa}_2-2\tilde{\kappa}_1)}, 2^{-k\tilde{\kappa}_1}, 2^{-k\tilde{\kappa}_2}, 2^{-k(n_1\tilde{\kappa}_1+\tilde{\kappa}_2-1)}, \dots, 2^{-k(n_{B-2}\tilde{\kappa}_1+(B-2)\tilde{\kappa}_2-1)}). \quad (8.3)$$

Observe also that if $A = 0$ (this is necessarily so if $B = 3$), then we have $Q(x) \equiv 1$, so that $\tilde{\kappa}_2 = 1/B$ and, consequently,

$$b(x_1, x_2, \delta) = b_B(\delta_1 x_1, \delta_2 x_2) \quad (8.4)$$

in (8.1). This indeed follows from our definition $\phi_k(x_1, x_2) := 2^k \tilde{\phi}^a(2^{-\tilde{\kappa}_1 k}x_1, 2^{-\tilde{\kappa}_2 k}x_2)$ of the rescaled phase function ϕ_k and the corresponding function ϕ_δ in Section 7.2, in combination with (7.15).

We finally recall that the complete phase corresponding to ϕ_δ is given by

$$\Phi(x, \delta, \xi) := \xi_1 x_1 + \xi_2(\delta_0 x_2 + x_1^m \omega(\delta_1 x_1)) + \xi_3 \phi_\delta(x_1, x_2).$$

Recall also that we are interested in the frequency domains where $|\xi_j| \sim \lambda_j,\ j = 1, 2, 3$, assuming that $\lambda_1 \sim \lambda_2 \sim \lambda_3$. For the sake of simplicity, we shall assume that

$$\lambda_1 = \lambda_2 = \lambda_3 \gg 1.$$

Changing notation in the same way as we did in Section 7.7, in the sequel we shall denote the common value of the λ_j by λ and accordingly write

$$\widehat{\nu_\delta^\lambda}(\xi) := \chi_1\left(\frac{\xi_1}{\lambda}\right)\chi_1\left(\frac{\xi_2}{\lambda}\right)\chi_1\left(\frac{\xi_3}{\lambda}\right)\int e^{-i\Phi(y,\delta,\xi)}\,\eta(y)\,dy,$$

that is,

$$\nu_\delta^\lambda(x) = \lambda^3 \int \check{\chi}_1(\lambda(x_1 - y_1)) \times \check{\chi}_1\left(\lambda(x_2 - \delta_0 y_2 - y_1^2\omega(\delta_1 y_1))\right) \\ \times \check{\chi}_1(\lambda(x_3 - \phi_\delta(y))\,\eta(y)\,dy.$$

Recall also that $\operatorname{supp}\eta \subset \{y_1 \sim 1,\ |y_2| < \varepsilon\}$, $(\varepsilon \ll 1)$.

As before, we change coordinates from $\xi = (\xi_1, \xi_2, \xi_3)$ to s_1, s_2, and $s_3 := \xi_3/\lambda$ by writing $\xi = \xi(s, \lambda)$, with

$$\xi_1 = s_1\xi_3 = \lambda s_1 s_3, \quad \xi_2 = s_2\xi_3 = \lambda s_2 s_3, \quad \xi_3 = \lambda s_3.$$

Accordingly, we write

$$\Phi(y, \delta, \xi) = \lambda s_3\left(\Phi_1(y_1, \delta_1, s) + s_2\delta_0 y_2 + y_2^B\, b(y_1, y_2, \delta) + r(y_1, y_2, \delta)\right), \quad (8.5)$$

where

$$\Phi_1(y_1, \delta_1, s) := s_1 y_1 + s_2 y_1^2\omega(\delta_1 y_1) + y_1^n \alpha(\delta_1 y_1).$$

Notice that here $|s_j| \sim 1,\ j = 1, 2, 3$, and that from now on we shall denote by s the pair $s := (s_1, s_2)$ (in order to simplify the notation, we deviate slightly here from our previous notation in Section 7.7, where this pair had been denoted by s'). By passing from s_j to $-s_j$, if necessary, we shall in the sequel always assume that

$$s_j \sim 1, \qquad j = 1, 2, 3$$

(notice that these changes of signs may cause a change of sign of the x_j and ω, respectively, of Φ).

Let us fix $s^0 = (s_1^0, s_2^0)$ such that $s_1^0 \sim 1 \sim s_2^0$, and consider the phase function $\Phi_1(x_1, 0, s^0)$ when $\delta_1 = 0$.

Assume first that this phase has at worst nondegenerate critical points $x_1^c \sim 1$. Then the same is true for sufficiently small δ_1, and the estimate (7.73) for $\widehat{\nu_\delta^\lambda}$ in Section 7.7 can be improved to

$$\|\widehat{\nu_\delta^\lambda}\|_\infty \lesssim \lambda^{-1/2-1/B},$$

and thus in Case A of Section 7.7 we obtain the better estimate

$$\|T_\delta^\lambda\|_{p_c \to p_c'} \lesssim \lambda^{-1/2-1/B+5\theta_c/2}$$

compared to (7.75). Moreover, by means of (7.89) and (7.90), one checks easily that the exponent of λ in this estimate is strictly negative if $B = 4$ and zero if $B = 3$. Thus we can sum these estimates over all dyadic $\lambda \gg 1$ if $B = 4$ and obtain at least a uniform estimate when $B = 3$. It is easy to see that this case can then still be treated by means of the real interpolation Proposition 2.6, since the relevant frequencies will here be restricted essentially to cuboids in the ξ-space.

In Case B, where $\lambda > \delta_0^{-B/(B-1)}$, we obtain the better estimate

$$\|T_\delta^\lambda\|_{p_c \to p_c'} \lesssim \lambda^{-1/2-1/B+(3/2+1/B)\theta_c} \delta_0^{-\theta_c},$$

compared to (7.76). The exponent of λ is strictly negative (compare the discussion leading to (7.88)), so summing over all $\lambda > \delta_0^{-B/(B-1)}$, we find that

$$\sum_{\lambda > \delta_0^{-B/(B-1)}} \|T_\delta^\lambda\|_{p_c \to p_c'} \lesssim \delta_0^{(B+2-5B\theta_c)/(2B-2)}.$$

And, since $\theta_c \le \tilde{\theta}_B$, one easily checks that the exponent of δ_0 in this estimate is nonnegative if $B \ge 3$.

8.2 REFINED AIRY-TYPE ANALYSIS

We are thus left with the more subtle situation where the phase function $\Phi_1(x_1, 0, s^0)$ has a degenerate critical point x_1^c of Airy type. Denoting by Φ_1', Φ_1'', ..., derivatives with respect to x_1 and arguing as in Section 5.1, we see by the implicit function theorem that for s sufficiently close to s^0 and δ sufficiently small, there is a unique, nondegenerate critical point $x_1^c = x_1^c(\delta_1, s_2) \sim 1$ of Φ_1', that is,

$$\Phi_1''(x_1^c(\delta_1, s_2), \delta_1, s) = 0, \quad |s - s^0| < \varepsilon, |\delta| < \varepsilon,$$

if ε is sufficiently small. We then shift this critical point to the origin, by putting

$$\Phi^\sharp(x, \delta, \xi) := \frac{1}{s_3\lambda}\Phi(x_1^c(\delta_1, s_2) + x_1, x_2, \delta, \xi), \quad |(x_1, x_2)| \ll 1 \tag{8.6}$$

(notice that we may indeed assume that $|(x_1, x_2)| \ll 1$, since away from x_1^c, we have at worst nondegenerate critical points, and the previous argument applies).

From Lemma 5.2 (with $\beta := \alpha, \sigma = 1, \delta_3 = 0$ and $b_0 = 0$), we then immediately get the following result, after scaling in x_1 so that we may assume that

$$-\frac{2\omega(0)}{n(n-1)\alpha(0)} = 1.$$

Lemma 8.1. *$\Phi^\sharp$ is of the form*

$$\Phi^\sharp(x, \delta, \xi) = x_1^3 B_3(s_2, \delta_1, x_1) - x_1 B_1(s, \delta_1) + B_0(s, \delta_1) + \phi^\sharp(x, \delta, s_2), \quad (8.7)$$

with

$$\phi^\sharp(x, \delta, s_2) := x_2^B b(x, \delta, s_2) + \sum_{j=2}^{B-2} \delta_{j+2} x_2^j \tilde{\alpha}_j(x_1, \delta, s_2)$$
$$+ x_2 \left(s_2\delta_0 + \delta_3 (x_1^c(\delta_1, s_2) + x_1)^{n_1} \alpha_1(\delta_1(x_1^c(\delta_1, s_2) + x_1))\right), \quad (8.8)$$

and where the following hold true:

The functions b (which will be different from b in (8.5)) and $\tilde{\alpha}_j$ are smooth, $b(x, \delta, s_2) \sim 1$, and also $|\tilde{\alpha}_j| \sim 1$, unless α_j is a flat function. Moreover, B_0, B_1, and B_3 are smooth functions, and

$$B_3(s_2, \delta_1, 0) = s_2^{(n-3)/(n-2)} G_4(\delta_1 s_2^{1/(n-2)}),$$

where

$$G_4(0) = \frac{n(n-1)(n-2)}{6} \alpha(0).$$

Furthermore, we may write

$$\begin{cases} x_1^c(\delta_1, s_2) &= s_2^{1/(n-2)} G_1(\delta_1 s_2^{1/(n-2)}), \\ B_0(s, \delta_1) &= s_1 s_2^{1/(n-2)} G_1(\delta_1 s_2^{1/(n-2)}) - s_2^{n/(n-2)} G_2(\delta_1 s_2^{1/(n-2)}), \\ B_1(s, \delta_1) &= -s_1 + s_2^{(n-1)/(n-2)} G_3(\delta_1 s_2^{1/(n-2)}), \end{cases} \quad (8.9)$$

where

$$\begin{cases} G_1(0) &= 1, \\ G_2(0) &= \dfrac{n^2 - n - 2}{2} \alpha(0), \\ G_3(0) &= n(n-2)\alpha(0). \end{cases} \quad (8.10)$$

Notice that all the numbers in (8.10) are nonzero, since we assume $n > 2B > 5$. Finally, if we put $G_5 := G_1 G_3 - G_2$, then we have

$$G_3(0) \neq 0, \quad G_5(0) \neq 0. \quad (8.11)$$

Observe that we here obtain a more specific dependency of x_1^c, B_0, B_1, and B_3 on δ_1 and s_2 than in Section 5.1, due to the fact that the equation for the critical point depends only on the parameter $\delta_1^{n-2} s_2$ in the coordinate $y_1 := \delta_1 x_1$.

Nevertheless, with a slight abuse of notation, we shall frequently also use the shorthand notation $G_j(s_2, \delta)$ in place of $G_j(\delta_1 s_2^{1/(n-2)})$, $j = 1, \ldots, 4$.

We also remark that the part of the measure ν_δ corresponding to the small neighborhood of the critical point x_1^c defined in (8.6) and which by some slight abuse of notation we shall again denote by ν_δ, is given by an expression for its Fourier transform of the form

$$\widehat{\nu_\delta}(\xi) = \int e^{-is_3\lambda\Phi^\sharp(x,\delta,\xi)} a(x,\delta,s)\,dx \qquad (\text{with } \xi = \xi(s,\lambda)), \tag{8.12}$$

where a is a smooth function with compact support in x such that $|x| \le \varepsilon$ on supp a.

Remark 8.2. *It will be important in the sequel to observe that every single δ_j is a fractional power of 2^{-k}, hence, also of δ_0; that is, there is some positive rational number $\mathfrak{r} > 0$ such that $\delta_j = \delta_0^{q_j\mathfrak{r}}$, with positive integers q_j (cf. (8.3)).*

In the sequel, we shall need more precise information on the structure of the last term of $\phi^\sharp$ in (8.8).

Lemma 8.3. *Let*

$$a_1(x_1,\delta,s_2) := s_2\delta_0 + \delta_3(x_1^c(\delta_1,s_2)+x_1)^{n_1}\alpha_1(\delta_1(x_1^c(\delta_1,s_2)+x_1))$$

be the coefficient of x_2 in the last term of $\phi^\sharp$ in (8.8). Then a_1 can be rewritten in the form

$$a_1(x_1,\delta,s_2) = \delta_{3,0}\,\tilde\alpha_1(x_1,\delta_0^{\mathfrak{r}},s_2) + \delta_0\,x_1\alpha_{1,1}(x_1,\delta_0^{\mathfrak{r}},s_2),$$

with smooth functions $\tilde\alpha_1$ and $\alpha_{1,1}$, where $\delta_{3,0}$ is of the form $\delta_{3,0} = \delta_0^{q_{3,0}\mathfrak{r}}$, with some positive integer $q_{3,0}$. Moreover, two possible cases may occur.

Case ND: The nondegenerate case. $\alpha_{1,1}\equiv 0$, $|\tilde\alpha_1|\sim 1$ *and* $\delta_{3,0} = \max\{\delta_0,\delta_3\} \ge \delta_0$.

Case D: The degenerate case. $|\alpha_{1,1}|\sim 1$, $\tilde\alpha_1 = \tilde\alpha_1(\delta_0^{\mathfrak{r}},s_2)$ *is independent of* x_1, $\delta_0 = \delta_3$, *and* $\delta_{3,0} \ll \delta_0$. *Moreover, either* $|\tilde\alpha_1|\sim 1$, *or we can choose* $q_{3,0}\in\mathbb{N}$ *as large as we wish.*

In particular, we may write

$$\phi^\sharp(x,\delta,s) = x_2^B\,b(x,\delta_0^{\mathfrak{r}},s_2) + \sum_{j=2}^{B-2}\delta_{j+2}x_2^j\,\tilde\alpha_j(x_1,\delta_0^{\mathfrak{r}},s_2)$$
$$+\delta_{3,0}\,x_2\,\tilde\alpha_1(x_1,\delta_0^{\mathfrak{r}},s_2) + \delta_0\,x_1x_2\,\alpha_{1,1}(x_1,\delta_0^{\mathfrak{r}},s_2), \tag{8.13}$$

with smooth functions $\tilde\alpha_j$ and b, where $|b|\sim 1$ and $\tilde\alpha_1$ and $\alpha_{1,1}$ are as in Case D (respectively, ND). Moreover, in both cases we have

$$\max\{\delta_0,\delta_3\} = \max\{\delta_0,\delta_{3,0}\}.$$

Proof. Recall that $\delta_0 = 2^{-k(\tilde\kappa_2-2\tilde\kappa_1)}$ and $\delta_3 = 2^{-k(n_1\tilde\kappa_1+\tilde\kappa_2-1)}$. We therefore distinguish two cases.

Case 1. $n_1 \ne n-2$. Then $\tilde\kappa_2 - 2\tilde\kappa_1 \ne n_1\tilde\kappa_1 + \tilde\kappa_2 - 1$, since $\tilde\kappa_1 = 1/n$, and thus either $\delta_0 \gg \delta_3$ or $\delta_0 \ll \delta_3$, for k sufficiently large. Notice that we may assume this to be true in particular if the function α_1 is flat, since we may then choose n_1 as large as we want. By putting

$$\delta_{3,0} := \max\{\delta_0, \delta_3\},$$

we then clearly may write a_1 as in Case ND.

Case 2. $\boldsymbol{n_1 = n-2}$. Then $\delta_0 = \delta_3$, and we may assume that $|\alpha_1| \sim 1$. Thus, expanding around $x_1 = 0$ and applying (8.9), we see that we may write

$$\begin{aligned}
a_1(x_1, \delta, s_2) &= \delta_0(s_2 + x_1^c(\delta_1, s_2)^{n_1}\alpha_1(\delta_1 x_1^c(\delta_1, s_2)) + x_1\alpha_{1,1}(x_1, \delta_0^\mathfrak{r}, s_2)) \\
&= \delta_0 s_2(1 + G_1(\delta_1 s_2^{1/(n-2)})\,\alpha_1(\delta_1 s_2^{1/(n-2)} G_1(\delta_1 s_2^{1/(n-2)}))) \\
&\quad + \delta_0 x_1 \alpha_{1,1}(x_1, \delta_0^\mathfrak{r}, s_2) \\
&= \delta_0 s_2(1 + g(\delta_1 s_2^{1/(n-2)})) + \delta_0 x_1\alpha_{1,1}(x_1, \delta_0^\mathfrak{r}, s_2),
\end{aligned}$$

with smooth functions $1+g$ and $\alpha_{1,1}$, where $|\alpha_{1,1}| \sim 1$. By means of a Taylor expansion of g around the origin, we thus find that

$$\begin{aligned}
a_1(x_1, \delta, s_2) &= \delta_0 s_2(\delta_1 s_2^{1/(n-2)})^N g_N(\delta_1 s_2^{1/(n-2)}) + \delta_0 x_1\alpha_{1,1}(x_1, \delta_0^\mathfrak{r}, s_2) \\
&= (\delta_0\delta_1^N)\tilde{\alpha}_1(\delta_0^\mathfrak{r}, s_2) + \delta_0 x_1 \alpha_{1,1}(x_1, \delta_0^\mathfrak{r}, s_2),
\end{aligned}$$

with $N \in \mathbb{N}$ and g_N smooth. Moreover, we may either assume that $|\tilde{\alpha}_1| \sim 1$ (if $1+g$ is a finite type N at the origin) or that we may choose N as large as we wish (if $1+g$ is flat). Notice that if $N = 0$, then we can include the second term into the first term and arrive again at Case ND. In all other cases we arrive at the situation described by Case D, where the second term cannot be included into the first term. Notice that then $\delta_{3,0} := \delta_0\delta_1^N \ll \delta_0$.

Q.E.D.

Let us next introduce the quantity

$$\rho := \begin{cases} \delta_{3,0}^{B/(B-1)} + \sum_{j=2}^{B-2} \delta_{j+2}^{B/(B-j)} & \text{in Case ND}, \\[2ex] \delta_0^{3B/(2B-3)} + \delta_{3,0}^{B/(B-1)} + \sum_{j=2}^{B-2} \delta_{j+2}^{B/(B-j)} & \text{in Case D}, \end{cases} \tag{8.14}$$

which we shall view as a function $\rho(\tilde{\delta})$ of the coefficients

$$\tilde{\delta} := \begin{cases} (\delta_{3,0}, \delta_4, \ldots, \delta_B) & \text{in Case ND}, \\[1ex] (\delta_0, \delta_{3,0}, \delta_4, \ldots, \delta_B) & \text{in Case D}. \end{cases}$$

Remark 8.4. *Observe that if we scale the complete phase $\Phi^\sharp$ from* (8.7) *in x_1 by the factor $r^{-1/3}$ and in x_2 by $r^{-1/B}$, $r > 0$, and multiply by r, that is, if we look at*

$$\Phi_r^\sharp(u_1, u_2, \delta, s) := r\Phi^\sharp(r^{-1/3}u_1, r^{-1/B}u_2, \delta, s),$$

then the effect is essentially that $\tilde{\delta}$ *in* (8.8) *is replaced by*

$$\tilde{\delta}^r := \begin{cases} (r^{(B-1)/B}\delta_{3,0}, \dots, r^{(B-j)/B}\delta_{j+2}, \dots, r^{2/B}\delta_B) & \text{in Case ND,} \\ (r^{(2B-3)/3B}\delta_0, r^{(B-1)/B}\delta_{3,0}, \dots, r^{(B-j)/B}\delta_{j+2}, \dots, r^{2/B}\delta_B) & \text{in Case D,} \end{cases} \tag{8.15}$$

whereas $B_1(s, \delta_1)$ *is replaced by* $r^{2/3}B_1(s_2, \delta_1)$ *and* $B_0(s, \delta_1)$ *by* $rB_0(s, \delta_1)$*. More precisely, if denote by* σ_r *the dilations* $\sigma_r(x_1, x_2) := (r^{1/3}x_1, r^{1/B}x_2)$*, so that* $\Phi_r^{\sharp}(u, \delta, s) = r\Phi^{\sharp}(\sigma_{1/r}u, \delta, s)$*, then we have*

$$\Phi_r^{\sharp}(u_1, u_2, \delta, s) = u_1^3\, B_3(s_2, \delta_1, r^{-1/3}u_1) - u_1\, r^{2/3}\, B_1(s, \delta_1) \\ + rB_0(s, \delta_1) + \phi_r^{\sharp}(u_1, u_2, \tilde{\delta}^r, s_2),$$

where

$$\phi_r^{\sharp}(u, \tilde{\delta}^r, s) := u_2^B\, b(\sigma_{r^{-1}}u, \delta_0^{\mathfrak{r}}, s_2) + \sum_{j=2}^{B-2} \tilde{\delta}_{j+2}^r u_2^j\, \tilde{\alpha}_j(r^{-1/3}u_1, \delta_0^{\mathfrak{r}}, s_2) \\ + \tilde{\delta}_{3,0}^r\, u_2\, \tilde{\alpha}_1(r^{-1/3}u_1, \delta_0^{\mathfrak{r}}, s_2) + \tilde{\delta}_0^r\, u_1u_2\, \alpha_{1,1}(r^{-1/3}u_1, \delta_0^{\mathfrak{r}}, s_2). \tag{8.16}$$

And, under these dilations, ρ *is homogeneous of degree* 1*, that is,*

$$\rho(\tilde{\delta}^r) = r\rho(\tilde{\delta}), \qquad r > 0.$$

In particular, after scaling $\Phi^{\sharp}$ *in this way by* $r := 1/\rho(\tilde{\delta})$*, we see that we have normalized the coefficients of* $\phi^{\sharp}$ *in such a way that* $\rho(\tilde{\delta}) = 1$.

This observation, which is based on ideas by Duistermaat [Dui74], will become important in the sequel.

8.3 THE CASE WHERE $\lambda\rho(\tilde{\delta}) \lesssim 1$

Assume now that $B \in \{3, 4\}$. Following the proof of Proposition 4.2 in Chapter 5, we define the functions $\nu_{\delta, Ai}^{\lambda}$ and $\nu_{\delta, l}^{\lambda}$ by

$$\widehat{\nu_{\delta, Ai}^{\lambda}}(\xi) := \chi_0(\lambda^{2/3}B_1(s, \delta_1))\widehat{\nu_\delta^\lambda}(\xi), \tag{8.17}$$

$$\widehat{\nu_{\delta, l}^{\lambda}}(\xi) := \chi_1((2^{-l}\lambda)^{2/3}B_1(s, \delta_1))\widehat{\nu_\delta^\lambda}(\xi), \quad M_0 \le 2^l \le \frac{\lambda}{M_1}, \tag{8.18}$$

so that

$$\nu_\delta^\lambda = \nu_{\delta, Ai}^\lambda + \sum_{M_0 \le 2^l \le \lambda/M_1} \nu_{\delta, l}^\lambda.$$

Here, $\chi_0, \chi_1 \in C_0^\infty(\mathbb{R})$, and $\chi_1(t)$ is supported where $2^{-4/3} \le |t| \le 2^{4/3}$, whereas $\chi_0(t) \equiv 1$ for $|t| \le M_0^{2/3}$. Thus, by choosing M_0 sufficiently large, we may assume that $2^{-l} \le 1/M_0 \ll 1$. Denote by $T_{\delta, Ai}^\lambda$ and $T_{\delta, l}^\lambda$ the corresponding operators of convolution with the Fourier transforms of these functions.

Our goal will be to adjust the proofs of the estimates in Lemma 5.3 and Lemma 5.4 to our present situation in order to derive the following estimates, which are analogous to the corresponding ones in those lemmas (formally, we have only to replace a factor $\lambda^{-1/2}$ by the factor $\lambda^{-1/B}$):

$$\|\widehat{\nu^\lambda_{\delta,\,Ai}}\|_\infty \le C_1\lambda^{-1/B-1/3}, \tag{8.19}$$

$$\|\nu^\lambda_{\delta,\,Ai}\|_\infty \le C_2\lambda^{5/3-1/B}, \tag{8.20}$$

as well as

$$\|\widehat{\nu^\lambda_{\delta,\,l}}\|_\infty \le C_1 2^{-l/6}\lambda^{-1/B-1/3}, \tag{8.21}$$

$$\|\nu^\lambda_{\delta,\,l}\|_\infty \le C_2 2^{l/3}\lambda^{5/3-1/B}. \tag{8.22}$$

8.3.1 Estimates for $\nu^\lambda_{\delta,\,Ai}$

Changing coordinates from x to u by putting $x = \sigma_{1/\lambda}u = (\lambda^{-1/3}u_1, \lambda^{-1/B}u_2)$ in the integral (8.12) and making use of Remark 8.4 (with $r := \lambda$), we find that

$$\begin{aligned}\widehat{\nu^\lambda_{\delta,Ai}}(\xi) = {}&\lambda^{-1/B-1/3}\chi_1(s,s_3)\,\chi_0(\lambda^{2/3}B_1(s,\,\delta_1))\,e^{-is_3\lambda B_0(s,\delta_1)}\\ &\times\iint e^{-is_3(u_1^3B_3(s_2,\delta_1,\lambda^{-1/3}u_1)-u_1\lambda^{2/3}B_1(s,\delta_1)+\phi^\sharp_\lambda(u_1,u_2,\tilde\delta^\lambda,s_2))}\\ &\times a(\sigma_{\lambda^{-1}}u,\delta,s)\,du_1\,du_2,\end{aligned} \tag{8.23}$$

where $\chi_1(s,s_3) := \chi_1(s_1s_3)\chi_1(s_2s_3)\chi_1(s_3)$ localizes to the region where $s_j\sim 1$, $j=1,2,3$. Observe that here we are integrating over the possibly large domain where $|u_1|\le\varepsilon\lambda^{1/3}$ and $|u_2|\le\varepsilon\lambda^{1/B}$. Recall also that $\phi^\sharp_\lambda$ is given by (8.16) and that $\rho(\tilde\delta^\lambda) = \lambda\rho(\tilde\delta)\lesssim 1$, and so we have

$$|\tilde\delta^\lambda|\lesssim 1 \quad\text{and}\quad \lambda^{2/3}|B_1(s,\,\delta_1)|\lesssim 1.$$

By means of this integral formula for $\widehat{\nu^\lambda_{\delta,Ai}}(\xi)$, we easily obtain the following.

Lemma 8.5. *If $\lambda\rho(\tilde\delta)\lesssim 1$, then we may write*

$$\begin{aligned}\widehat{\nu^\lambda_{\delta,Ai}}(\xi) = {}&\lambda^{-1/B-1/3}\chi_1(s,s_3)\,\chi_0\left(\lambda^{2/3}B_1(s,\,\delta_1)\right)\,e^{-is_3\lambda B_0(s,\delta_1)}\\ &\times a\left(\lambda^{2/3}B_1(s,\,\delta_1),\,\tilde\delta^\lambda,\,s,\,s_3,\,\delta_0^{\mathfrak{r}},\,\lambda^{-1/3B}\right),\end{aligned} \tag{8.24}$$

where a is again a smooth function of all its (bounded) variables.

Proof. Given $L\gg 1$, we decompose the integral in (8.23) by means of suitable smooth cutoff functions into the integral I_1 over the region where $|(u_1,u_2)|\le L$, the integral I_2 over the region where $|(u_1,u_2)|>L$ and $|u_2|^{B-1}\gg|u_1|$, and the integral I_3 over the region where $|(u_1,u_2)|>L$ and $|u_2|^{B-1}\lesssim|u_1|$. For each of these contributions I_j, we then show that it is of the form $a_j(\lambda^{2/3}B_1(s,\,\delta_1),\,\tilde\delta^\lambda,\,s,\,s_3,\,\delta_0^{\mathfrak{r}},\,\lambda^{-1/3B})$, with a suitable smooth function a_j, provided L is sufficiently large. For I_1, this claim is obvious.

On the remaining region where $|(u_1, u_2)| \geq L$, we may use iterated integrations by parts with respect to u_1 or u_2 in order to convert the integral into an absolutely convergent integral, to which we then may apply the standard rules for differentiation with respect to parameters (such as s_j, etc.). Denote to this end the complete phase function appearing in this integral by Φ_c. It is then easily seen that we may estimate

$$|\partial_{u_2}\Phi_c| \gtrsim |u_2|^{B-1} - c|u_1|, \tag{8.25}$$

$$|\partial_{u_1}\Phi_c| \gtrsim u_1^2 - c\lambda^{-1/3}|u_2|^B - c|u_2|, \tag{8.26}$$

with a fixed constant $c > 0$. The last terms appear only in the degenerate case D, due to the presence of the term $\tilde{\delta}_0^\lambda\, u_1 u_2\, \alpha_{1,1}(\lambda^{-1/3}u_1, \delta_0^\mathfrak{r}, s_2)$ in $\phi_\lambda^\sharp$.

Denote by I_2 the contribution by the subregion on which $|u_2|^{B-1} \gg |u_1|$. On this region, we may gain factors of order $|u_2|^{-2N(B-1)}$ (for any $N \in \mathbb{N}$) in the amplitude by means of iterated integrations by parts in u_2. And, since $|u_2|^{-2N(B-1)} \lesssim |u_1|^{-N}|u_2|^{-N(B-1)}$, we see that we arrive at an absolutely convergent integral.

Similarly, denote by I_3 the contribution by the subregion on which $|u_2|^{B-1} \lesssim |u_1|$. Observe that $|u_2|^B \lesssim |u_1|^{B/(B-1)} \ll u_1^2$, since $B \geq 3$, and since we may assume that $|u_1| \gg 1$. This shows that in this region, we have $|\partial_{u_1}\Phi_c| \gtrsim u_1^2$, and thus we may gain factors of order $|u_1|^{-2N}$ (for any $N \in \mathbb{N}$) in the amplitude, by means of iterated integrations by parts in u_1. And, since $|u_1|^{-2N} \lesssim |u_1|^{-N}$ $|u_2|^{-N(B-1)}$, we arrive again at an absolutely convergent integral. Q.E.D.

Lemma 8.5 implies, in particular, estimate (8.19). As for the more involved estimate (8.20), with Lemma 8.5 at hand, we can basically follow the arguments from Section 5.2.2, only with the factor $\lambda^{-1/2}$ appearing there replaced by the factor $\lambda^{-1/B}$ here and with the amplitude $g(\lambda^{2/3}B_1(s', \delta, \sigma), \lambda, \delta, \sigma, s)$ replaced by the amplitude $a(\lambda^{2/3}B_1(s, \delta_1), \tilde{\delta}^\lambda, s, s_3, \delta_0^\mathfrak{r}, \lambda^{-1/3B})$ here (compare with (5.17)).

8.3.2 Estimates for $\nu_{\delta,l}^\lambda$

Changing coordinates from x to u by letting $x = \sigma_{2^l/\lambda}u$ in the integral (8.12), and making use of Remark 8.4 (with $r := \lambda/2^l$), we find that

$$\begin{aligned}\widehat{\nu_{\delta,l}^\lambda}(\xi) = {} & (2^{-l}\lambda)^{-1/B-1/3}\chi_1(s, s_3)\,\chi_1\left((2^{-l}\lambda)^{2/3}B_1(s, \delta_1)\right)e^{-is_3\lambda B_0(s,\delta_1)} \\ & \times \iint e^{-is_3 2^l\Phi(u_1,u_2,s,\delta,\lambda,l)}a(\sigma_{2^l\lambda^{-1}}u, \delta, s)\,du_1 du_2,\end{aligned} \tag{8.27}$$

with phase function

$$\begin{aligned}&\Phi(u_1, u_2, s, \delta, \lambda, l) \\ &:= u_1^3\,B_3(s_2, \delta_1, (2^{-l}\lambda)^{-1/3}u_1) - u_1(2^{-l}\lambda)^{2/3}B_1(s, \delta_1) + \phi_{2^{-l}\lambda}^\sharp(u_1, u_2, \tilde{\delta}^{2^{-l}\lambda}, s_2).\end{aligned} \tag{8.28}$$

Observe that here we are integrating over the possibly large domain where $|u_1| \leq (\varepsilon 2^{-l}\lambda)^{1/3}$ and $|u_2| \leq \varepsilon(2^{-l}\lambda)^{1/B}$. Recall also that $\phi_{2^{-l}\lambda}^\sharp$ is given by (8.16) and that $\rho(\tilde{\delta}^{2^{-l}\lambda}) = 2^{-l}\lambda\rho(\tilde{\delta}) \lesssim 2^{-l}$, so that we have

$$|\tilde{\delta}^{2^{-l}\lambda}| \ll 1 \quad \text{and} \quad (2^{-l}\lambda)^{2/3}|B_1(s, \delta_1)| \sim 1.$$

Notice that this implies that

$$\phi^{\sharp}_{2^{-l}\lambda}(u_1, u_2, \tilde{\delta}^{2^{-l}\lambda}, s_2) = u_2^B\, b\left(\sigma_{(2^{-l}\lambda)^{-1}}u, \delta_0^{\mathfrak{r}}, s_2\right) + \text{ small error.} \tag{8.29}$$

Arguing in a somewhat similar way as in Section 5.2, we first decompose

$$\nu^{\lambda}_{\delta,l} = \nu^{\lambda}_{l,0} + \nu^{\lambda}_{l,\infty},$$

where

$$\begin{aligned}\widehat{\nu^{\lambda}_{l,0}}(\xi) &:= (2^{-l}\lambda)^{-1/B-1/3}\chi_1(s, s_3)\,\chi_1\left((2^{-l}\lambda)^{2/3}B_1(s,\,\delta_1)\right)\,e^{-is_3\lambda B_0(s,\delta_1)}\\ &\quad\times\iint e^{-is_3 2^l\Phi(u_1,u_2,s,\delta,\lambda,l)}a(\sigma_{2^l\lambda^{-1}}u, \delta, s)\,\chi_0(u)\,du_1du_2,\\ \widehat{\nu^{\lambda}_{l,\infty}}(\xi) &:= (2^{-l}\lambda)^{-1/B-1/3}\chi_1(s, s_3)\,\chi_1\left((2^{-l}\lambda)^{2/3}B_1(s,\,\delta_1)\right)\,e^{-is_3\lambda B_0(s,\delta_1)}\\ &\quad\times\iint e^{-is_3 2^l\Phi(u_1,u_2,s,\delta,\lambda,l)}a(\sigma_{2^l\lambda^{-1}}u, \delta, s)\,(1-\chi_0(u))\,du_1du_2.\end{aligned}$$

Here, we choose $\chi_0 \in C_0^\infty(\mathbb{R}^2)$ such that $\chi_0(u) \equiv 1$ for $|u| \le L$, where L is supposed to be a sufficiently large positive constant.

Let us first consider the contribution given by the $\nu^{\lambda}_{l,\infty}$: Arguing as in the proof of Lemma 8.5, we can easily see by means of integrations by parts that, given $k \in \mathbb{N}$, then for every $N \in \mathbb{N}$ we may write

$$\begin{aligned}\widehat{\nu^{\lambda}_{l,\infty}}(\xi) &= 2^{-lN}\,(2^{-l}\lambda)^{-1/B-1/3}\chi_1(s, s_3)\,\chi_1((2^{-l}\lambda)^{2/3}B_1(s,\,\delta_1))\,e^{-is_3\lambda B_0(s,\delta_1)}\\ &\quad\times a_{N,l}((2^{-l}\lambda)^{2/3}B_1(s,\,\delta_1), s, s_3, \tilde{\delta}^{2^{-l}\lambda}, \delta_0^{\mathfrak{r}}, (2^{-l}\lambda)^{-1/3}, \lambda^{-1/3B}),\end{aligned} \tag{8.30}$$

where $a_{N,l}$ is a smooth function of all its (bounded) variables such that $\|a_{N,l}\|_{C^k}$ is uniformly bounded in l. In particular, we see that

$$\|\widehat{\nu^{\lambda}_{l,\infty}}\|_\infty \lesssim 2^{-lN}\lambda^{-1/B-1/3} \qquad \forall N \in \mathbb{N}. \tag{8.31}$$

Next, applying the Fourier inversion formula and changing coordinates from s_1 to

$$z := (2^{-l}\lambda)^{2/3}B_1(s, \delta_1), \qquad \text{that is,} \quad s_1 = s_2^{(n-1)/(n-2)}G_3(s_2, \delta_1) - (2^{-l}\lambda)^{-2/3}z, \tag{8.32}$$

as in Section 5.2, we find that

$$\begin{aligned}\nu^{\lambda}_{l,\infty}(x) &= \lambda^3 2^{-lN}\,(2^{-l}\lambda)^{-1/B-1}\int e^{-is_3\lambda\Psi(z,s_2,\delta)}\chi_1(s_2^{(n-1)/(n-2)}G_3(s_2, \delta_1)\\ &\quad -(2^{-l}\lambda)^{-2/3}z, s_2, s_3)\,\chi_1(z)\\ &\quad\times a_{N,l}(z, s_2, s_3, \delta_0^{\mathfrak{r}}, \tilde{\delta}^{2^{-l}\lambda}, (2^{-l}\lambda)^{-1/3}, \lambda^{-1/3B})\,dz\,ds_2\,ds_3,\end{aligned} \tag{8.33}$$

where the phase function Ψ is given by

$$\begin{aligned}\Psi(z, s_2, \delta) &:= s_2^{n/(n-2)}G_5(s_2, \delta) - x_1 s_2^{(n-1)/(n-2)}G_3(s_2, \delta) - s_2x_2 - x_3\\ &\quad + (2^l\lambda^{-1})^{2/3}z\,(x_1 - s_2^{1/(n-2)}G_1(s_2, \delta)).\end{aligned} \tag{8.34}$$

Applying the van der Corput–type Lemma 2.1 (of order $M = 3$) to the integration in s_2, which allows for the gain of a factor $\lambda^{-1/3}$, this easily implies that

$$\|\nu^\lambda_{l,\infty}\|_\infty \lesssim 2^{-lN}\lambda^{5/3-1/B} \qquad \forall N \in \mathbb{N}. \tag{8.35}$$

Notice that estimates (8.33) and (8.35) are stronger than the desired estimates, (8.21) and (8.22).

Let us next look at the contribution given by the $\nu^\lambda_{l,0}$: observe first that on a region where $|u_1|$ is sufficiently small, iterated integrations by parts with respect to u_1 allow to gain factors 2^{-lN} in the integral defining $\nu^\lambda_{l,0}$, for every $N \in \mathbb{N}$. Afterwords, freezing u_1, we can reduce this to the integration in u_2 alone. A similar argument applies whenever we are allowed to integrate by parts in u_1 (this is also the case when B_1 and B_3 have opposite signs). We shall therefore assume from now on that B_1 and B_3 have the same sign. Moreover, let us assume without loss of generality that $u_1 > 0$. Then there is a unique nondegenerate critical point $u_1^c = u_1^c((2^{-l}\lambda)^{-1/B}u_2, \tilde{\delta}^{2^{-l}\lambda}, ...)$ of the phase Φ in (8.28), of size $|u_1^c| \sim 1$, and we may restrict the integration in u_1 to a small neighborhood of u_1^c. That is, we may replace the cutoff function $\chi_0(u_1)$ by a cutoff function $\chi_1(u_1)$ supported in a sufficiently small neighborhood of a point u_1^0 containing u_1^c. Then the method of stationary phase shows that the corresponding integral in u_1 will be of order $2^{-l/2}$, so that this term will give the main contribution.

For the sake of simplicity, we shall therefore restrict ourselves in the sequel to the discussion of this main term $\nu^\lambda_{l,1}$, given by

$$\begin{aligned}\widehat{\nu^\lambda_{l,1}}(\xi) :=& (2^{-l}\lambda)^{-1/B-1/3}\chi_1(s, s_3)\, \chi_1\left((2^{-l}\lambda)^{2/3}B_1(s,\, \delta_1)\right)\, e^{-is_3\lambda B_0(s,\delta_1)}\\ &\times \iint e^{-is_3 2^l\Phi(u_1,u_2,s,\delta,\lambda,l)}a(\sigma_{2^l\lambda^{-1}}u, \delta, s)\, \chi_0(u)\chi_1(u_1)\, du_1\, du_2,\end{aligned} \tag{8.36}$$

where $|u_1| \sim 1$ on the support of $\chi_1(u_1)$.

First applying the method of stationary phase to the integration in u_1 and, subsequently, van der Corput's estimate of order B to the integration in u_2, we easily arrive at the estimate

$$\|\widehat{\nu^\lambda_{l,1}}\|_\infty \lesssim (2^{-l}\lambda)^{-1/B-1/3}2^{-l/2-l/B},$$

which is exactly what we need to verify (8.21) (also recall here estimate (8.31)).

As for the more involved estimation of $\nu^\lambda_{l,1}(x)$, Fourier inversion allows to write

$$\begin{aligned}\nu^\lambda_{l,1}(x) =& \lambda^3\, (2^{-l}\lambda)^{-1/B-1/3}\int e^{-is_3\Psi(u,s,x,\delta,\lambda,l)}\chi_1(s,\, s_3)\, \chi_1((2^{-l}\lambda)^{2/3}B_1(s,\, \delta_1))\\ &\times a(\sigma_{2^l\lambda^{-1}}u, \delta, s)\, \chi_0(u)\chi_1(u_1)\, du_1\, du_2\, ds,\end{aligned} \tag{8.37}$$

where the complete phase Ψ is given by

$$\Psi(u, s, x, \delta, \lambda, l) := 2^l\Phi(u_1, u_2, s, \delta, \lambda, l) + \lambda\,(B_0(s, \delta_1) - s_1x_1 - s_2x_2 - x_3)\,,$$

with Φ given by (8.28). Changing again from the coordinate s_1 to z as in (8.32), we may also write

$$\nu_{l,1}^{\lambda}(x) = \lambda^3 (2^{-l}\lambda)^{-1/B-1} \int e^{-is_3\tilde{\Psi}(u,z,s_2,x,\delta,\lambda,l)} \chi_1(s, s_3)\, \chi_1(z) \\ \times \tilde{a}\left(\sigma_{2^l\lambda^{-1}}u, (2^l\lambda^{-1})^{2/3}z, s_2, \delta\right) \chi_0(u)\chi_1(u_1)\, du_1\, du_2\, dz\, ds_2\, ds_3, \quad (8.38)$$

with phase

$$\begin{aligned}\tilde{\Psi}(u, z, s_2, x, \delta, \lambda, l) := {} & \lambda(2^l\lambda^{-1})^{2/3} z\,(x_1 - s_2^{1/(n-2)} G_1(s_2, \delta)) \\ & + \lambda(s_2^{n/(n-2)} G_5(s_2, \delta) - x_1 s_2^{(n-1)/(n-2)} \\ & \times G_3(s_2, \delta) - s_2 x_2 - x_3) \\ & + 2^l(u_1^3\, B_3(s_2, \delta_1, (2^{-l}\lambda)^{-1/3} u_1) - z u_1 \\ & + \phi^{\sharp}_{2^{-l}\lambda}(u_1, u_2, \tilde{\delta}^{2^{-l}\lambda}, s_2)). \end{aligned} \quad (8.39)$$

We shall prove the following estimate:

$$|\nu_{l,1}^{\lambda}(x)| \le C 2^{l/3} \lambda^{5/3-1/B}, \quad (8.40)$$

with a constant C that is independent of λ, l, x, and δ.

As in Section 5.2.2, this estimate is easily verified if $|x| \gg 1$, simply by means of integrations by parts in the variables s_2, s_3 and z, in combination with the method of stationary phase in u_1 and van der Corput's estimate of order B in u_2. Similarly, if $|x| \lesssim 1$ and $|x_1| \ll 1$, we can arrive at the same conclusion by first integrating by parts in z. Indeed, in these situations we may gain factors $2^{-2Nl/3}\lambda^{-N/3}$ through integrations by parts, so that the corresponding estimates can be summed in a trivial way.

Let us thus assume that $|x| \lesssim 1$ and $|x_1| \sim 1$. Following Section 5.2, we then decompose

$$\nu_{l,1}^{\lambda} = \nu_{l,I}^{\lambda} + \nu_{l,II}^{\lambda} + \nu_{l,III}^{\lambda},$$

where $\nu_{l,I}^{\lambda}$, $\nu_{l,II}^{\lambda}$ and $\nu_{l,III}^{\lambda}$ correspond to the contributions to the integral in (8.38) given by the regions where $\lambda(2^l\lambda^{-1})^{2/3}|x_1 - s_2^{1/(n-2)} G_1(s_2, \delta)| \gg 2^l$, $\lambda(2^l\lambda^{-1})^{2/3} \times |x_1 - s_2^{1/(n-2)} G_1(s_2, \delta)| \sim 2^l$ and $\lambda(2^l\lambda^{-1})^{2/3}|x_1 - s_2^{1/(n-2)} G_1(s_2, \delta)| \ll 2^l$, respectively. The first and the last term can easily be handled as in Section 5.2 by means of integrations by parts in z, with subsequent exploitation of the oscillations with respect to u_1 and u_2, which leads to the following estimate:

$$|\nu_{l,I}^{\lambda}(x)| + |\nu_{l,III}^{\lambda}(x)| \le C 2^{-l/3} \lambda^{5/3-1/B},$$

which is better than (8.40) by a factor $2^{-2l/3}$, so that summation in l is no problem for these terms. Nevertheless, summation in λ still will require an interpolation argument if $B = 3$.

Let us next concentrate on $\nu_{l,II}^{\lambda}(x)$, which is of the form

$$\nu_{l,II}^{\lambda}(x) = \lambda^3 (2^{-l}\lambda)^{-1/B-1} \int e^{-is_3\tilde{\Psi}(u,z,s_2,x,\delta,\lambda,l)} \tilde{a}(\sigma_{2^l\lambda^{-1}}u, (2^l\lambda^{-1})^{2/3}z, s_2, \delta) \\ \times \chi_1(s_2, s_3)\, \chi_1((2^l\lambda^{-1})^{-1/3}(x_1 - s_2^{1/(n-2)} G_1(s_2, \delta)))\, \chi_0(u) \\ \times \chi_1(u_1)\, \chi_1(z)\, du_1\, du_2\, dz\, ds_2\, ds_3. \quad (8.41)$$

Writing

$$\begin{aligned}\tilde{\Psi}(u, z, s_2, x, \delta, \lambda, l) = {} & \lambda(s_2^{n/(n-2)} G_5(s_2, \delta) - x_1 s_2^{(n-1)/(n-2)} G_3(s_2, \delta) - s_2 x_2 - x_3) \\ & + 2^l \Big[z\,(2^{-l}\lambda)^{1/3}(x_1 - s_2^{1/(n-2)} G_1(s_2, \delta)) - z u_1 \\ & + u_1^3\, B_3(s_2, \delta_1, (2^{-l}\lambda)^{-1/3} u_1) + \phi^{\sharp}_{2^{-l}\lambda}(u_1, u_2, \tilde{\delta}^{2^{-l}\lambda}, s_2)\Big],\end{aligned}$$

and observing that here $|(2^{-l}\lambda)^{1/3}(x_1 - s_2^{1/(n-2)} G_1(s_2, \delta))| \sim 1$ and $|u_1| \sim 1$, we see that the phase $\tilde{\Psi}$ may have a critical point (u_1^c, z^c) within the support of the amplitude as a function of u_1 and z. Moreover, in a similar way as in Section 5.2, we see that this critical point will be nondegenerate. Of course, if there is no critical point, we may obtain even better estimates by means of integrations by parts. Thus, let us assume in the sequel that there is such a critical point.

Applying the method of stationary phase to the integration in (u_1, z), we see that we essentially may write

$$\begin{aligned}\nu^{\lambda}_{l,II}(x) = {} & \lambda^2\,(2^{-l}\lambda)^{-1/B} \int e^{-is_3 \Psi_2(u_2, s_2, x, \delta, \lambda, l)} a_2((2^l\lambda^{-1})^{1/3} u_2, s_2, x, (2^l\lambda^{-1})^{1/3}, \delta) \\ & \times \chi_1(s_2, s_3)\, \chi_1((2^l\lambda^{-1})^{-1/3}(x_1 - s_2^{1/(n-2)} G_1(s_2, \delta)))\, \chi_0(u_2)\, du_2\, ds_2\, ds_3,\end{aligned}$$

where the phase Ψ_2 arises from $\tilde{\Psi}$ by replacing (u_1, z) by the critical point (u_1^c, z^c) (which, of course, also depends on the other variables).

In order to compute Ψ_2 more explicitly, we go back to our original coordinates, in which our complete phase is given by (compare (8.5))

$$\begin{aligned}& \lambda(s_1 y_1 + s_2 y_1^2 \omega(\delta_1 y_1) + y_1^n \alpha(\delta_1 y_1) + s_2 \delta_0 y_2 + y_2^B\, b(y_1, y_2, \delta) \\ & + r(y_1, y_2, \delta) - s_1 x_1 - s_2 x_2 - x_3).\end{aligned}$$

Recall also that we had passed from the coordinates (y_1, s_1) to the coordinates (u_1, z) by means of a smooth change of coordinates (depending on the remaining variables (y_2, s_2)). Since the value of a function at a critical point does not depend on the chosen coordinates, arguing by means of Lemma 5.6, we find that in the coordinates (y_2, s_2) the phase Ψ_2 is given by

$$\lambda(s_2 x_1^2 \omega(\delta_1 x_1) + x_1^n \alpha(\delta_1 x_1) + s_2 \delta_0 y_2 + y_2^B\, b(x_1, y_2, \delta) + r(x_1, y_2, \delta) - s_2 x_2 - x_3). \tag{8.42}$$

And, since $y_2 = (2^{-l}\lambda)^{-1/B} u_2$, this means that

$$\begin{aligned}\Psi_2(u_2, s_2, x, \delta, \lambda, l) = {} & \lambda(s_2 x_1^2 \omega(\delta_1 x_1) + x_1^n \alpha(\delta_1 x_1) - s_2 x_2 - x_3) \\ & + 2^l (u_2^B\, b(x_1, (2^{-l}\lambda)^{-1/B} u_2, \delta) + \sum_{j=2}^{B-2} u_2^j\, \tilde{\delta}_{j+2}^{2^{-l}\lambda}\, x_1^{n_j} \alpha_j(\delta_1 x_1) \\ & + (2^{-l}\lambda)^{(B-1)/B}\, u_2(\delta_0 s_2 + \delta_3 x_1^{n_1} \alpha_1(\delta_1 x_1)))\end{aligned}$$

(compare (8.2)). Note that $\partial_{s_2}(s_2^{1/(n-2)} G_1(s_2, \delta)) \sim 1$ because $s_2 \sim 1$ and $G_1(s_2, 0) = 1$. Therefore, the relation $|x_1 - s_2^{1/(n-2)} G_1(s_2, \delta)| \sim (2^l\lambda^{-1})^{1/3}$ can be rewritten as $|s_2 - \tilde{G}_1(x_1, \delta)| \sim (2^l\lambda^{-1})^{1/3}$, where $\tilde{G}_1$ is again a smooth function such that $|\tilde{G}_1| \sim 1$. If we write

$$s_2 = (2^l\lambda^{-1})^{1/3} v + \tilde{G}_1(x_1, \delta),$$

then this means that $|v| \sim 1$. We shall therefore change variables from s_2 to v, which leads to

$$\begin{aligned} \nu^\lambda_{l,II}(x) = \lambda^2 \, (2^{-l}\lambda)^{-1/B-1/3} \int & e^{-is_3\Psi_3(u_2, v, x, \delta, \lambda, l)} a_3((2^l\lambda^{-1})^{1/3} u_2, v, x, \\ & (2^l\lambda^{-1})^{1/3}, \delta)\chi_1(s_3)\, \chi_1(v)\, \chi_0(u_2)\, du_2\, dv\, ds_3, \end{aligned} \tag{8.43}$$

with a smooth amplitude a_3 and the new phase function

$$\begin{aligned} \Psi_3(u_2, v, x, \delta, \lambda, l) = {} & \lambda(v\, (2^{-l}\lambda)^{-1/3}(x_1^2\omega(\delta_1 x_1) - x_2) \\ & + (2^{-l}\lambda)^{-1/B-1/3}\delta_0\, v u_2 + Q_A(x, \delta)) \\ & + 2^l (u_2^B\, b(x_1, (2^{-l}\lambda)^{-1/B} u_2, \delta) + \sum_{j=2}^{B-2} u_2^j\, \tilde{\delta}^{2^{-l}\lambda}_{j+2}\, x_1^{n_j}\alpha_j(\delta_1 x_1) \\ & + u_2\, (2^{-l}\lambda)^{(B-1)/B}(\delta_0\tilde{G}_1(x_1, \delta) + \delta_3 x_1^{n_1}\alpha_1(\delta_1 x_1))), \end{aligned} \tag{8.44}$$

where

$$Q_A(x, \delta) := \tilde{G}_1(x_1, \delta)(x_1^2\omega(\delta_1 x_1) - x_2) + x_1^n\alpha(\delta_1 x_1) - x_3.$$

Applying van der Corput's estimate of order B to the integration in u_2 in (8.43), we find that

$$|\nu^\lambda_{l,II}(x)| \le C\lambda^2\, (2^{-l}\lambda)^{-1/B-1/3} 2^{-l/B} = 2^{l/3}\lambda^{5/3-1/B}.$$

This proves (8.40), which completes the proof of estimate (8.22).

8.3.3 Consequences of the estimates (8.19)–(8.22)

By interpolation, these estimates imply

$$\|T^\lambda_{\delta,\,Ai}\|_{p_c\to p_c'} \lesssim \lambda^{-1/3-1/B+2\theta_c}, \tag{8.45}$$

$$\|T^\lambda_{\delta,\,l}\|_{p_c\to p_c'} \lesssim 2^{-l(1-3\theta_c)/6}\lambda^{-1/3-1/B+2\theta_c}. \tag{8.46}$$

But, by Lemma 7.4, we have $\theta_c \le \tilde{\theta}_B = 3/(2B+3)$, and this easily implies that the exponents of λ and 2^l in the preceding estimates are strictly negative if $B = 4$ and zero if $B = 3$ (where $\theta_c = \frac{1}{3}$). We can therefore sum these estimates over all l, as well as $\lambda \gg 1$ if $B = 4$, and the desired estimate follow, whereas if $B = 3$, then we get only uniform estimates

$$\|T^\lambda_{\delta,\,Ai}\|_{p_c\to p_c'} \le C, \quad \|T^\lambda_{\delta,\,l}\|_{p_c\to p_c'} \le C, \qquad (\lambda\rho(\tilde{\delta}) \lesssim 1),$$

with a constant C not depending on δ. The case where $B = 3$ will therefore again require a complex interpolation argument in order to capture the endpoint, as in the proof of Proposition 4.2 (c).

In particular, we have proven the following.

Proposition 8.6. *If $B = 4$, then under the assumptions in this section,*

$$\sum_{\{\lambda \geq 1 : \lambda\rho(\tilde{\delta}) \lesssim 1\}} \|T_\delta^\lambda\|_{p_c \to p_c'} \lesssim 1,$$

where the constant in this estimate will not depend on δ.

In combination with the following proposition, which will be proved in the next two sections, this will complete the discussion of the remaining case where $B = 4$ in (7.89) and, hence, also the proof of Proposition 7.1 for this case.

Proposition 8.7. *If $B = 4$, then under the assumptions in this section,*

$$\sum_{\{\lambda \geq 1 : \lambda\rho(\tilde{\delta}) \gg 1\}} \|T_\delta^\lambda\|_{p_c \to p_c'} \lesssim 1,$$

where the constant in this estimate will not depend on δ.

8.4 THE CASE WHERE $m = 2$, $B = 4$, AND $A = 1$

According to Proposition 8.6, we are left with controlling the operators T_δ^λ with $\lambda\rho \gg 1$, where we have used the abbreviation $\rho = \rho(\tilde{\delta})$. If $A = 1$, then according to (7.90), we have $h^r + 1 = 4$; hence, $\theta_c = \frac{1}{4}$ and $p_c' = 8$. We shall then use the first estimate in (7.74), that is,

$$\|\nu_\delta^\lambda\|_\infty \lesssim \lambda^{7/4}. \tag{8.47}$$

The crucial observation is that under the assumption $\lambda\rho \gg 1$, we can here improve on estimate (7.73) for $\widehat{\nu_\delta^\lambda}$. Indeed, we shall prove that

$$\|\widehat{\nu_\delta^\lambda}\|_\infty \lesssim \rho^{-1/12}\lambda^{-2/3}. \tag{8.48}$$

Under the assumption that this estimate is valid, we obtain by interpolation from (8.47) and (8.48) that

$$\|T_\delta^\lambda\|_{p_c \to p_c'} \lesssim (\lambda\rho)^{-1/16},$$

hence the desired remaining estimate

$$\sum_{\lambda\rho \gg 1} \|T_\delta^\lambda\|_{p_c \to p_c'} \lesssim 1.$$

In order to prove (8.48), recall from (8.12) and Lemma 8.3 that

$$\widehat{\nu_\delta}(\xi) = \iint e^{-i\lambda s_3 \Phi^\sharp(x,\delta,s)}\, a(x,\delta,s)\, dx_2\, dx_1,$$

where the complete phase $\Phi^\sharp$ is given by

$$\Phi^\sharp(x,\delta,s) = x_1^3\, B_3(s_2,\delta_1,x_1) - x_1 B_1(s,\delta_1) + B_0(s,\delta_1) + \phi^\sharp(x,\delta,s_2),$$

with

$$\begin{aligned}\phi^\sharp(x,\delta,s_2) = {} & x_2^4\, b(x,\delta_0^\natural,s_2) + \delta_4 x_2^2\, \tilde{\alpha}_2(x_1,\delta_0^\natural,s_2) \\ & + \delta_{3,0} x_2\, \tilde{\alpha}_1(x_1,\delta_0^\natural,s_2) + \delta_0 x_1 x_2\, \alpha_{1,1}(x_1,\delta_0^\natural,s_2),\end{aligned} \tag{8.49}$$

and where a is a smooth amplitude supported in a small neighborhood of the origin in x. Recall also from Lemma 8.3 that $|\alpha_{1,1}| \equiv 0$ and $|\tilde{\alpha}_1| \sim 1$ in Case ND, whereas $|\alpha_{1,1}| \sim 1$ and $\tilde{\alpha}_1$ is independent of x_1 in Case D. Recall also that $s_j \sim 1$ for $j = 1, 2, 3$.

Moreover, in Case ND we have

$$\rho = \delta_{3,0}^{4/3} + \delta_4^2,$$

whereas in Case D

$$\rho = \delta_0^{12/5} + \delta_{3,0}^{4/3} + \delta_4^2,$$

where $\delta_{3,0} \ll \delta_0$. We shall treat both cases ND and D at the same time, assuming implicitly that $\delta_0 = 0$ in Case ND.

Estimate (8.48) will thus be a direct consequence of the following lemma, which can be derived from more general results by Duistermaat (cf. Proposition 4.3.1 in [Dui74]). For the convenience of the reader, we shall give a more elementary, direct proof for our situation, which requires only C^2-smoothness of the amplitude. Our approach will be based on arguments similar to the ones used on pages 196–205 in [IKM10].

Lemma 8.8. *Denote by $J(\lambda, \delta, s)$ the oscillatory integral*

$$J(\lambda, \delta, s) = \chi_1(s_1, s_2) \iint e^{-i\lambda\Phi(x,\delta,s)}\, a(x, \delta, s)\, dx_1\, dx_2,$$

with phase

$$\Phi(x, \delta, s) = x_1^3\, B_3(s_2, \delta_1, x_1) - x_1 B_1(s, \delta_1) + \phi^{\sharp}(x_1, x_2, \delta, s_2),$$

where $\phi^{\sharp}$ is given by (8.49) and where $\chi_1(s_1, s_2)$ localizes to the region where $s_j \sim 1$, $j = 1, 2$. Let us also put

$$\tilde{\rho} := \rho + |B_1(s, \delta_1)|^{3/2}.$$

Then the estimate

$$|J(\lambda, \delta, s)| \le \frac{C}{\tilde{\rho}^{1/12}\lambda^{2/3}} \tag{8.50}$$

holds true, provided the amplitude a is supported in a sufficiently small neighborhood of the origin. The constant C in this estimate is independent of δ and s.

Proof. Note that $\tilde{\rho} \lesssim 1$. We may even assume that $\tilde{\rho} \ll 1$.

For, if $|B_1(s, \delta_1)| \sim 1$ and if we choose the support of a sufficiently small in x, then it is easily seen that the phase Φ has no critical point with respect to x_1 on the support of the amplitude, and thus an integration by parts in x_1 allows to estimate $|J(\lambda, \delta, s)| \le C\lambda^{-1}$, which is better than what is needed for (8.50). And, if $|B_1(s, \delta_1)| \ll 1$, then we also have $\tilde{\rho} \ll 1$.

We begin with the case where $\tilde{\rho}\lambda \lesssim 1$. Here we can argue as in the proof of Lemma 8.5: changing coordinates from x to u by putting $x = \sigma_{1/\lambda} u = (\lambda^{-1/3} u_1,$

$\lambda^{-1/4}u_2)$ and making use of Remark 8.4 (with $r := \lambda$), we find that

$$\begin{aligned} J(\lambda, \delta, s) = {}& \lambda^{-1/4-1/3}\, \chi_1(s_1, s_2) \\ & \times \iint e^{-is_3\left(u_1^3 B_3(s_2,\delta_1,\lambda^{-1/3}u_1) - u_1\lambda^{2/3}B_1(s,\delta_1) + \phi_\lambda^{\sharp}(u_1,u_2,\tilde{\delta}^\lambda, s_2)\right)} \\ & \times a(\sigma_{\lambda^{-1}}u, \delta, s)\, du_1\, du_2 . \end{aligned}$$

Observe that here we are integrating over the possibly large domain where $|u_1| \le \varepsilon\lambda^{1/3}$ and $|u_2| \le \varepsilon\lambda^{1/4}$. Recall also that $\phi_\lambda^{\sharp}$ is given by (8.16) and that $\rho(\tilde{\delta}^\lambda) = \lambda\rho(\tilde{\delta}) \lesssim 1$, and so we have

$$|\tilde{\delta}^\lambda| \lesssim 1 \quad \text{and} \quad \lambda^{2/3}|B_1(s, \delta_1)| \lesssim 1.$$

It is then easily seen by means of integrations by parts in u_1, respectively, u_2 (whenever these quantities are large), that the double integral in this expression is uniformly bounded in δ and s, and thus we arrive at the uniform estimate

$$|J(\lambda, \delta, s)| \le \frac{C}{\lambda^{7/12}}.$$

This estimate is stronger than estimate (8.50) when $\tilde{\rho}\lambda \lesssim 1$.

From now on, we may thus assume that $\Lambda := \tilde{\rho}\lambda \gtrsim 1$. We then apply the change of coordinates $x = \sigma_{\tilde{\rho}}u = (\tilde{\rho}^{1/3}u_1, \tilde{\rho}^{1/4}u_2)$ and find that

$$J(\lambda, \delta, s) = \tilde{\rho}^{7/12} I(\lambda\tilde{\rho}, \delta, s),$$

where we have put

$$I(\Lambda, \delta, s) = \chi_1(s_1, s_2) \iint e^{-i\Lambda\Phi_1(u,\delta,s)} a(\sigma_{\tilde{\rho}}u, \delta, s)\, du_1\, du_2, \qquad (\Lambda \gtrsim 1), \tag{8.51}$$

with

$$\Phi_1(u, \delta, s) := u_1^3\, B_3(s_2, \delta_1, \tilde{\rho}^{1/3}u_1) - u_1 B_1'(s, \delta_1) + \phi(u, \tilde{\rho}, \delta, s_2)$$

and

$$\begin{aligned} \phi(u, \tilde{\rho}, \delta, s_2) :={}& u_2^4\, b(\sigma_{\tilde{\rho}}u, \delta_0^{r}, s_2) + \delta_4' u_2^2\, \tilde{\alpha}_2(\tilde{\rho}^{1/3}u_1, \delta_0^{\mathfrak{r}}, s_2) \\ & + \delta_{3,0}' u_2\, \tilde{\alpha}_1(\tilde{\rho}^{1/3}u_1, \delta_0^{\mathfrak{r}}, s_2) + \delta_0' u_1 u_2\, \alpha_{1,1}(\tilde{\rho}^{1/3}u_1, \delta_0^{\mathfrak{r}}, s_2). \end{aligned} \tag{8.52}$$

Here,

$$B_1'(s, \delta_1) := \frac{B_1(s, \delta_1)}{\tilde{\rho}^{2/3}}, \quad \delta_0' := \frac{\delta_0}{\tilde{\rho}^{5/12}}, \quad \delta_{3,0}' := \frac{\delta_{3,0}}{\tilde{\rho}^{3/4}}, \quad \delta_4' := \frac{\delta_4}{\tilde{\rho}^{1/2}},$$

so that, in analogy with Remark 8.4, we have

$$(\delta_0')^{12/5} + (\delta_{3,0}')^{4/3} + (\delta_4')^2 + |B_1'(s, \delta_1)|^{3/2} = 1. \tag{8.53}$$

Note that, in particular,

$$\delta_0' + \delta_{3,0}' + \delta_4' + |B_1'(s, \delta_1)| \sim 1.$$

In order to prove (8.50), we have thus to verify the following estimate:

$$|I(\Lambda, \delta, s)| \le C\Lambda^{-2/3}. \tag{8.54}$$

Again, take a smooth cutoff function $\chi_0 \in C_0^\infty(\mathbb{R}^2)$ such that $\chi_0(u) = 1$ for $|u| \le L$, where L is a sufficiently large, fixed positive number, and decompose

$$I(\Lambda, \delta, s) = I_0(\Lambda, \delta, s) + I_\infty(\Lambda, \delta, s),$$

with

$$I_0(\Lambda, \delta, s) := \chi_1(s_1, s_2) \iint e^{-i\Lambda\Phi_1(u,\delta,s)} a(\sigma_{\tilde\rho} u, \delta, s)\, \chi_0(u)\, du_1\, du_2,$$

and

$$I_\infty(\Lambda, \delta, s) := \chi_1(s_1, s_2) \iint e^{-i\Lambda\Phi_1(u,\delta,s)} a(\sigma_{\tilde\rho} u, \delta, s)\, (1 - \chi_0(u))\, du_1\, du_2.$$

Note that on the support of $1 - \chi_0$ we have $|u_1| \gtrsim L$ or $|u_2| \gtrsim L$. Thus, by choosing L sufficiently large, we see by (8.53) that the phase Φ_1 has no critical point on the support of $1 - \chi_0$, and in fact we may use integrations by parts in u_1, respectively, u_2, in order to prove that the double integral in the expression for $I_\infty(\Lambda, \delta, s)$ is of order $O(\Lambda^{-1})$, uniformly in δ and s. This is stronger than what is required for (8.54).

There remains the integral $I_0(\Lambda, \delta, s)$. Here we use arguments from [IKM10] (compare pp. 203–5). Recall from (8.53) that $(B_1'(s, \delta_1), \delta_0', \delta_{3,0}', \delta_4')$ lies on the "unit sphere"

$$\Sigma := \{(B_1', \delta_0', \delta_{3,0}', \delta_4') \in \mathbb{R}^4 : |B_1'|^{3/2} + (\delta_0')^{12/5} + (\delta_{3,0}')^{4/3} + (\delta_4')^2 = 1\}.$$

Following [IKM10] let us fix a point $((B_1')^0, (\delta_0')^0, (\delta_{3,0}')^0, (\delta_4')^0) \in \Sigma$, a point s^0 in the support of $\chi_1(s_1, s_2)$, and a point $u^0 = (u_1^0, u_2^0) \in \operatorname{supp} \chi_0$ and denote by η a smooth cutoff function supported near u^0. By I_0^η we denote the corresponding oscillatory integral containing η as a factor in the amplitude:

$$I_0^\eta(\Lambda, \delta, s) := \chi_1(s_1, s_2) \iint e^{-i\Lambda\Phi_2(u, B_1', \delta_0', \delta_{3,0}', \delta_4', \tilde\rho, s)} a(\sigma_{\tilde\rho} u, \delta, s)\, \chi_0(u) \times \eta(u)\, du_1\, du_2,$$

where

$$\Phi_2(u, B_1', \delta_0', \delta_{3,0}', \delta_4', \tilde\rho, s) := u_1^3\, B_3(s_2, \delta_1, \tilde\rho^{1/3} u_1) - u_1 B_1' + \phi(u, \tilde\rho, \delta, s_2), \tag{8.55}$$

with ϕ as before.

We shall prove that I_0^η satisfies the estimate

$$|I_0^\eta(\Lambda, \delta, s)| \le C \|a(\cdot, \delta, s)\|_{C^2}\, \Lambda^{-2/3}, \tag{8.56}$$

provided η is supported in a sufficiently small neighborhood of U of u^0, s lies in a sufficiently small neighborhood S of s^0, and $(B_1', \delta_0', \delta_{3,0}', \delta_4')$ is in a sufficiently small neighborhood V of the point $((B_1')^0, (\delta_0')^0, (\delta_{3,0}')^0, (\delta_4')^0)$ in Σ. The constant C in these estimates may depend on the "base points" u^0, s^0, and $((B_1')^0, (\delta_0')^0, (\delta_{3,0}')^0, (\delta_4')^0)$, as well as on the chosen neighborhoods, but not on Λ, δ and s.

By means of a partition-of-unity argument, this will imply the same type of estimate for I_0 and, hence, for I, which will conclude the proof of Lemma 8.8.

Now, if $\nabla_u \Phi_2(u^0, (B_1')^0, (\delta_0')^0, (\delta_{3,0}')^0, (\delta_4')^0, \tilde{\rho}, s^0) \neq 0$, then by using an integration-by-parts argument in a similar way as for I_∞, we arrive at the same type of estimate for I_0^η as for I_∞, which is better than what is required.

We may therefore assume from now on that $\nabla_u \Phi_2(u^0, (B_1')^0, (\delta_0')^0, (\delta_{3,0}')^0, (\delta_4')^0, \tilde{\rho}, s^0) = 0$, and shall distinguish two cases.

Case 1: $u_1^0 \neq 0$. In this case, it is easy to see from (8.55) and (8.52) that

$$\partial_{u_1}^2 \Phi_2(u^0, (B_1')^0, (\delta_0')^0, (\delta_{3,0}')^0, (\delta_4')^0, \tilde{\rho}, s^0) \neq 0,$$

provided $\tilde{\rho}$ is sufficiently small. Then, by the implicit function theorem, the phase Φ_2 has a unique critical point $u_1^c(u_2, B_1', \delta_0', \delta_{3,0}', \delta_4', \tilde{\rho}, s)$ with respect to u_1, which is a smooth function of its variables, provided we choose the neighborhoods U, and so on, sufficiently small. Indeed, when $\tilde{\rho} = 0$, then by (8.55) and (8.52),

$$u_1^c(u_2, B_1', \delta_0', \delta_{3,0}', \delta_4', 0, s) = \left(\frac{B_1' - \delta_0' u_2 \alpha_{11}(0, \delta_0^{\mathfrak{r}}, s_2)}{3B_3(s_2, \delta_1, 0)} \right)^{1/2}. \tag{8.57}$$

We may thus apply the method of stationary phase to the integration with respect to the variable u_1 in the integral defining I_0^η. Let us denote by Ψ the phase function

$$\begin{aligned}\Psi(u_2, B_1', \delta_0', \delta_{3,0}', \delta_4', \tilde{\rho}, s) := \Phi_2(&u_1^c(u_2, B_1', \delta_0', \delta_{3,0}', \delta_4', \tilde{\rho}, s),\\ &B_1', \delta_0', \delta_{3,0}', \delta_4', \tilde{\rho}, s),\end{aligned}$$

which arises through this application of the method of stationary phase. We claim that then

$$\max_{j=4,5} |\partial_{u_2}^j \Psi(u_2^0, B_1', \delta_0', \delta_{3,0}', \delta_4', \tilde{\rho}, s)| \neq 0. \tag{8.58}$$

Notice that it suffices to prove this for $\tilde{\rho} = 0$, since then the result also follows for $\tilde{\rho}$ sufficiently small.

In order to prove (8.58) when $\tilde{\rho} = 0$, we make use of (8.57). Since $|B_3| \sim 1$, (8.57) shows that we may assume that

$$|B_1' - \delta_0' u_2 \alpha_{11}(0, \delta_0^{\mathfrak{r}}, s_2)| \sim |u_1^c| \sim |u_1^0|. \tag{8.59}$$

Note also that by (8.57) we have

$$\begin{aligned}\Psi(u_2, B_1', \delta_0', \delta_{3,0}', \delta_4', 0, s) = &\Gamma(u_2) + u_2^4 b(0, \delta_0^{\mathfrak{r}}, s_2) + \delta_4' u_2^2 \tilde{\alpha}_2(0, \delta_0^{\mathfrak{r}}, s_2)\\ &+ \delta_{3,0}' u_2 \tilde{\alpha}_1(0, \delta_0^{\mathfrak{r}}, s_2),\end{aligned}$$

where we have put

$$\Gamma(u_2) := -2 \cdot 3^{-3/2} B_3(s_2, \delta_1, 0)^{-1/2} \left(B_1' - \delta_0' u_2 \alpha_{11}(0, \delta_0^{\mathfrak{r}}, s_2)\right)^{3/2}.$$

In Case ND we have $\alpha_{11} \equiv 0$, and thus $|\partial_{u_2}^4 \Psi(u_2^0, B_1', \delta_0', \delta_{3,0}', \delta_4', \tilde{\rho}, s)| \neq 0$.

Next, if we are in Case D, then $|\alpha_{11}| \sim 1$, and (8.59) implies that $|\Gamma^{(j)}(u_2)| \sim |u_1^0|^{3/2} |\delta_0'/u_1^0|^j$. Therefore, if $\delta_0' \ll |u_1^0|$, then we find that $|\partial_{u_2}^4 \Psi(u_2^0, B_1', \delta_0', \delta_{3,0}', \delta_4', \tilde{\rho}, s)| \neq 0$, and if $\delta_0' \gtrsim |u_1^0|$, then

$$|\partial_{u_2}^5 \Psi(u_2^0, B_1', \delta_0', \delta_{3,0}', \delta_4', \tilde{\rho}, s)| \gtrsim |u_1^0|^{3/2} \neq 0.$$

This verifies (8.58) in this case too. But, (8.58) allows us to apply the van der Corput–type Lemma 2.1 to the integration in u_2 (after the application of the method of stationary phase in u_1), and altogether we obtain the estimate

$$|I_0^\eta(\Lambda, \delta, s)| \le C\Lambda^{-1/2-1/5},$$

which is, again, even stronger than what is required by (8.56).

Case 2: $u_1^0 = 0$. Assume first that $(\delta_0')^0 \neq 0$ and $|\alpha_{11}| \sim 1$ (this situation can occur only in Case D). Then

$$\partial_{u_1}\partial_{u_2}\Phi_2(u^0, (B_1')^0, (\delta_0')^0, (\delta_{3,0}')^0, (\delta_4')^0, 0, s^0) = \delta_0'\alpha_{11}(0, 0, s_2^0) \neq 0,$$
$$\partial_{u_1}^2\Phi_2(u^0, (B_1')^0, (\delta_0')^0, (\delta_{3,0}')^0, (\delta_4')^0, 0, s^0) = 0.$$

Therefore, we can apply the method of stationary phase to the integration in both variables (u_1, u_2) and again obtain an estimate of order $O(\Lambda^{-1})$, which is again stronger than what we need.

From now on, we may thus assume that $(\delta_0')^0 = 0$ (recall that in Case ND, we have $\alpha_{11} \equiv 0$ and are assuming that $\delta_0 = 0$ and, hence, also $\delta_0' = 0$, so that this assumption is automatically satisfied).

Then, necessarily, $(B_1')^0 = 0$, for otherwise, in view of (8.57) we would have $|u_1^c| \sim |(B_1')^0| \neq 0$ when $\tilde{\rho} = 0$, which would contradict our assumption that $u_1^0 = 0$. Since $((B_1')^0, (\delta_0')^0, (\delta_{3,0}')^0, (\delta_4')^0) \in \Sigma$, we thus see that $((\delta_{3,0}')^0)^{4/3} + ((\delta_4')^0)^2 = 1$.

Therefore, at the "base point" $((B_1')^0, (\delta_0')^0, (\delta_{3,0}')^0, (\delta_4')^0)$, the function ϕ satisfies, for $\tilde{\rho} = 0$, the inequality

$$\sum_{j=2}^{3} |\partial_{u_2}^j \phi(0, u_2^0, 0, \delta, s_2^0)| \neq 0,$$

and this inequality will persist for parameters sufficiently close to this base point.

Assume first that we have $\partial_{u_2}^2\phi(0, u_2^0, 0, \delta, s_2^0) \neq 0$. Then we can first apply the method of stationary phase to the u_2 integration and, subsequently, van der Corput's estimate in u_1 (with $M = 3$), which results in the estimate $|I_0^\eta(\Lambda, \delta, s)| \le C\Lambda^{-1/2-1/3}$. This is again stronger than what we need.

There remains the case where $\partial_{u_2}^2\phi(0, u_2^0, 0, \delta, s_2^0) = 0$ and $\partial_{u_2}^3\phi(0, u_2^0, 0, \delta, s_2^0) \neq 0$. In this case the phase function Φ_2 is a small smooth perturbation of a function Φ_2^0 of the form

$$\Phi_2^0(u_1, u_2) = c_3u_1^3 + (u_2 - u_2^0)^3 b_3(u_2) + c_0,$$

where $c_3 := B_3(s_2^0, \delta_1, 0) \neq 0$ and where $b_3(u_2)$ is a smooth function such that $b_3(u_2^0) \neq 0$. This means that Φ_2 has a so-called D_4^+-type singularity in the sense of [AGV88] and the distance between the associated Newton polyhedron and the origin is $\frac{3}{2}$. Estimate (8.56) therefore follows in this situation from the particular case of D_4^+-type singularities in Proposition 4.3.1 of [Dui74].

Alternatively, one could also first treat the integration with respect to u_1 by means of Lemma 2.2, with $B = 3$, and subsequently estimate the integration in u_2 by means of van der Corput's lemma (we leave the details to the interested reader).

This concludes the proof of Lemma 8.8. Q.E.D.

REMARK 8.9. *Notice that our phase* Φ *in Lemma 8.8 is a small perturbation of a phase of the form* $c_1 x_1^3 + c_2 x_2^4$, *with* $c_1 \neq 0 \neq c_2$, *at least if we assume that* $|B_1(s, \delta_1)| \ll 1$ *(this has been the interesting case in the preceding proof). It is, however, not true that for arbitrary small perturbations of such a phase function, depending, say, on a small perturbation parameter* $\delta > 0$, *an estimate analogous to* (8.50) *of order* $O\left(c(\delta)\lambda^{-2/3}\right)$ *as* $\lambda \to \infty$ *holds true. A counterexample is given by the function*

$$\Phi(x, \delta) := x_1^3 + (x_2 - \delta)^4 + 4\delta(x_2 - \delta)^3 - 3\sqrt[3]{4\delta^2}x_1(x_2 - \delta)^2 + C(\delta),$$

where $C(\delta)$ *is chosen such that* $\Phi(0, \delta) \equiv 0$. *Note that* $\Phi(x, 0) = x_1^3 + x_2^4$.

To see this, consider an oscillatory integral $J(\lambda, \delta) := \int e^{i\lambda\Phi(x,\delta)}a(x)\,dx$ with phase function Φ, whose amplitude is supported in a sufficiently small neighborhood of the origin and such that $a(0) = 1$.

When $\delta > 0$ is sufficiently small, then Φ has exactly two critical points, namely, the degenerate critical point $x_d := (0, \delta)$ and the nondegenerate critical point $x_{nd} := (6\sqrt[3]{2}\delta^{4/3}, -6\delta)$.

Let us consider the contribution of the degenerate critical point x_d to the oscillatory integral. The linear change of variables

$$z_1 = x_1 - \sqrt[3]{2\delta}(x_2 - \delta), \qquad z_2 = x_1 + 2\sqrt[3]{2\delta}(x_2 - \delta)$$

transforms x_d into $z_d = (0, 0)$ and the phase function into $\tilde{\Phi}(z) + C(\delta)$, where

$$\tilde{\Phi}(z) := z_1^2 z_2 + \left(\frac{z_2 - z_1}{3\sqrt[3]{2\delta}}\right)^4.$$

A look at the Newton polyhedron of $\tilde{\Phi}$ reveals that the principal face of $\mathcal{N}(\tilde{\Phi})$ is given by the compact edge $[(0, 4), (2, 1)]$ that lies on the line given by $\kappa_1 t_1 + \kappa_2 t_2 = 1$, with associated weight $\kappa = (\kappa_1, \kappa_2) := (\frac{3}{8}, \frac{1}{4})$, and the principal part of $\tilde{\Phi}$ is given by

$$\tilde{\Phi}_{\text{pr}}(z) = z_1^2 z_2 + \frac{z_2^4}{81\sqrt[3]{16\delta^4}}.$$

Moreover, the Newton distance is given by $d = \frac{8}{5}$, whereas the nontrivial roots of $\tilde{\Phi}_{\text{pr}}$ have multiplicity 1. Therefore, by Theorem 3.3 in [IM11a], the coordinates (z_1, z_2) are adapted to $\tilde{\Phi}$ in a sufficiently small neighborhood of the origin, so that the height h of $\tilde{\Phi}$ in the sense of Varchenko is also given by $h = d = \frac{8}{5}$. This implies that for every sufficiently small, fixed $\delta > 0$, we have that

$$J(\lambda, \delta) = C(\delta)\lambda^{-5/8} + O\left(\lambda^{-7/8}\right) \qquad \text{as} \quad \lambda \to \infty,$$

with a nontrivial constant $C(\delta)$, because the contribution of the nondegenerate critical point x_{nd} is of order $O(\lambda^{-1})$ (compare, for instance, [IM11b]). This shows that an estimate of the type $|J(\lambda, \delta)| \leq C(\delta)\lambda^{-2/3}$ cannot hold in this example.

8.5 THE CASE WHERE $m = 2$, $B = 4$ AND $A = 0$

Again, we are assuming that $\lambda\rho \gg 1$, where $\rho = \rho(\tilde{\delta})$ is given by (8.14). Observe that if $A = 0$, then according to (7.90) we have $h^r + 1 = 4(n+3)/(n+4)$, where $n \geq 9$, so that

$$\theta_c \leq \tfrac{13}{48}.$$

Here we shall again perform a frequency decomposition near the Airy cone by defining functions $\nu^\lambda_{\delta,\,Ai}$ and $\nu^\lambda_{\delta,l}$ as follows:

$$\widehat{\nu^\lambda_{\delta,\,Ai}}(\xi) := \chi_0\left(\rho^{-2/3} B_1(s,\,\delta_1)\right) \widehat{\nu^\lambda_\delta}(\xi),$$

$$\widehat{\nu^\lambda_{\delta,l}}(\xi) := \chi_1\left((2^l\rho)^{-2/3} B_1(s,\,\delta_1)\right) \widehat{\nu^\lambda_\delta}(\xi), \qquad M_0 \leq 2^l \leq \frac{\rho^{-1}}{M_1},$$

so that

$$\nu^\lambda_\delta = \nu^\lambda_{\delta,\,Ai} + \sum_{\{l:M_0 \leq 2^l \leq \rho^{-1}/M_1\}} \nu^\lambda_{\delta,l}. \tag{8.60}$$

Denote by $T^\lambda_{\delta,\,Ai}$ and $T^\lambda_{\delta,l}$ the corresponding operators of convolution with the Fourier transforms of these functions.

8.5.1 Estimation of $T^\lambda_{\delta,\,Ai}$

Here we have $|\rho^{-2/3}B_1(s,\,\delta_1)| \lesssim 1$. In this case, we use the change of variables $x =: \sigma_\rho u := (\rho^{1/3}u_1,\ \rho^{1/4}u_2)$ in the integral (8.12) defining $\widehat{\nu_\delta}$ and obtain

$$\widehat{\nu_\delta}(\xi) = \rho^{7/12} e^{-i\lambda s_3 B_0(s,\delta_1)} \int_{|\sigma_\rho u|<\varepsilon} e^{-i\lambda\rho\, s_3 \Phi_1(u,s,\delta)} a(\sigma_\rho u,\,\delta,\,s)\, du, \tag{8.61}$$

where the phase Φ_1 has the form

$$\begin{aligned}\Phi_1(u,\,s,\,\delta) &= u_1^3\, B_3(s_2,\,\delta_1,\,\rho^{1/3}u_1) - u_1\,\rho^{-2/3} B_1(s,\,\delta_1)\\ &\quad + u_2^4\, b(\sigma_\rho u,\,\delta_0^\mathfrak{r},\,s_2) + \delta_4' u_2^2\,\tilde{\alpha}_2(\rho^{1/3}u_1,\,\delta_0^\mathfrak{r},\,s_2) + \delta_{3,0}' u_2\,\tilde{\alpha}_1(\delta_0^\mathfrak{r},\,s_2)\\ &\quad + \delta_0' u_1 u_2\,\alpha_{1,1}(\rho^{1/3}u_1,\,\delta_0^\mathfrak{r},\,s_2)\end{aligned} \tag{8.62}$$

and where, according to Remark 8.4 (in which we choose $r := 1/\rho$), $\rho(\widetilde{\delta'}) = 1$, so that

$$\delta_0' + \delta_{3,0}' + \delta_4' \sim 1$$

(recall that the coefficient δ_0' does not appear in Case ND, where $\alpha_{1,1} = 0$). We have also indicated that the amplitude $a(\sigma_\rho u,\,\delta,\,s)$ is supported where $|\sigma_\rho u| < \varepsilon$, where we may assume that $\varepsilon > 0$ is sufficiently small since this will become important soon.

We shall proceed in a somewhat similar way as in Section 5.2, by choosing a cutoff function $\chi_0 \in C_0^\infty(\mathbb{R}^2)$ such that $\chi_0(u) = 1$ for $|u| \leq R$, where R will be chosen sufficiently large, and further decomposing

$$\widehat{\nu_\delta}(\xi) = \widehat{\nu_{\delta,0}}(\xi) + \widehat{\nu_{\delta,\infty}}(\xi),$$

where

$$\widehat{\nu_{\delta,0}}(\xi) := \rho^{7/12} e^{-i\lambda s_3 B_0(s,\delta_1)} \int e^{-i\lambda\rho\, s_3 \Phi_1(u,s,\delta)} a(\sigma_\rho u, \delta, s)\chi_0(u)\, du,$$

and

$$\widehat{\nu_{\delta,\infty}}(\xi) := \rho^{7/12} e^{-i\lambda s_3 B_0(s,\delta_1)} \int e^{-i\lambda\rho\, s_3 \Phi_1(u,s,\delta,s)} a(\sigma_\rho u, \delta, s)(1 - \chi_0(u))\, du.$$

Accordingly, we decompose

$$\nu_{\delta,\,Ai}^{\lambda} = \nu_{\delta,0}^{\lambda} + \nu_{\delta,\infty}^{\lambda},$$

where we have let

$$\widehat{\nu_{\delta,0}^{\lambda}}(\xi) := \chi_0\left(\rho^{-2/3} B_1(s,\ \delta_1)\right) \chi_1(s, s_3)\, \widehat{\nu_{\delta,0}}(\xi),$$

$$\widehat{\nu_{\delta,\infty}^{\lambda}}(\xi) := \chi_0\left(\rho^{-2/3} B_1(s,\ \delta_1)\right) \chi_1(s, s_3)\, \widehat{\nu_{\delta,\infty}}(\xi).$$

Recall from (8.24) that $\chi_1(s, s_3) = \chi_1(s_1, s_2, s_3)$ localizes to the region where $s_j \sim 1$, $j = 1, 2, 3$. The corresponding operators of convolution with $\widehat{\nu_{\delta,0}^{\lambda}}$ and $\widehat{\nu_{\delta,\infty}^{\lambda}}$ will be denoted by $T_{\delta,0}^{\lambda}$ and $T_{\delta,\infty}^{\lambda}$, respectively.

Let us first consider the operators $T_{\delta,\infty}^{\lambda}$: By means of integrations by parts, we easily see that if R is chosen sufficiently large, then the phase will have no critical point, and thus for every $N \in \mathbb{N}$, we have

$$\|\widehat{\nu_{\delta,\infty}^{\lambda}}\|_\infty \lesssim \rho^{7/12}(\lambda\rho)^{-N}. \tag{8.63}$$

Moreover, by Fourier inversion we find that

$$\nu_{\delta,\infty}^{\lambda}(x) = \lambda^3 \int_{\mathbb{R}^3} e^{i\lambda s_3(s_1 x_1 + s_2 x_2 + x_3)} \widehat{\nu_{\delta,\infty}^{\lambda}}(\xi)\, ds \tag{8.64}$$

(with $\xi = \lambda s_3(s_1, s_2, 1)$). We then use the change of variables from $s = (s_1, s_2)$ to (z, s_2), where

$$z := \rho^{-2/3} B_1(s, \delta_1),$$

and find that (compare (8.9))

$$s_1 = s_2^{(n-1)/(n-2)} G_3(s_2, \delta) - \rho^{2/3} z$$

and, in particular,

$$B_0(s, \delta, \sigma) = -\rho^{2/3} z\, s_2^{1/(n-2)} G_1(s_2, \delta) + s_2^{n/(n-2)} G_5(s_2, \delta).$$

And, if we plug the previous formula for $\widehat{\nu_{\delta,\infty}^{\lambda}}$ into (8.64), we see that we may write $\nu_{\delta,\infty}^{\lambda}(x)$ as an oscillatory integral

$$\begin{aligned}\nu_{\delta,\infty}^{\lambda}(x) = \rho^{7/12+2/3}\lambda^3 \int e^{-i\lambda s_3 \Phi_2(u,z,s_2,\delta)} \chi_0(z)(1 - \chi_0(u)) a(\sigma_\rho u, \rho^{2/3} z, s, \delta) \\ \times \tilde{\chi}_1(s_2, s_3)\, du\, dz\, ds_2\, ds_3\end{aligned} \tag{8.65}$$

with respect to the variables u_1, u_2, z, s_2, s_3, where the complete phase is given by

$$\Phi_2(u, z, s_2, \delta) := s_2^{n/(n-2)} G_5(s_2, \delta) - x_1 s_2^{(n-1)/(n-2)} G_3(s_2, \delta)$$
$$-s_2 x_2 - x_3 + \rho^{2/3} z \left(x_1 - s_2^{1/(n-2)} G_1(s_2, \delta)\right) + \rho\, \Phi_1(u, z, s_2, \delta), \tag{8.66}$$

where, according to (8.62), the phase Φ_1 is given in the new coordinates by

$$\Phi_1(u, z, s_2, \delta) = u_1^3\, B_3(s_2, \delta_1, \rho^{1/3} u_1) - u_1 z$$
$$+ u_2^4\, b(\sigma_\rho u, \delta_0^{\mathfrak{r}}, s_2) + \delta_4' u_2^2\, \tilde{\alpha}_2(\rho^{1/3} u_1, \delta_0^{\mathfrak{r}}, s_2)$$
$$+ \delta_{3,0}' u_2\, \tilde{\alpha}_1(\delta_0^{\mathfrak{r}}, s_2) + \delta_0' u_1 u_2\, \alpha_{1,1}(\rho^{1/3} u_1, \delta_0^{\mathfrak{r}}, s_2). \tag{8.67}$$

Recall also from (8.11) that $|G_5| \sim 1 \sim |G_3|$ (where $G_5 := G_1 G_3 - G_2$). The new amplitude a is again a smooth function of its arguments, and $\tilde{\chi}_1(s_2, s_3)$ localizes to the region where $|s_2| \sim 1 \sim |s_3|$. Observe also that here $|z| \lesssim 1$ and $|\rho^{1/3} u_1| \le \varepsilon$, $|\rho^{1/4} u_2| \le \varepsilon$, so that the sum of the last two terms in Φ_2 can be viewed as a small error term of order $O(\rho^{2/3} + \varepsilon)$, provided $|x| \lesssim 1$.

First applying N integrations by parts with respect to the variables u_1, u_2 and then van der Corput's lemma to the integration in s_2, we find that

$$|\nu_{\delta,\infty}^\lambda(x)| \lesssim \rho^{7/12+2/3} \lambda^3 (\lambda\rho)^{-N} \lambda^{-1/3}$$

if $|x| \lesssim 1$.

However, if $|x| \gg 1$, then we may argue as in Section 5.2: if $|x_1| \ll |(x_2, x_3)|$, then we easily see by means a further integration by parts with respect to the variables s_2 or s_3 that $|\nu_{\delta,\infty}^\lambda(x)| \lesssim \rho^{7/12+2/3} \lambda^3 (\lambda\rho)^{-N} \lambda^{-1}$, and if $|x_1| \gtrsim |(x_2, x_3)|$, then an integration by parts in z leads to $|\nu_{\delta,\infty}^\lambda(x)| \lesssim \rho^{7/12+2/3} \lambda^3 (\lambda\rho)^{-N} (\lambda\rho^{2/3})^{-1}$. Both these estimates are stronger than the previous one, and so altogether we have shown that

$$\|\nu_{\delta,\infty}^\lambda\|_\infty \lesssim \rho^{7/12+2/3} \lambda^{8/3} (\lambda\rho)^{-N}. \tag{8.68}$$

Interpolating between this estimate and (8.63), we obtain

$$\|T_{\delta,\infty}^\lambda\|_{p_c \to p_c'} \lesssim \rho^{7/12} (\lambda\rho)^{-N} \rho^{2\theta_c/3} \lambda^{8\theta_c/3},$$

which implies the desired estimate

$$\sum_{\lambda\rho \gg 1} \|T_{\delta,\infty}^\lambda\|_{p_c \to p_c'} \lesssim \rho^{7/12 - 2\theta_c} \le 1,$$

since $\theta_c \le \frac{13}{48}$.

We next turn to the main terms $\nu_{\delta,0}^\lambda$ and the corresponding operators $T_{\delta,0}^\lambda$. First, we claim that

$$\|\widehat{\nu_{\delta,0}^\lambda}\|_\infty \lesssim \rho^{7/12} (\lambda\rho)^{-2/3}. \tag{8.69}$$

This is an immediate consequence of Lemma 8.8. Indeed our phase Φ_1 in (8.62) is of the form as required in this lemma, if we choose the ρ in the lemma of size 1 and replace λ in the lemma by $\lambda\rho$. Therefore, the oscillatory integral in the definition of $\widehat{\nu_{\delta,0}^\lambda}$ can be estimated by $C/(1^{1/12}(\lambda\rho)^{2/3})$.

Finally, we want to estimate $\|\nu_{\delta,0}^\lambda\|_\infty$. In analogy with (8.65), we may write $\nu_{\delta,0}^\lambda(x)$ as an oscillatory integral of the form

$$\nu_{\delta,0}^\lambda(x) = \rho^{7/12+2/3}\lambda^3 \int e^{-i\lambda s_3 \Phi_2(u,z,s_2,\delta)} \\ \times \chi_0(z)\chi_0(u)\tilde{a}(\sigma_\rho u, \rho^{2/3}z, s, \delta)\tilde{\chi}_1(s_2, s_3)\, du\, dz\, ds_2\, ds_3,$$

with Φ_2 given by (8.66). We can reduce this to the following situation, in which the amplitude is independent of z, that is, where

$$\nu_{\delta,0}^\lambda(x) = \rho^{7/12+2/3}\lambda^3 \int e^{-i\lambda s_3 \Phi_2(u,z,s_2,\delta)} \\ \times \chi_0(z)\chi_0(u)a(\sigma_\rho u, s, \rho^{1/3}, \delta)\tilde{\chi}_1(s_2, s_3)\, du\, dz\, ds_2\, ds_3. \tag{8.70}$$

In fact, we may develop the amplitude $\tilde{a}$ into a convergent series of smooth functions, each of which is a tensor product of a smooth function of the variable z with a smooth function depending on the remaining variables only. Thus, by considering each of the corresponding terms separately, we can reduce to the situation (8.70) (the function $\chi_0(z)$ will of course have to be different from the previous one).

We claim that

$$\|\nu_{\delta,0}^\lambda\|_\infty \lesssim \rho^{7/12}\lambda^2(\lambda\rho)^{-1/4}. \tag{8.71}$$

Indeed, if $|x| \gg 1$, arguing in a similar way as for $\nu_{\delta,\infty}^\lambda(x)$, we see that $|\nu_{\delta,0}^\lambda(x)| \lesssim \rho^{7/12+2/3}\lambda^3(\lambda\rho^{2/3})^{-N}$ for every $N \in \mathbb{N}$, which is stronger than what is needed for (8.71).

Therefore, from now on we shall assume that $|x| \lesssim 1$. For such x fixed, we can argue in a similar way as in in Section 5.2 (compare also with the discussion in Subsection 8.7.2): we decompose

$$\nu_{\delta,0}^\lambda = \nu_{0,I}^\lambda + \nu_{0,II}^\lambda, \tag{8.72}$$

where $\nu_{0,I}^\lambda$ and $\nu_{0,II}^\lambda$ denote the contributions to the integral (8.70) by the region L_I given by

$$|x_1 - s_2^{1/(n-2)}G_1(s_2,\delta)| \gg \rho^{1/3},$$

and the region L_{II} where

$$|x_1 - s_2^{1/(n-2)}G_1(s_2,\delta)| \lesssim \rho^{1/3},$$

respectively. Recall from (8.66) and (8.67) that

$$\begin{aligned}\Phi_2(u,z,s_2,\delta) = {} & s_2^{n/(n-2)}G_5(s_2,\delta) - x_1 s_2^{(n-1)/(n-2)}G_3(s_2,\delta) - s_2x_2 - x_3 \\ & + z\rho(\rho^{-1/3}(x_1 - s_2^{1/(n-2)}G_1(s_2,\delta)) - u_1) \\ & + \rho u_1^3\, B_3(s_2,\delta_1,\rho^{1/3}u_1) + \rho\,\big(u_2^4\, b(\sigma_\rho u, \delta_0^\mathfrak{r}, s_2) \\ & + \delta_4' u_2^2\, \tilde{\alpha}_2(\rho^{1/3}u_1, \delta_0^\mathfrak{r}, s_2) + \delta_{3,0}' u_2\, \tilde{\alpha}_1(\delta_0^\mathfrak{r}, s_2) \\ & + \delta_0' u_1 u_2\, \alpha_{1,1}(\rho^{1/3}u_1, \delta_0^\mathfrak{r}, s_2)\big).\end{aligned} \tag{8.73}$$

Let us change variables from s_2 first to $v := x_1 - s_2^{1/(n-2)} G_1(s_2, \delta)$ and then to $w := \rho^{-1/3} v = \rho^{-1/3}(x_1 - s_2^{1/(n-2)} G_1(s_2, \delta))$. In these new coordinates, Φ_2 can be written as

$$\Phi_2 = z\rho(w - u_1) + \Phi_3,$$

with Φ_3 of the form

$$\begin{aligned}\Phi_3(u, w, x, \delta) = {} & \Psi_3(\rho^{1/3} w, x, \delta) + \rho u_1^3 \, B_3(\rho^{1/3} w, \rho^{1/3} u_1, x, \delta) \\ & + \rho \left(u_2^4 \, b(\sigma_\rho u, \rho^{1/3} w, x, \delta_0^{\mathfrak{r}}) + \delta_4' u_2^2 \, \tilde{\alpha}_2(\rho^{1/3} u_1, \rho^{1/3} w, x, \delta_0^{\mathfrak{r}}) \right. \\ & \left. + \, \delta_{3,0}' u_2 \, \tilde{\alpha}_1(\rho^{1/3} w, x, \delta_0^{\mathfrak{r}}) + \delta_0' u_1 u_2 \, \alpha_{1,1}(\rho^{1/3} u_1, \rho^{1/3} w, x, \delta_0^{\mathfrak{r}}) \right), \end{aligned} \tag{8.74}$$

where Ψ_3 is a smooth, real-valued function. With a slight abuse of notation, we have here used the same symbols $B_3, b, \ldots, \alpha_{1,1}$ as before, since these functions will have the same basic properties here as the corresponding ones in (8.73). A similar remark will apply to the amplitudes, which we shall always denote by the letter a, even though they may change from line to line. Moreover, we may write

$$\begin{aligned}\nu_{0,I}^\lambda(x) = {} & (\rho^{7/12+2/3} \lambda^3) \, \rho^{1/3} \int e^{-i\lambda s_3 \Phi_3(u,w,x,\delta)} \, \widehat{\chi_0}(\lambda\rho s_3(w - u_1)) \, (1 - \chi_0(w)) \, \chi_0(u) \\ & \times a(\sigma_\rho u, \rho^{1/3} w, s_1, \rho^{1/3}, x, \delta) \tilde{\chi}_1(s_3) \, dw \, du \, ds_3. \end{aligned} \tag{8.75}$$

$$\begin{aligned}\nu_{0,II}^\lambda(x) = {} & (\rho^{7/12+2/3} \lambda^3) \, \rho^{1/3} \int e^{-i\lambda s_3 \Phi_3(u,w,x,\delta)} \, \widehat{\chi_0} \, (\lambda\rho s_3(w - u_1)) \, \chi_0(w) \, \chi_0(u) \\ & \times a(\sigma_\rho u, \rho^{1/3} w, s_1, \rho^{1/3}, x, \delta) \tilde{\chi}_1(s_3) \, dw \, du \, ds_3. \end{aligned} \tag{8.76}$$

Here, $\chi_0(w)$ will again denote a smooth function with compact support, which is identically 1 on a sufficiently large neighborhood of the origin.

Observe that in (8.75) we have $|w| \gg 1 \gtrsim |u_1|$, so that $|\widehat{\chi_0} \, (\lambda\rho s_3(w - u_1))| \le C_N(\lambda\rho|w|)^{-(N+1)}$ for every $N \in \mathbb{N}$, and we immediately obtain the estimate

$$\|\nu_{0,I}^\lambda(x)\|_\infty \le C_N(\rho^{7/12+2/3} \lambda^3) \, \rho^{1/3} (\lambda\rho)^{-(N+1)} = C_N \rho^{7/12} \lambda^2 \, (\lambda\rho)^{-N}, \tag{8.77}$$

which is even stronger than (8.71).

In order to estimate the second term, we perform yet another change of variables from u_1 to y_1 so that $u_1 = w - (\lambda\rho)^{-1} y_1$, that is, $y_1 = \lambda\rho(w - u_1)$. This leads to the following expression for $\nu_{0,II}^\lambda(x)$:

$$\begin{aligned}\nu_{0,II}^\lambda(x) = {} & (\rho^{7/12+2/3} \lambda^3) \, \rho^{1/3} \, (\lambda\rho)^{-1} \int e^{-i\lambda s_3 \Phi_4(y_1, u_2, w, x, \delta)} \, \widehat{\chi_0}(s_3 y_1) \\ & \times \chi_0(w) \, \chi_0(w - (\lambda\rho)^{-1} y_1) \\ & \times \chi_0(u_2) \, a_4 \left((\lambda\rho^{2/3})^{-1} y_1, \rho^{1/4} u_2, w, s_1, \rho^{1/3}, x, \delta \right) \\ & \times \tilde{\chi}_1(s_3) \, dy_1 \, du_2 \, dw \, ds_3, \end{aligned} \tag{8.78}$$

with phase Φ_4 of the form

$$\begin{aligned}\Phi_4(y_1, u_2, w, x, \delta) = {} & \Psi_3\left(\rho^{1/3}w, x, \delta\right) + \rho\left(w - (\lambda\rho)^{-1}y_1\right)^3 \\ & \times \tilde{B}_3\left(\rho^{1/3}w, (\lambda\rho^{2/3})^{-1}y_1, x, \delta\right) \\ & + \rho\left(u_2^4\, b\left(\rho^{1/3}w, (\lambda\rho^{2/3})^{-1}y_1, \rho^{1/4}u_2, x, \delta_0^{\mathfrak{r}}\right)\right. \\ & + \delta_4' u_2^2\, \tilde{\alpha}_2\left(\rho^{1/3}w, (\lambda\rho^{2/3})^{-1}y_1, \rho^{1/4}u_2, x, \delta_0^{\mathfrak{r}}\right) \\ & \times\ \delta_{3,0}' u_2\, \tilde{\alpha}_1\left(\rho^{1/3}w, x, \delta_0^{\mathfrak{r}}\right) + \delta_0' u_2\left(w - (\lambda\rho)^{-1}y_1\right) \\ & \left.\times\, \alpha_{1,1}\left(\rho^{1/3}w, (\lambda\rho^{2/3})^{-1}y_1, x, \delta_0^{\mathfrak{r}}\right)\right). \end{aligned} \tag{8.79}$$

Observe that in this integral, $|u_2| + |w| \lesssim 1$ and $|y_1| \lesssim \lambda\rho$. Moreover, the factor $\widehat{\chi_0}(s_3 y_1)$ guarantees the absolute convergence of this integral with respect to the variable y_1. We can thus first apply van der Corput's estimate of order $M = 4$ for the integration in u_2, which leads to an additional factor of order $(\lambda\rho)^{-1/4}$, and then perform the remaining integrations in w, y_1, and s_3. Altogether, this leads to the estimate

$$\|\nu_{0,l}^{\lambda}(x)\|_\infty \le C(\rho^{7/12+2/3}\lambda^3)\,\rho^{1/3}(\lambda\rho)^{-1}(\lambda\rho)^{-1/4} = \rho^{7/12}\lambda^2\,(\lambda\rho)^{-1/4}.$$

In combination with (8.77), this proves (8.71).

Finally, interpolating between the estimates (8.69) and (8.71), we obtain

$$\|T_{\delta,0}^{\lambda}\|_{p_c \to p_c'} \lesssim \rho^{7/12 - 2\theta_c}(\lambda\rho)^{29\theta_c/12 - 2/3}.$$

But, since $\theta_c \le \frac{13}{48}$, we have $\frac{29}{12}\theta_c - \frac{2}{3} < 0$ and $\frac{7}{12} - 2\theta_c > 0$, which implies the desired estimate,

$$\sum_{\lambda\rho \gg 1} \|T_{\delta,0}^{\lambda}\|_{p_c \to p_c'} \lesssim \rho^{7/12 - 2\theta_c} \le 1.$$

Altogether, we have thus proved that

$$\sum_{\lambda\rho \gg 1} \|T_{\delta,Ai}^{\lambda}\|_{p_c \to p_c'} \lesssim 1. \tag{8.80}$$

8.5.2 Estimation of $T_{\delta,l}^{\lambda}$

Here we have $|(2^l\rho)^{-2/3}B_1(s,\delta_1)| \sim 1$. Recall also that $2^l\rho \le 1/M_1 \ll 1$. In this case, we use the change of variables $x =: \sigma_{2^l\rho}u := ((2^l\rho)^{1/3}u_1, (2^l\rho)^{1/4}u_2)$ in the integral (8.12) defining $\widehat{\nu_\delta}$, and obtain

$$\widehat{\nu_\delta}(\xi) = (2^l\rho)^{7/12}\, e^{-i\lambda s_3 B_0(s,\delta_1)} \int_{|\sigma_{2^l\rho}u| < \varepsilon} e^{-i\lambda 2^l\rho\, s_3 \Phi_1(u,s,\delta)}\, a(\sigma_{2^l\rho}u, \delta, s)\, du,$$

where now the phase $\Phi_1 = \Phi_{1,l}$ has the form

$$\begin{aligned}\Phi_1(u,s,\delta) = {} & u_1^3\, B_3(s_2, \delta_1, (2^l\rho)^{1/3}u_1) - u_1\,(2^l\rho)^{-2/3}B_1(s,\delta_1) + u_2^4\, b(\sigma_{2^l\rho}u, \delta_0^{\mathfrak{r}}, s_2) \\ & + \delta_4' u_2^2\, \tilde{\alpha}_2((2^l\rho)^{1/3}u_1, \delta_0^{\mathfrak{r}}, s_2) + \delta_{3,0}' u_2\, \tilde{\alpha}_1(\delta_0^{\mathfrak{r}}, s_2) \\ & + \delta_0' u_1 u_2\, \alpha_{1,1}((2^l\rho)^{1/3}u_1, \delta_0^{\mathfrak{r}}, s_2); \end{aligned} \tag{8.81}$$

according to Remark 8.4 (in which we choose $r := 1/(2^l\rho)$), $\rho(\widetilde{\delta'}) = 2^{-l}$, so that

$$(\delta_4')^2 + (\delta_{3,0}')^{4/3} + (\delta_0')^{12/5} = 2^{-l} \le \frac{1}{M_0} \ll 1$$

(recall that the coefficient δ_0' does not appear in Case ND, where $\alpha_{1,1} = 0$). We have also indicated that the amplitude $a(\sigma_{2^l\rho}u, \delta, s)$ is supported where $|\sigma_{2^l\rho}u| < \varepsilon$ and that we may assume that $\varepsilon > 0$ is sufficiently small.

Observe that the second and third rows in (8.81) are a small perturbation of the leading term, given by the first row.

Again, we choose a cutoff function $\chi_0 \in C_0^\infty(\mathbb{R}^2)$ such that $\chi_0(u) = 1$ for $|u| \le R$, where R will be chosen sufficiently large, and further decompose

$$\widehat{\nu_\delta}(\xi) = \widehat{\nu_{\delta,0}}(\xi) + \widehat{\nu_{\delta,\infty}}(\xi),$$

where now

$$\widehat{\nu_{\delta,0}}(\xi) := (2^l\rho)^{7/12} e^{-i\lambda s_3 B_0(s,\delta_1)} \int_{|\sigma_{2^l\rho}u|<\varepsilon} e^{-i\lambda 2^l\rho\, s_3 \Phi_1(u,s,\delta)} a(\sigma_{2^l\rho}u, \delta, s)\chi_0(u)\, du,$$

and

$$\widehat{\nu_{\delta,\infty}}(\xi) := (2^l\rho)^{7/12} e^{-i\lambda s_3 B_0(s,\delta_1)} \times \int_{|\sigma_{2^l\rho}u|<\varepsilon} e^{-i\lambda 2^l\rho\, s_3 \Phi_1(u,s,\delta)} a(\sigma_{2^l\rho}u, \delta, s)(1-\chi_0(u))\, du.$$

Accordingly, we decompose

$$\nu_{\delta,l}^\lambda = \nu_{l,0}^\lambda + \nu_{l,\infty}^\lambda,$$

where we have put

$$\widehat{\nu_{l,0}^\lambda}(\xi) := \chi_1\big((2^l\rho)^{-2/3} B_1(s, \delta_1)\big)\, \chi_1(s, s_3)\, \widehat{\nu_{\delta,0}}(\xi),$$
$$\widehat{\nu_{l,\infty}^\lambda}(\xi) := \chi_1\big((2^l\rho)^{-2/3} B_1(s, \delta_1)\big)\, \chi_1(s, s_3)\, \widehat{\nu_{\delta,\infty}}(\xi).$$

The corresponding operators of convolution with $\widehat{\nu_{l,0}^\lambda}$ and $\widehat{\nu_{l,\infty}^\lambda}$ will be denoted by $T_{l,0}^\lambda$ and $T_{l,\infty}^\lambda$, respectively.

Again, let us first consider the operators $T_{l,\infty}^\lambda$: by means of integrations by parts, we easily see that if R is chosen sufficiently large, then the phase will have no critical point, and thus for every $N \in \mathbb{N}$ we have

$$\|\widehat{\nu_{l,\infty}^\lambda}\|_\infty \lesssim (2^l\rho)^{7/12} (\lambda 2^l\rho)^{-N}. \tag{8.82}$$

Moreover, Fourier inversion again leads to (8.64), and performing the change of variables from $s = (s_1, s_2)$ to (z, s_2), where here

$$z := (2^l\rho)^{-2/3} B_1(s, \delta_1),$$

we find that

$$s_1 = s_2^{(n-1)/(n-2)} G_3(s_2, \delta) - (2^l\rho)^{2/3} z$$

and, in particular,

$$B_0(s, \delta, \sigma) = -(2^l\rho)^{2/3} z\, s_2^{1/(n-2)} G_1(s_2, \delta) + s_2^{n/(n-2)} G_5(s_2, \delta).$$

In a similar way as before, this leads to

$$\begin{aligned} \nu_{l,\infty}^{\lambda}(x) &= (2^l\rho)^{7/12+2/3}\lambda^3 \\ &\times \int e^{-i\lambda s_3\Phi_2(u,z,s_2,\delta)}\chi_1(z)(1-\chi_0(u))a(\sigma_{2^l\rho}u,(2^l\rho)^{2/3}z,s,\delta) \\ &\times \tilde{\chi}_1(s_2,s_3)\,du\,dz\,ds_2\,ds_3 \end{aligned} \tag{8.83}$$

with respect to the variables u_1, u_2, z, s_2, s_3, where the complete phase is now given by

$$\begin{aligned} \Phi_2(u,z,s_2,\delta) := {} & s_2^{n/(n-2)}G_5(s_2,\delta) - x_1 s_2^{(n-1)/(n-2)}G_3(s_2,\delta) \\ & - s_2x_2 - x_3 + (2^l\rho)^{2/3}z\big(x_1 - s_2^{1/(n-2)}G_1(s_2,\delta)\big) \\ & + 2^l\rho\,\Phi_1(u,z,s_2,\delta). \end{aligned} \tag{8.84}$$

According to (8.81), the phase Φ_1 is given in the new coordinates by

$$\begin{aligned} \Phi_1(u,z,s_2,\delta) = {} & u_1^3\,B_3(s_2,\delta_1,(2^l\rho)^{1/3}u_1) - u_1z + u_2^4\,b(\sigma_{2^l\rho}u,\delta_0^{\mathfrak{r}},s_2) \\ & + \delta_4' u_2^2\,\tilde{\alpha}_2((2^l\rho)^{1/3}u_1,\delta_0^{\mathfrak{r}},\, s_2) + \delta_{3,0}'u_2\,\tilde{\alpha}_1(\delta_0^{\mathfrak{r}},s_2) \\ & + \delta_0'u_1u_2\,\alpha_{1,1}((2^l\rho)^{1/3}u_1,\delta_0^{\mathfrak{r}},s_2). \end{aligned} \tag{8.85}$$

Arguing as in the preceding section, by first applying N integrations by parts with respect to the variables u_1, u_2, and then van der Corput's lemma (of order $M=3$) to the integration in s_2, we thus find that

$$|\nu_{l,\infty}^{\lambda}(x)| \lesssim (2^l\rho)^{7/12+2/3}\lambda^3(\lambda 2^l\rho)^{-N}\lambda^{-1/3},$$

first if $|x|\lesssim 1$ and then also for $|x|\gg 1$, by the same kind of arguments. We obtain

$$\|\nu_{l,\infty}^{\lambda}\|_\infty \lesssim (2^l\rho)^{7/12+2/3}\lambda^{8/3}(\lambda 2^l\rho)^{-N}.$$

Interpolating between this estimate and (8.82), we find that

$$\|T_{l,\infty}^{\lambda}\|_{p_c\to p_c'} \lesssim (2^l\rho)^{7/12}(\lambda 2^l\rho)^{-N}(2^l\rho)^{2\theta_c/3}\lambda^{8\theta_c/3},$$

which implies that

$$\sum_{\lambda\rho\gg 1}\|T_{l,\infty}^{\lambda}\|_{p_c\to p_c'} \lesssim 2^{(7/12+2\theta_c/3-N)l}\rho^{7/12-2\theta_c} \le 1;$$

this yields the desired estimate

$$\sum_{\{l:M_0\le 2^l\le\rho^{-1}/M_1\}}\ \sum_{\lambda\rho\gg 1}\|T_{l,\infty}^{\lambda}\|_{p_c\to p_c'} \lesssim \rho^{7/12-2\theta_c} \le 1$$

since $\theta_c \le \frac{13}{48}$.

We next turn to the main terms $\nu_{l,0}^{\lambda}$ and the corresponding operators $T_{l,0}^{\lambda}$. First we show that

$$\|\widehat{\nu_{l,0}^{\lambda}}\|_\infty \lesssim (2^l\rho)^{7/12}(\lambda 2^l\rho)^{-3/4}. \tag{8.86}$$

Indeed, (8.81) in combination with (8.4) shows that the phase Φ_1 is a small perturbation of the phase

$$\Phi_{1,0}(u,s) := u_1^3\,B_3(s_2,0,0) - c_1u_1 + u_2^4\,b_4(0,0),$$

where c_1 corresponds to a fixed value of $(2^l\rho)^{-2/3}B_1(s,\delta_1)$, so that $|c_1|\sim 1$. Now, this phase will either have a critical point with respect to the variable u_1 (recall that $u\in \operatorname{supp}\chi_0$), and then we can apply the method of stationary phase to the u_1-integration, or, otherwise, we may use integrations by parts in u_1 (for instance, if c_1 and B_3 have opposite signs). Subsequently applying van der Corput's estimate (of order $M=4$) to the u_1-integration, we immediately get (8.86).

We finally want to estimate $\|\nu^\lambda_{l,0}\|_\infty$. In analogy with (8.83), we may write $\nu^\lambda_{l,0}(x)$ as an oscillatory integral of the form

$$\begin{aligned}\nu^\lambda_{l,0}(x) &= (2^l\rho)^{7/12+2/3}\lambda^3\\ &\times \int e^{-i\lambda s_3\Phi_2(u,z,s_2,\delta)}\chi_1(z)\chi_0(u)a(\sigma_{2^l\rho}u,(2^l\rho)^{2/3}z,s,\delta)\\ &\times \tilde\chi_1(s_2,s_3)\,du\,dz\,ds_2\,ds_3,\end{aligned}$$

with Φ_2 given by (8.84). We can then basically argue as we did for $\nu^\lambda_{\delta,Ai}$ in the previous section, only with ρ replaced by $2^l\rho$, and arrive at the following analogue of estimate (8.71):

$$\|\nu^\lambda_{l,0}\|_\infty \lesssim (2^l\rho)^{7/12}\lambda^2(\lambda 2^l\rho)^{-1/4}. \tag{8.87}$$

Finally, interpolating between estimates (8.86) and (8.87), we obtain

$$\|T^\lambda_{l,0}\|_{p_c\to p_c'} \lesssim (2^l\rho)^{7/12}\lambda^{2\theta_c}(\lambda 2^l\rho)^{\theta_c/2-3/4} = 2^{(3\theta_c-1)/6}(\lambda\rho)^{(10\theta_c-3)/4}\rho^{7/12-2\theta_c}.$$

But, since $\theta_c\le \frac{13}{48}$, we have $10\theta_c-3<0$ and $3\theta_c-1<0$, which implies the desired estimate

$$\sum_{\{l:M_0\le 2^l\le\rho^{-1}/M_1\}}\ \sum_{\lambda\rho\gg 1}\|T^\lambda_{l,0}\|_{p_c\to p_c'}\lesssim 1.$$

Altogether, we have thus proved that

$$\sum_{\{l:M_0\le 2^l\le\rho^{-1}/M_1\}}\ \sum_{\lambda\rho\gg 1}\|T^\lambda_{\delta,l}\|_{p_c\to p_c'}\lesssim 1.$$

This completes the proof of Proposition 8.7, as well as the case where $A=0$.

8.6 THE CASE WHERE $m=2$, $B=3$, AND $A=0$: WHAT STILL NEEDS TO BE DONE

What remains open in (7.89)—hence, also in the proof of Proposition 7.1—is the case where $B=3$ and $A=0$, in which $\theta_c=\frac13$ and $p_c=\frac65$. Notice also that in this case (8.14) means that

$$\rho := \begin{cases}\delta_{3,0}^{3/2} & \text{in Case ND},\\ \delta_0^3+\delta_{3,0}^{3/2} & \text{in Case D},\end{cases}$$

where in Case ND, $\delta_{3,0}\ge\delta_0$. This shows that $\rho\ge\delta_0^3$ in both cases.

Let us first observe that estimate (7.76) shows that we "trivially" have

$$\sum_{\lambda \gtrsim \delta_0^{-3}} \|T_\delta^\lambda\|_{p_c \to p_c'} \lesssim 1$$

since $B = 3$ and $\theta_c = \frac{1}{3}$. In the sequel, we shall therefore always assume that $\lambda \ll \delta_0^{-3}$.

According to our discussion in Section 8.3, if $\lambda\rho \lesssim 1$ (where $\rho = \rho(\tilde{\delta})$), the endpoint $p = p_c$ is still left open.

On the other hand, if $\lambda\rho \gg 1$, we can basically follow our approach from the previous section, with only minor modifications. Let us describe some more details.

Again, we perform the "Airy-cone decomposition" (8.60). In order to estimate $T_{\delta,Ai}^\lambda$, we may here use the scaling $x =: \sigma_\rho u := (\rho^{1/3}u_1, \rho^{1/3}u_2)$, which leads to

$$\widehat{\nu_\delta}(\xi) = \rho^{2/3} e^{-i\lambda s_3 B_0(s,\delta_1)} \int_{|\sigma_\rho u|<\varepsilon} e^{-i\lambda\rho\, s_3 \Phi_1(u,s,\delta)} a(\sigma_\rho u, \delta, s)\, du,$$

where the phase Φ_1 now has the form

$$\begin{aligned} \Phi_1(u,s,\delta) = {}& u_1^3\, B_3(s_2, \delta_1, \rho^{1/3}u_1) - u_1\, \rho^{-2/3} B_1(s,\delta_1) \\ & + u_2^3\, b(\sigma_\rho u, \delta_0^{\mathfrak{r}}, s_2) + \delta_{3,0}' u_2\, \tilde{\alpha}_1(\delta_0^{\mathfrak{r}}, s_2) + \delta_0' u_1 u_2\, \alpha_{1,1}(\rho^{1/3}u_1, \delta_0^{\mathfrak{r}}, s_2) \end{aligned} \tag{8.88}$$

in place of (8.61) and (8.62) and where

$$\delta_0' + \delta_{3,0}' \sim 1.$$

Recall that the coefficient δ_0' does not appear in Case ND, in which $\alpha_{1,1} = 0$. If we again decompose

$$\nu_{\delta,Ai}^\lambda = \nu_{\delta,0}^\lambda + \nu_{\delta,\infty}^\lambda,$$

then we find here that, in place of (8.63) and (8.68), we obtain $\|\widehat{\nu_{\delta,\infty}^\lambda}\|_\infty \lesssim \rho^{2/3}(\lambda\rho)^{-N}$ and $\|\nu_{\delta,\infty}^\lambda\|_\infty \lesssim \rho^{4/3}\lambda^{8/3}(\lambda\rho)^{-N}$, for every $N \in \mathbb{N}$. By interpolation, this leads to

$$\|T_{\delta,\infty}^\lambda\|_{p_c \to p_c'} \lesssim (\lambda\rho)^{8/9-N},$$

which implies the desired estimate,

$$\sum_{\lambda\rho \gg 1} \|T_{\delta,\infty}^\lambda\|_{p_c \to p_c'} \lesssim 1. \tag{8.89}$$

As for the main term $\nu_{\delta,0}^\lambda$, a similar type of discussion that led to (8.69) here yields the following estimate:

$$\|\widehat{\nu_{\delta,0}^\lambda}\|_\infty \lesssim \rho^{2/3}(\lambda\rho)^{-1/2-1/3} = \rho^{2/3}(\lambda\rho)^{-5/6}. \tag{8.90}$$

Indeed, recall that we are assuming that $\lambda\rho \gg 1$ and $\lambda \ll \delta_0^{-3}$, so that $\rho \gg \delta_0^3$. The estimate (8.90) therefore follows from the following analogue of Lemma 8.8 for the case where $B = 3$.

LEMMA 8.10. *Denote by $J(\lambda, \delta, s)$ the oscillatory integral*

$$J(\lambda, \delta, s) = \chi_1(s_1, s_2) \iint e^{-i\lambda\Phi(x,\delta,s)}\, a(x, \delta, s)\, dx_1\, dx_2,$$

with phase

$$\Phi(x, \delta, s) = x_1^3\, B_3(s_2, \delta_1, x_1) - x_1 B_1(s, \delta_1) + \phi^\sharp(x_1, x_2, \delta,\ s_2),$$

where

$$\phi^\sharp(x, \delta,\ s_2) := x_2^3\, b(x, \delta_0^r, s_2) + \delta_{3,0} x_2\, \tilde{\alpha}_1(x_1, \delta_0^{\mathfrak{r}}, s_2) + \delta_0 x_1 x_2\, \alpha_{1,1}(x_1, \delta_0^{\mathfrak{r}}, s_2)$$

and where $\chi_1(s_1, s_2)$ localizes to the region where $s_j \sim 1$, $j = 1, 2$. Also, let

$$\tilde{\rho} := \rho + |B_1(s, \delta_1)|^{3/2}$$

and assume that $\tilde{\rho} \geq M\delta_0^3$. Then the estimate

$$|J(\lambda, \delta, s)| \leq \frac{C}{\tilde{\rho}^{1/6}\lambda^{5/6}} \tag{8.91}$$

holds true, provided the amplitude a is supported in a sufficiently small neighborhood of the origin and $M \geq 1$ is sufficiently large. The constant C in this estimate is independent of δ and s.

Proof. The proof is analogous to the proof of Lemma 8.8. In the more difficult case where $\Lambda := \tilde{\rho}\lambda \gtrsim 1$, after applying the change of coordinates $x = \tilde{\rho}^{1/3} u$, we see that it suffices to prove an estimate of the form $|I(\Lambda, \delta, s)| \leq C\Lambda^{-5/6}$, where the oscillatory integral $I(\Lambda, \delta, s)$ is as in (8.51), only with ϕ in the phase Φ_1 replaced by

$$\begin{aligned}\phi(u, \tilde{\rho}, \delta,\ s_2) :=\, & u_2^3\, b(\tilde{\rho}^{1/3} u, \delta_0^r, s_2) + \delta'_{3,0} u_2\, \tilde{\alpha}_1(\tilde{\rho}^{1/3} u_1, \delta_0^{\mathfrak{r}}, s_2)\\ & + \delta'_0 u_1 u_2\, \alpha_{1,1}(\tilde{\rho}^{1/3} u_1, \delta_0^{\mathfrak{r}}, s_2)\end{aligned}$$

(compare (8.52)). Here,

$$B'_1(s, \delta_1) := \frac{B_1(s, \delta_1)}{\tilde{\rho}^{2/3}},\quad \delta'_0 := \frac{\delta_0}{\tilde{\rho}^{1/3}},\quad \delta'_{3,0} := \frac{\delta_{3,0}}{\tilde{\rho}^{2/3}}.$$

Notice that our assumption implies that $\delta'_0 \ll 1$, and then

$$(\delta'_{3,0})^{3/2} + |B'_1(s, \delta_1)|^{3/2} \sim 1.$$

This shows that the phase Φ_1 will have at worst an Airy-type singularity in at most one of the variables u_1 or u_2. Applying first the method of stationary phase to the integration in one of these variables and subsequently van der Corput's estimates of order 3 to the integration in the second variable, we arrive at the desired estimate for $I(\Lambda, \delta, s)$. Q.E.D.

REMARK 8.11. *Without the assumption $\tilde{\rho} \gg \delta_0^3$, estimate (8.91) may fail. Indeed, in the worst case, it may happen that, after rescaling, all the quantities $B'_1(s, \delta_1)$, δ'_0 and $\delta'_{3,0}$ are of size 1, and a degenerate critical point of order 4 arises for the integration in u_2, after we have applied the method of stationary phase in u_1. In this case, we will obtain only an estimate $|J(\lambda, \delta, s)| \leq C/(\tilde{\rho}^{1/2}\lambda^{1/4})$, and this estimate will be sharp.*

Next, in order to estimate $\nu^\lambda_{\delta,0}(x)$, recall that

$$\nu^\lambda_{\delta,0}(x) = \rho^{2/3+2/3}\lambda^3 \int e^{-i\lambda s_3 \Phi_2(u,z,s_2,\delta)} \chi_0(z)\chi_0(u)a(\sigma_\rho u, \rho^{2/3}z, s, \delta) \times \tilde{\chi}_1(s_2, s_3)\, du\, dz\, ds_2\, ds_3, \tag{8.92}$$

with Φ_2 given by (8.66), and in place of (8.71) we now get

$$\|\nu^\lambda_{\delta,0}\|_\infty \lesssim \rho^{2/3}\lambda^2(\lambda\rho)^{-1/3} = \rho^{-4/3}(\lambda\rho)^{5/3}. \tag{8.93}$$

By means of interpolation, we arrive from (8.90) and (8.90) at a uniform estimate

$$\|T^\lambda_{\delta,0}\|_{p_c \to p'_c} \lesssim 1. \tag{8.94}$$

The corresponding estimate for $p < p_c$ allows for summation over all dyadic $\lambda \gg 1$, but in order also to reach the endpoint $p = p_c$, similar to our discussion in Subsection 5.2.1 for the case $B = 2$, we shall have to apply a complex interpolation argument.

For the estimation of the operators $T^\lambda_{\delta,l}$, very similar statements hold true (compare the analogous discussion in Subsection 8.5.2).

Scaling $x = \sigma_{2^l\rho} u := ((2^l\rho)^{1/3}u_1, (2^l\rho)^{1/3}u_2)$ leads to

$$\widehat{\nu_\delta}(\xi) = (2^l\rho)^{2/3}\, e^{-i\lambda s_3 B_0(s,\delta_1)} \int_{|\sigma_{2^l\rho}u|<\varepsilon} e^{-i\lambda 2^l\rho\, s_3 \Phi_1(u,s,\delta)} a(\sigma_{2^l\rho}u, \delta, s)\, du,$$

where now the phase $\Phi_1 = \Phi_{1,l}$ has the form

$$\Phi_1(u,s,\delta) = u_1^3\, B_3(s_2, \delta_1, (2^l\rho)^{1/3}u_1) - u_1\,(2^l\rho)^{-2/3} B_1(s,\delta_1) + u_2^3\, b(\sigma_{2^l\rho}u, \delta_0^\mathfrak{r}, s_2) + \delta'_{3,0}u_2\,\tilde{\alpha}_1(\delta_0^\mathfrak{r}, s_2) + \delta'_0 u_1 u_2\, \alpha_{1,1}((2^l\rho)^{1/3}u_1, \delta_0^\mathfrak{r}, s_2), \tag{8.95}$$

and where

$$(\delta'_{3,0})^{3/2} + (\delta'_0)^3 = 2^{-l} \le \frac{1}{M_0} \ll 1.$$

Moreover, in analogy with (8.89), one easily verifies that also

$$\sum_{\{l: M_0 \le 2^l \le \rho^{-1}/M_1\}} \sum_{\lambda\rho \gg 1} \|T^\lambda_{l,\infty}\|_{p_c \to p'_c} \lesssim 1. \tag{8.96}$$

As for the main terms $\nu^\lambda_{l,0}$ of

$$\nu^\lambda_{\delta,l} = \nu^\lambda_{l,0} + \nu^\lambda_{l,\infty},$$

which is here given by

$$\widehat{\nu_{l,0}}(\xi) := (2^l\rho)^{2/3} e^{-i\lambda s_3 B_0(s,\delta_1)} \int_{|\sigma_{2^l\rho}u|<\varepsilon} e^{-i\lambda 2^l\rho\, s_3 \Phi_1(u,s,\delta)} a(\sigma_{2^l\rho}u, \delta, s)\chi_0(u)\, du,$$

we find that in place of (8.86) we now have

$$\|\widehat{\nu^\lambda_{l,0}}\|_\infty \lesssim (2^l\rho)^{2/3}(\lambda 2^l\rho)^{-5/6}. \tag{8.97}$$

Moreover, after changing coordinates, we may write

$$\begin{aligned} \nu^{\lambda}_{l,0}(x) = {}& (2^l\rho)^{2/3+2/3}\lambda^3 \\ & \times \int e^{-i\lambda s_3 \Phi_2(u,z,s_2,\delta)} \chi_1(z)\chi_0(u) a(\sigma_{2^l\rho}u, (2^l\rho)^{2/3}z, s, \delta) \\ & \times \tilde{\chi}_1(s_2, s_3)\, du\, dz\, ds_2\, ds_3, \end{aligned} \tag{8.98}$$

with Φ_2 given by

$$\begin{aligned} \Phi_2(u, z, s_2, \delta) := {}& s_2^{n/(n-2)} G_5(s_2, \delta) - x_1 s_2^{(n-1)/(n-2)} G_3(s_2, \delta) \\ & - s_2 x_2 - x_3 + (2^l\rho)^{2/3} z \big(x_1 - s_2^{1/(n-2)} G_1(s_2, \delta)\big) \\ & + 2^l\rho\, \Phi_1(u, z, s_2, \delta), \end{aligned} \tag{8.99}$$

and

$$\begin{aligned} \Phi_1(u, z, s_2, \delta) = {}& u_1^3\, B_3(s_2, \delta_1, (2^l\rho)^{1/3} u_1) - u_1 z + u_2^3\, b(\sigma_{2^l\rho}u, \delta_0^{\mathfrak{r}}, s_2) \\ & + \delta'_{3,0} u_2\, \tilde{\alpha}_1(\delta_0^{\mathfrak{r}}, s_2) + \delta'_0 u_1 u_2\, \alpha_{1,1}((2^l\rho)^{1/3} u_1, \delta_0^{\mathfrak{r}}, s_2). \end{aligned} \tag{8.100}$$

This leads to the estimate

$$\|\nu^{\lambda}_{l,0}\|_\infty \lesssim (2^l\rho)^{2/3}\lambda^2(\lambda 2^l\rho)^{-1/3}.$$

Interpolating between this estimate and (8.97), we obtain again only a uniform estimate

$$\|T^{\lambda}_{l,0}\|_{p_c\to p_c'} \lesssim 1$$

(whereas the corresponding estimate for $p < p_c$ allows for summation over all dyadic $\lambda \gg 1$ and all l.) So, again, we shall have to apply a complex interpolation argument in order to include the endpoint $p = p_c$.

For the operators $T^{\lambda}_{l,\infty}$, we get a better estimate of the form $\|T^{\lambda}_{l,\infty}\|_{p_c\to p_c'} \le C_N(\lambda 2^l\rho)^{-N}$, which allows us to sum absolutely in l and λ.

Recall also that we have seen that we may restrict ourselves to those λ for which $\lambda \ll \delta_0^{-3}$. This assumption has the additional advantage that we shall have to deal only with finite sums in Proposition 8.12. Notice also that $\rho^{-1} \le \delta_0^{-3}$ since we have seen that $\rho \ge \delta_0^3$.

Taking into account these observations, the following proposition puts together those estimates that still need to be established in order to complete the proof of Proposition 7.1, and, hence, also that of our main result, Theorem 1.14.

Proposition 8.12. *Assume that $m = 2$, $B = 3$, and $A = 0$ in (7.89), so that $\theta_c = \frac{1}{3}$, $p_c = \frac{6}{5}$ and $n \ge 7$. Then the following hold true, provided $M \in \mathbb{N}$ is sufficiently large and δ is sufficiently small:*

(a) *If $\lambda\rho \lesssim 1$ and if $\nu^{\lambda}_{\delta,Ai}$ and $\nu^{\lambda}_{\delta,l}$ are given by (8.17) respectively, (8.18), then let*

$$\nu^{I}_{\delta,Ai} := \sum_{2^M \le \lambda \le 2^M\rho^{-1}} \nu^{\lambda}_{\delta,Ai} \quad \text{and} \quad \nu^{II}_{\delta} := \sum_{2^M \le \lambda \le 2^M\rho^{-1}} \ \sum_{\{l: M_0 \le 2^l \le \lambda/M_1\}} \nu^{\lambda}_{\delta,l},$$

and denote by $T^{I}_{\delta,Ai}$ and T^{II}_{δ} the convolution operators $\varphi \mapsto \varphi * \widehat{\nu^{I}_{\delta,Ai}}$ and $\varphi \mapsto \varphi * \widehat{\nu^{II}_{\delta}}$, respectively. Then

$$\|T^{I}_{\delta,Ai}\|_{p_c\to p_c'} \le C \quad \text{and} \quad \|T^{II}_{\delta}\|_{p_c\to p_c'} \le C. \tag{8.101}$$

(b) *If $\lambda\rho \gg 1$ and if $\nu^\lambda_{\delta,0}$ and $\nu^\lambda_{l,0}$ denote the main terms of of $\nu^\lambda_{\delta,Ai}$, respectively, $\nu^\lambda_{\delta,l}$ (cf. (8.92), (8.98)), then let*

$$\nu^{III}_{\delta,Ai} := \sum_{2^M\rho^{-1}<\lambda\le 2^{-M}\delta_0^{-3}} \nu^\lambda_{\delta,0} \quad \text{and} \quad \nu^{IV}_\delta := \sum_{\{l:M_0\le 2^l\le\rho^{-1}/M_1\}} \sum_{2^M\rho^{-1}<\lambda\le 2^{-M}\delta_0^{-3}} \nu^\lambda_{l,0},$$

and denote by $T^{III}_{\delta,Ai}$ and T^{IV}_δ the convolution operators $\varphi \mapsto \varphi * \widehat{\nu^{III}_{\delta,Ai}}$ and $\varphi \mapsto \varphi * \widehat{\nu^{IV}_\delta}$, respectively. Then

$$\|T^{III}_{\delta,Ai}\|_{p_c\to p_c'} \le C \quad \text{and} \quad \|T^{IV}_\delta\|_{p_c\to p_c'} \le C. \tag{8.102}$$

Here, the constant C is independent of δ.

8.7 PROOF OF PROPOSITION 8.12(a): COMPLEX INTERPOLATION

In this section, we assume that $B = 3$ and $A = 0$ in (7.89) and that $\lambda\rho \lesssim 1$.

8.7.1 Estimation of $T^I_{\delta,Ai}$

Recall formula (8.24) for $\widehat{\nu^\lambda_{\delta,Ai}}$. Applying the Fourier inversion formula to this expression and performing the change of variables $z := \lambda^{2/3}B_1(s,\delta)$, we find that we may write

$$\nu^\lambda_{\delta,Ai}(x) = \lambda^{5/3}\int e^{-is_3\lambda\Phi(z,s_2,x,\delta)}\, a(z, s_2, \tilde\delta^\lambda, \delta_0^{\mathfrak{r}}, \lambda^{-1/9}) \times \chi_0(z)\tilde\chi_1(s_2)\chi_1(s_3)\, dz\, ds_2\, ds_3, \tag{8.103}$$

where a is again a smooth function of all its (bounded) variables and where

$$\Phi(z, s_2, x, \delta) := \phi(s_2, x, \delta) + \lambda^{-2/3} z(x_1 - s_2^{1/(n-2)} G_1(s_2, \delta)),$$

with

$$\phi(s_2, x, \delta) := s_2^{n/(n-2)} G_5(s_2, \delta) - s_2^{(n-1)/(n-2)} G_3(s_2, \delta)x_1 - s_2 x_2 - x_3$$

(compare with a similar discussion in Section 5.1, in particular with (5.21)). Recall also that $G_3(0) \ne 0$ and $G_5(0) \ne 0$.

In fact, a priori we have to assume that the density a also depends on the variable s_3. However, arguing as in Subsection 8.5.1, we may develop this function into a convergent series of smooth functions, each of which is a tensor product of a smooth function of the variable s_3 with a smooth function depending only on the remaining variables. Thus, by considering each of the corresponding terms separately, we can reduce the function to situation (8.103), provided we properly choose the functions $\tilde\chi_1$ and χ_1, which localize to the regions where $|s_2| \sim 1$ and $|s_3| \sim 1$.

We next embed $\nu^I_{\delta,Ai}$ into an analytic family of measures

$$\nu^I_{\delta,\zeta} := \gamma(\zeta) \sum_{2^M\le\lambda\le 2^M\rho^{-1}} \lambda^{2(1-3\zeta)/3} \nu^\lambda_{\delta,Ai},$$

where ζ lies again in the complex strip Σ given by $0 \le \operatorname{Re}\zeta \le 1$ and where $\gamma(\zeta) := (1 - 2^{2(1-\zeta)})/(1 - 2^{4/3})$.

Here, summation is again over dyadic $\lambda = 2^j$, $j \in \mathbb{N}$. Observe that indeed $\nu^I_{\delta, Ai} = \nu^I_{\delta,\theta_c}$, since $\theta_c = \frac{1}{3}$.

Since the supports of the $\widehat{\nu^\lambda_{\delta,\,Ai}}$ are almost disjoint, (8.19) implies that

$$\|\widehat{\nu^I_{\delta,\,it}}\|_\infty \lesssim 1 \qquad \forall t \in \mathbb{R}.$$

We shall also prove that

$$|\nu^I_{\delta,\,1+it}(x)| \le C \qquad \forall t \in \mathbb{R}, x \in \mathbb{R}^3. \tag{8.104}$$

Again, Stein's interpolation theorem will then imply that the operator $T^I_{\delta,Ai}$ is bounded from L^{p_c} to $L^{p'_c}$, which will complete the proof of the first part of Proposition 8.12(a).

In order to prove (8.104), we first consider the case where $|x| \gg 1$. In this case, we can use a formula similar to (5.17) in order to argue as in Section 5.1 and find that

$$|\nu^\lambda_{\delta,Ai}(x)| \le C_N \lambda^{-N}, \quad N \in \mathbb{N}, \text{ if } |x| \gg 1.$$

This estimate allows us, in a trivial way, to sum in λ and thus to obtain (8.104).

We may therefore assume from now on that $|x| \lesssim 1$. We then write

$$\nu^I_{\delta,1+it}(x) = \gamma(1+it) \sum_{2^M \le \lambda \le 2^M \rho^{-1}} \lambda^{-2it} \mu_\lambda(x), \tag{8.105}$$

where

$$\mu_\lambda(x) := \lambda^{1/3} \int e^{-is_3\lambda\Phi(z,\,s_2,\,x,\,\delta)}\, a(z, s_2, \tilde{\delta}^\lambda, \delta^\mathfrak{r}_0, \lambda^{-1/9}) \chi_0(z) \tilde{\chi}_1(s_2) \chi_1(s_3)\, dz\, ds_2\, ds_3.$$

Let us look at the contribution to this integral given by a small neighborhood of a given point $s_2^0 \sim 1$. If $|\partial^2_{s_2}\phi(s_2^0, x, \delta)| \sim 1$, then van der Corput's estimate applied to the integration in s_2 shows that $|\mu_\lambda(x)| \lesssim \lambda^{1/3}\lambda^{-1/2} = \lambda^{-1/6}$, which clearly implies (8.104). This situation arises in particular when $|x_1| \ll 1$ or when $|x_1| \sim 1$ and G_5 and $x_1\, G_3$ have opposite signs.

So, let us assume from now on that $|x_1| \sim 1$ and that G_5 and $x_1 G_3$ have the same sign. Notice that if $\delta = 0$, then

$$\phi(s_2,\, x,\, 0) = s_2^{n/(n-2)} G_5(0) - s_2^{(n-1)/(n-2)} G_3(0) x_1 - s_2 x_2 - x_3.$$

Since the exponents $n/(n-1)$ and $(n-1)/(n-2)$ are different, this shows that there is a unique $s_2^c(0) \sim 1$ so that $\partial^2_{s_2}\phi(s_2^c(0), x, 0) = 0$, whereas $|\partial^3_{s_2}\phi(s_2^c(0), x, 0)| \sim 1$. By the implicit function theorem, we then find a smooth function $s_2^c(x_1, \delta)$ such that

$$\partial^2_{s_2}\phi(s_2^c(x_1, \delta), x, \delta) \equiv 0.$$

We let

$$\phi^\sharp(v, x, \delta) := \phi(s_2^c(x_1, \delta) + v, x, \delta), \quad \Phi^\sharp(z, v, x, \delta) := \Phi(z, s_2^c(x_1, \delta) + v, x, \delta).$$

By means of Taylor expansion around $v = 0$, we may write

$$\phi^{\daleth}(v, x, \delta) = v^3 Q_3(v, x, \delta) - vQ_1(x, \delta) + Q_0(x, \delta),$$

where the Q_j are smooth functions of all their variables and where we may assume that $|Q_3(v, x, \delta)| \sim 1$, since we had $|\partial_{s_2}^3 \phi(s_2^c(0), x, 0)| \sim 1$. Moreover, developing

$$x_1 - (s_2^c(x_1, \delta) + v)^{1/(n_2-2)} G_1(s_2^c(x_1, \delta) + v, \ \delta_1) = Q_5(x, \delta) + vQ_6(v, x, \delta),$$

after scaling $v \mapsto \lambda^{-1/3} v$, we find that

$$\begin{aligned}\lambda \Phi^{\daleth}(z, \lambda^{-1/3} v, x, \delta) = {}& v^3 Q_3(\lambda^{-1/3} v, x, \delta) - v\lambda^{2/3} Q_1(x, \delta) + \lambda Q_0(x, \delta) \\ &+ zv\, Q_6(\lambda^{-1/3} v, x, \delta) + \lambda^{1/3} z\, Q_5(x, \delta).\end{aligned}$$

This allows to rewrite

$$\mu_\lambda(x) = \int \int_{-\lambda^{1/3}}^{\lambda^{1/3}} \chi_0(z) F_\delta(\lambda, x, z, v)\, dv\, dz,$$

where F_δ is of the form

$$\begin{aligned}F_\delta(\lambda, x, z, v) = {}& \chi_0(\lambda^{-1/3} v)\, \widehat{\chi_1}\, \big(A + Dz - Bv - v^3 Q_3(\lambda^{-1/3} v, x, \delta) \\ &+ \ zv\, Q_6(\lambda^{-1/3} v, x, \delta)\big)\, \tilde{a}(z, \lambda^{-1/3} v, \ \tilde{\delta}^\lambda, \ x_1, \ \delta_0^{\mathfrak{r}}, \lambda^{-1/9}).\end{aligned}$$

Here, $\tilde{a}$ is again a smooth function of all its variables with compact support, and

$$\begin{aligned}&A = A(x, \lambda, \delta) := \lambda Q_0(x, \delta), \qquad B := B(x, \lambda, \delta) := \lambda^{2/3} Q_1(x, \delta), \\ &D = D(x, \lambda, \delta) := \lambda^{1/3} Q_5(x, \delta).\end{aligned}$$

Now we can argue in a very similar way is in previous proofs and will, therefore, only briefly sketch the arguments.

We first consider the contribution to the sum in (8.105) given by the λs satisfying $|A| = \lambda |Q_0(x, \delta)| \gg 1$. Observe that for z fixed, we may estimate $\int_{-\lambda^{1/3}}^{\lambda^{1/3}} |F_\delta(\lambda, x, z, v)|\, dv$ by means of Lemma 2.4, where we choose $y_2 = v$, $T := \lambda^{1/3}$, $\epsilon = 0$, $r_i \equiv 0$ (so that the integral in y_1 just yields a positive constant), and $Q(y_2) = zQ_6(y_2, x, \delta)$. The condition (2.1) is here also satisfied, since $(\phi^{\daleth})''(0, x, \delta) = 0$ and $|(\phi^{\daleth})'''(0, x, \delta)| \sim 1$. Thus, Lemma 2.4 implies that for L sufficiently large, we may estimate

$$|\mu_\lambda(x)| \leq C\Big(\int_I |A + Dz|^{-1/6}\, |\chi_0(z)|\, dz + |J|\Big),$$

where I and J denote the sets of all $z \in \operatorname{supp} \chi_0$ for which $|A + Dz| \geq L$, respectively, $|A + Dz| < L$. The integral can easily be estimated by

$$|D|^{-1/6} \int_I \left|z + \frac{A}{D}\right|^{-1/6} |\chi_0(z)|\, dz \lesssim |D|^{-1/6} \left(1 + \left|\tfrac{A}{D}\right|\right)^{-1/6} \lesssim |A|^{-1/6}.$$

Moreover, if $|D| \ll L$, then the set J is empty, if we assume that $|A| \geq 2L$. So, let us assume that $|D| \gtrsim L$. Then, on the set J we have $|z + A/D| < L/|D|$, and since $|z| \lesssim 1$, this implies that $|A/D| \lesssim 1$, and we see that $|J| \leq L/|D| \lesssim L/|A|$. Putting all this together, we find that

$$|\mu_\lambda(x)| \leq C|A|^{-1/6},$$

which allows us to sum over those λ for which $\lambda|Q_0(x,\delta)| \gg 1$ in such a way that the resulting estimates are independent of x.

Next, we consider the λs for which $|A| \lesssim 1$. If, in addition, $|D| \gg 1$, then we can argue as before and obtain an estimate of the form $|\mu_\lambda(x)| \le C|D|^{-1/6}$, which again allows us to sum. Similarly, if $|A| \lesssim 1$ and $|D| \lesssim 1$ but $|B| \gg 1$, we may apply Lemma 2.4 once more and obtain that $|\mu_\lambda(x)| \le C|B|^{-1/4}$. This allows us again to sum.

We are thus left with the oscillatory sum (8.105) over only those λs for which, say, $\max\{|A|, |B|, |D|\} \le L$. However, this can again easily be handled by means of Lemma 2.7, and we arrive at (8.104). Recall here that by (8.15) the components of $\tilde{\delta}^\lambda$ are of the form $\lambda^{\beta_i}\delta_i$, where we may assume that $|\lambda^{\beta_i}\delta_i| \le C$, since we are assuming that $\rho(\tilde{\delta}^\lambda) = \lambda\rho(\tilde{\delta}) \lesssim 1$.

8.7.2 Estimation of T_δ^{II}

We next come to the proof of the second estimate in Proposition 8.12(a), where we still assume that $\lambda\rho \lesssim 1$. Recall from Section 8.3 that we have decomposed

$$\nu_{\delta,l}^\lambda = \nu_{l,\infty}^\lambda + \nu_{l,0,0}^\lambda + \nu_{l,I}^\lambda + \nu_{l,II}^\lambda + \nu_{l,III}^\lambda. \qquad (8.106)$$

Here, we denote by $\nu_{l,0,0}^\lambda$ the contribution to $\nu_{l,0}^\lambda$ by the domain where $|u_1| \ll 1$.

Again, we embed the measure ν_δ^{II} into an analytic family of measures

$$\nu_{\delta,\zeta}^{II} := \gamma(\zeta) \sum_{2^M \le \lambda \le 2^M\rho^{-1}} \sum_{\{l: M_0 \le 2^l \le \lambda/M_1\}} 2^{l(1-3\zeta)/6}\lambda^{2(1-3\zeta)/3}\, \nu_{\delta,l}^\lambda,$$

where ζ again lies in the complex strip Σ. The analytic function $\gamma(\zeta)$ will be a finite product of factors $\gamma_i(\zeta)$, which will be specified in the course of the proof.

In view of (8.21), by following our standard approach it will suffice to prove the following estimate in order to establish the second inequality in (8.101):

$$|\nu_{\delta,1+it}^{II}(x)| \le C \qquad \forall t \in \mathbb{R}, x \in \mathbb{R}^3. \qquad (8.107)$$

By putting $\mu_{l,\lambda} := 2^{-l/3}\lambda^{-4/3}\nu_{\delta,l}^\lambda$, we may write

$$\nu_{\delta,1+it}^{II}(x) = \gamma(1+it) \sum_{2^M \le \lambda \le 2^M\rho^{-1}} \sum_{\{l: M_0 \le 2^l \le \lambda/M_1\}} 2^{-itl/2}\lambda^{-2it}\, \mu_{l,\lambda}(x).$$

8.7.3 Contribution by the $\nu_{l,II}^\lambda$

Let us begin with the contribution of the main terms $\nu_{l,II}^\lambda$ in (8.106), that is, let us look at

$$\nu_{II,1+it} := \gamma(1+it) \sum_{2^M \le \lambda \le 2^M\rho^{-1}} \sum_{\{l: M_0 \le 2^l \le \lambda/M_1\}} 2^{-itl/2}\lambda^{-2it}\, \mu_{l,\lambda,II}(x), \qquad (8.108)$$

where we have set $\mu_{l,\lambda,II} := 2^{-l/3}\lambda^{-4/3}\,\nu_{l,II}^\lambda$. As a major step in the proof of (8.107), we want to prove that

$$|\nu_{II,1+it}(x)| \le C \qquad \forall t \in \mathbb{R}, x \in \mathbb{R}^3. \qquad (8.109)$$

Recall from Section 8.3 that we may restrict ourselves to those x for which $|x| \lesssim 1$ and $|x_1| \sim 1$. Making use of (8.43) and (8.44), after scaling the variable u_2 by the factor $2^{-l/3}$, we find that

$$\begin{aligned}\mu_{l,\lambda,II}(x) &= \int e^{-is_3\tilde{\Psi}_3(y_2,v,x,\delta,\lambda,l)} a_3\big(\lambda^{-1/3} y_2, v, x, 2^{-l/3}, (2^l\lambda^{-1})^{1/3}, \delta\big) \\ &\quad \times \chi_1(s_3)\, \chi_1(v)\, \chi_0(2^{-l/3} y_2)\, dy_2\, dv\, ds_3 \\ &= \int \widehat{\chi_1}\big(\tilde{\Psi}_3(y_2, v, x, \delta, \lambda, l)\big) a_3\big(\lambda^{-1/3} y_2, v, x, 2^{-l/3}, (2^l\lambda^{-1})^{1/3}, \delta\big) \\ &\quad \times \chi_1(v)\, \chi_0(2^{-l/3} y_2)\, dy_2\, dv, \end{aligned} \tag{8.110}$$

with

$$\begin{aligned}\tilde{\Psi}_3(y_2, v, x, \delta, \lambda, l) &:= v\,\lambda(2^{-l}\lambda)^{-1/3}(x_1^2\omega(\delta_1 x_1) - x_2) + \lambda Q_A(x, \delta) \\ &\quad + y_2^3\, b(x_1, \lambda^{-1/3} y_2, \delta) \\ &\quad + y_2\,(\lambda^{2/3}(\delta_0 \tilde{G}_1(x_1, \delta) + \delta_3 x_1^{n_1} \alpha_1(\delta_1 x_1)) + (2^l\lambda)^{1/3}\delta_0\, v).\end{aligned}$$

We shall write this as

$$\tilde{\Psi}_3(y_2, v, x, \delta, \lambda, l) = A + Bv + y_2^3\, b(x_1, \lambda^{-1/3} y_2, \delta) + y_2(D + Ev), \tag{8.111}$$

with

$$\begin{aligned} A &= A(x, \lambda, \delta) := \lambda Q_A(x, \delta), & B &= B(x, \lambda, l, \delta) := 2^{l/3}\lambda^{2/3} Q_B(x, \delta), \\ D &= D(x, \lambda, \delta) := \lambda^{2/3} Q_D(x, \delta), & E &= E(\lambda, l, \delta) = 2^{l/3}\lambda^{1/3}\delta_0, \end{aligned} \tag{8.112}$$

and

$$\begin{aligned} Q_A(x, \delta) &:= \tilde{G}_1(x_1, \delta)(x_1^2\omega(\delta_1 x_1) - x_2) + x_1^n \alpha(\delta_1 x_1) - x_3, \\ Q_B(x, \delta) &:= x_1^2 \omega(\delta_1 x_1) - x_2, \quad Q_D(x, \delta) := \delta_0 \tilde{G}_1(x_1, \delta) + \delta_3 x_1^{n_1} \alpha_1(\delta_1 x_1). \end{aligned}$$

Now, applying Lemma 2.5 to the integration in y_2, with $T := 2^{l/3}$ and $\delta := \lambda^{-1/3}$ (so that $\delta T = (2^l\lambda^{-1})^{1/3} \ll 1$), we see that we may estimate

$$\begin{aligned} &\Big|\int \widehat{\chi_1}\big(\tilde{\Psi}_3(y_2, v, x, \delta, \lambda, l)\big) a_3\big(\lambda^{-1/3} y_2, v, x, 2^{-l/3}, (2^l\lambda^{-1})^{1/3}, \delta\big)\, \chi_0(2^{-l/3} y_2)\, dy_2\Big| \\ &\qquad \lesssim \big(1 + \max\{|A + Bv|, |D + Ev|\}\big)^{-1/6}, \end{aligned} \tag{8.113}$$

with a constant that does not depend on A, B, D, E, and v. The following simple lemma, which is essentially a special case of a more general result on integrals of sublevel type for polynomial functions by E. M. Stein and F. Ricci [RS87], will thus be useful.

Lemma 8.13. *Let $0 < \varepsilon < 1$ and consider for $A, B, D, E \in \mathbb{R}$ the integral*

$$J(A, B, D, E) := \int \big(1 + \max\{|A + Bv|, |D + Ev|\}\big)^{-\varepsilon} \chi_0(v)\, dv,$$

where, as usual, χ_0 is a smooth, nonnegative bump function with compact support. Then

$$|J(A, B, D, E)| \le C\big(\max\{|A|, |B|, |D|, |E|\}\big)^{-\varepsilon},$$

where the constant C is independent of A, B, D, and E.

Proof. If $|A| \gg |B|$, then $|A + Bv| \gtrsim |A|$, and we see that $|J(A, B, D, E)| \leq C|A|^{-\varepsilon}$. Next, if $|A| \lesssim |B|$ then we apply the change of variables $v \mapsto w := Bv$ and find that we can estimate

$$|J(A, B, D, E)| \lesssim \frac{1}{|B|} \int_{|w| \lesssim |B|} (1 + |A + w|)^{-\varepsilon}\, dw$$

$$\lesssim \frac{1}{|B|} \int_{|y| \lesssim |B|} (1 + |y|)^{-\varepsilon}\, dw \lesssim |B|^{-\varepsilon},$$

provided $|B| \geq 1$. Of course, if $|B| < 1$, then we can always use the trivial estimate $|J(A, B, D, E)| \lesssim 1$.

We may now conclude the proof by interchanging the roles of A, B and D, E.

Q.E.D.

In combination with (8.113) and (8.110) this lemma implies that

$$|\mu_{l,\lambda,II}(x)| \lesssim \left(\max\{|A|, |B|, |D|, |E|\}\right)^{-1/6}, \tag{8.114}$$

with A, B, D and E given by (8.112).

In order to estimate $\nu_{II,1+it}(x)$, we shall again distinguish various cases in a similar way as in preceding arguments of this type and shall thus only briefly sketch the ideas.

Let us first consider the contribution by those terms in (8.108) for which $|D| \gtrsim 1$ and $|E| \gtrsim 1$. Since by (8.114) we may estimate $|\mu_{l,\lambda,II}(x)| \lesssim |D|^{-1/12}|E|^{-1/12}$, we can first sum these contributions absolutely over all l for which $|E| = 2^{l/3}\lambda^{1/3}\delta_0 \gtrsim 1$, and subsequently over all dyadic $\lambda = 2^j$ for which $|D| = \lambda^{2/3}|Q_D(x, \delta)| \gtrsim 1$, and arrive at a bound that is uniform in x and δ.

In essentially the same way, we can sum (absolutely) the contributions by those terms in (8.108) for which $|A| \gtrsim 1$ and $|B| \gtrsim 1$.

Consider next the terms for which $|E| \ll 1$ and $|D| \ll 1$. For these terms, we rewrite

$$\mu_{l,\lambda,II}(x) = \int e^{-is_3(A+Bv)}\, J(v, s_3)\, \chi_1(s_3)\, \chi_1(v)\, dv\, ds_3, \tag{8.115}$$

with

$$J(v, s_3) := \int e^{-is_3(y_2^3 b(x_1, \lambda^{-1/3} y_2, \delta) + y_2(D+Ev))} a_3\big(\lambda^{-1/3} y_2, v, x, 2^{-l/3}, (2^l\lambda^{-1})^{1/3}, \delta\big)$$
$$\times\, \chi_0(2^{-l/3} y_2)\, dy_2. \tag{8.116}$$

But, since $|D + Ev| \lesssim 1$, the proof of Lemma 2.2(a) shows that $J(v, s_3) = g(D + Ev, s_3, x, 2^{-l/3}, (2^l\lambda^{-1})^{1/3}, \delta)$, with a smooth function g, and thus

$$\mu_{l,\lambda,II}(x) = \int e^{-is_3(A+Bv)} g(D + Ev, s_3, x, 2^{-l/3}, (2^l\lambda^{-1})^{1/3}, \delta)\, \chi_1(s_3)\, \chi_1(v)\, dv\, ds_3.$$

Arguing in a similar way as in Subsection 8.7.1, without loss of generality we may and shall assume that $g(D+Ev, s_3, x, 2^{-l/3}, (2^l\lambda^{-1})^{1/3}, \delta) = g(D+Ev, x, 2^{-l/3}, (2^l\lambda^{-1})^{1/3}, \delta)$ is independent of s_3. Then we may write

$$\mu_{l,\lambda,II}(x) = \int g(D+Ev, x, 2^{-l/3}, (2^l\lambda^{-1})^{1/3}, \delta)\, \widehat{\chi_1}(A+Bv)\, \chi_1(v)\, dv \tag{8.117}$$

(alternatively, one could also use integrations by parts in s_3 in the previous formula, but the other approach appears a bit clearer).

Recall that we assume that either $|A| \gtrsim 1$ and $|B| \ll 1$, or $|A| \ll 1$ and $|B| \gtrsim 1$, or $|A| \ll 1$ and $|B| \ll 1$.

If $|A| \gtrsim 1$ and $|B| \ll 1$, then we can treat the summation in l by means of Lemma 2.7, where we choose, for λ fixed,

$$H_{\lambda,x}(u_1, u_2, u_3, u_4) := \int g(D+u_1 v, x, u_3, u_4, \delta)\, \widehat{\chi_1}(A+u_2 v)\, \chi_1(v)\, dv.$$

Then clearly $\|H_{\lambda,x}\|_{C^1(Q)} \lesssim |A|^{-1}$, and so after summation in those l for which $|E| \ll 1$ and $|B| \ll 1$, we can also sum (absolutely) in the λs for which $|A| \gtrsim 1$. Observe that this requires that $\gamma(\zeta)$ contains a factor

$$\gamma_1(\zeta) := \frac{2^{(1-\zeta)/2} - 1}{2^{1/3} - 1}.$$

Consider next the case where $|B| \gtrsim 1$ and $|A| \ll 1$. If we write $\lambda = 2^j$, then $2^{l/3}\lambda^{2/3} = 2^{k/3}$, where we put $k := l + 2j$. We therefore pass from the summation variables j and l to the variables j and k, which allows us to write $B = 2^{k/3} Q_B(x)$. For k fixed, we then sum first in j by means of Lemma 2.7, which gives an estimate of order $O(|B|^{-1})$; in return, this allows us to sum (absolutely) in those k for which $|B| = 2^{k/3}|Q_B(x)| \gtrsim 1$. Since $-l/2 - 2j = -k/2 - j$, the application of Lemma 2.7 requires in this case that $\gamma(\zeta)$ contains a factor

$$\gamma_2(\zeta) := \frac{2^{1-\zeta} - 1}{2^{2/3} - 1}.$$

There remains the case where $|A| + |B| \ll 1$ and $|D| + |E| \lesssim 1$. The summation over all ls and λs for which these conditions are satisfied can easily be treated by means of the double summation Lemma 2.9, in a very similar way as in the last part of the proof of Proposition 4.2(a). The corresponding vector (α_1, α_2) to be used in Lemma 8.1 will here be given by $(\alpha_1, \alpha_2) = (2, \frac{1}{2})$, and the vectors (β_1^k, β_2^k), by $(1, 0), (\frac{2}{3}, \frac{1}{3}), (\frac{2}{3}, 0), (\frac{1}{3}, \frac{1}{3}), (-1, 1)$ and $(0, -\frac{1}{3})$; they obviously satisfy the assumptions of Lemma 2.9. For the application of this lemma, we need to assume that $\gamma(\zeta)$ contains also a factor $\gamma_3(\zeta)$ given by Remark 2.10.

What remains are the contributions by those l and λ for which either $|D| \gtrsim 1$ and $|E| \ll 1$ or $|D| \ll 1$ and $|E| \gtrsim 1$.

We begin with the case where $|E| \gtrsim 1$ and $|D| \ll 1$. Then we may assume in addition that $|B| \ll 1$, for otherwise by (8.114) we have $|\mu_{l,\lambda,II}(x)| \lesssim |E|^{-1/12}|B|^{-1/12}$, which allows us to sum absolutely in j and l, as can easily be seen by means of a change of the summation variables from j and l to $k := 2j + l$

and $m := j + l$ (compare (8.112)). In a very similar way, we may also assume that $|A| \ll 1$.

Recall from (8.110) and (8.111) that

$$\mu_{l,\lambda,II}(x) = \int e^{-is_3(A+y_2^3 b(x_1,\lambda^{-1/3}y_2,\delta)+Dy_2+v(B+Ey_2))} \times a_3\left(\lambda^{-1/3}y_2, v, x, 2^{-l/3}, (2^l\lambda^{-1})^{1/3}, \delta\right) \times \chi_1(s_3)\,\chi_1(v)\,\chi_0(2^{-l/3}y_2)\,dy_2\,dv\,ds_3.$$

Again, by our usual argument, we may assume without loss of generality that a_3 is independent of v. Then we find that

$$\mu_{l,\lambda,II}(x) = \int e^{-is_3\left(A+y_2^3 b(x_1,\lambda^{-1/3}y_2,\delta)\right)}\,\widehat{\chi_1}\,(s_3(B+Ey_2)) \times a_3\left(\lambda^{-1/3}y_2, x, 2^{-l/3}, (2^l\lambda^{-1})^{1/3}, \delta\right) \times \chi_1(s_3)\,\chi_1(v)\,\chi_0(2^{-l/3}y_2)\,dy_2\,ds_3. \tag{8.118}$$

We then change the summation variables from j, l to j, k, where $k := j + l$, so that $E = 2^{k/3}\delta_0$. Then, for k fixed, we can treat the summation in j by means of Lemma 2.7, where we choose

$$H_{k,x}(u_1, u_2, u_3, u_4, u_5) := \int e^{-is_3\left(u_1+y_2^3 b(x_1,u_2y_2,\delta)\right)}\,\widehat{\chi_1}\,(s_3(u_3 + Ey_2)) \times a_3(u_2y_2, x, u_4, u_5, \delta)\chi_1(s_3)\,\chi_1(v)\,\chi_0(u_3y_2)\,dy_2\,ds_3.$$

By means of the change of variables $y_2 \mapsto y_2/E$, we thus see that $|H_{k,x}(u_1, \dots, u_5)| \lesssim |E|^{-1}$ on the natural cuboid Q arising in this context, since $\widehat{\chi_1}$ is a Schwartz function. Next, observe that by the product rule, $\partial_{u_i} H_{k,x}(u_1, \dots, u_5)$ can be written as a finite sum of integrals of a similar form, where the amplitude may carry additional factors of the form y_2^n, with $n = 0, \dots, 4$. Again, the change of variables $y_2 \mapsto y_2/E$ shows that these can be estimated by $C|E|^{-1}$ (or even higher powers of $|E|^{-1}$). We thus find that $\|H_{k,x}\|_{C^1(Q)} \lesssim |E|^{-1}$, and after the summation over the $\lambda = 2^j$, this allows us to subsequently also sum over the k for which $|E| = 2^{k/3}\delta_0 \gtrsim 1$.

There remains the contribution by those l and λ for which $|D| \gtrsim 1$ and $|E| \ll 1$. Observe that here we have $|D + Ev| \gtrsim 1$ in (8.116).

Applying the change of variables $y_2 = \lambda^{1/3}t$ in the integral defining $J(v, s_3)$, we obtain

$$J(v, s_3) = \lambda^{1/3} \int e^{-is_3\lambda\left(t^3 b(x_1,t,\delta)+t(\lambda^{-2/3}(D+Ev))\right)} \times a_3\left(t, v, x, 2^{-l/3}, (2^l\lambda^{-1})^{1/3}, \delta\right)\,\chi_0((\lambda 2^{-l})^{1/3}t)\,dt. \tag{8.119}$$

It is important to observe that here the phase function is independent of l. Notice also that by (8.112)

$$\lambda^{-2/3}D = Q_D(x, \delta), \quad \lambda^{-2/3}E = (2^l\lambda^{-1})^{1/3}\delta_0 \ll 1,$$

so that, in particular, $\lambda^{-2/3}|D + vE| \lesssim 1$.

We may then argue in a similar way as in the proof of Lemma 2.2(b) to see that for every $N \in \mathbb{N}$,

$$\begin{aligned} J(v, s_3) = & |D + Ev|^{-1/4} a_+ \left(\lambda^{-1/3}|D + Ev|^{1/2}, v, x, 2^{-l/3}, (2^l\lambda^{-1})^{1/3}, \delta\right) \\ & \times \chi_0 \left(2^{-l/3}|D + Ev|^{1/2}\right) e^{-is_3|D+Ev|^{3/2} q_+(\lambda^{-1/3}|D+Ev|^{1/2}, x, \delta)} \\ & + |D + Ev|^{-1/4} a_- \left(\lambda^{-1/3}|D + Ev|^{1/2}, v, x, 2^{-l/3}, (2^l\lambda^{-1})^{1/3}, \delta\right) \\ & \times \chi_0 \left(2^{-l/3}|D + Ev|^{1/2}\right) e^{-is_3|D+Ev|^{3/2} q_-(\lambda^{-1/3}|D+Ev|^{1/2}, x, \delta)} \\ & + (D + Ev)^{-N} \\ & \times F_N \left(|D + vE|^{3/2}, \lambda^{-1/3}|D + Ev|^{1/2}, v, x, 2^{-l/3}, (2^l\lambda^{-1})^{1/3}, \delta\right), \end{aligned} \tag{8.120}$$

where $a_\pm$, $q_\pm$, and F_N are smooth functions of their (bounded) variables. Moreover, $|q_\pm(0, x, (2^l\lambda^{-1})^{1/3}, \delta)| \sim 1$.

Indeed, notice that the phase in (8.119) has a critical point in the support of the amplitude only if $2^{-l/3}|D + vE|^{1/2} \lesssim 1$, and so we obtain the first two terms in (8.120) by applying the method of stationary phase, whereas the last term arises from integrations by parts on intervals in t on which there is no stationary point.

We shall concentrate on the first term only. The second term can be treated in the same way as the first one, and the last term can be handled in an even easier way by a similar method, since it is of order $= O(|D|^{-N})$ and, unlike the first term, carries no oscillatory factor.

We denote by

$$\begin{aligned} \mu^1_{l,\lambda}(x) := \int & e^{-is_3\left[(A+Bv)+|D+Ev|^{3/2} q_+(\lambda^{-1/3}|D+Ev|^{1/2}, x, \delta)\right]} |D + Ev|^{-1/4} \chi_1(s_3)\, \chi_1(v) \\ & \times a_+ \left(\lambda^{-1/3}|D + Ev|^{1/2}, v, x, 2^{-l/3}, (2^l\lambda^{-1})^{1/3}, \delta\right) \\ & \times \chi_0 \left(2^{-l/3}|D + Ev|^{1/2}\right) dv\, ds_3 \end{aligned}$$

the contribution by the first term in (8.120) to $\mu_{l,\lambda,II}(x)$ and by $\nu^1_{II,1+it}(x)$ the contribution of the $\mu^1_{l,\lambda}(x)$ to the sum defining $\nu_{II,1+it}(x)$.

Assuming for instance that $D > 0$, and keeping in mind that, according to (8.112), $\lambda^{-1/3}D^{1/2} = Q_D(x, \delta)^{1/2}$ depends only on x and δ, a Taylor expansion then shows that

$$|D + Ev|^{3/2} = D^{3/2} + \tfrac{3}{2} D^{1/2} Ev + D^{-1/2} E^2 r_1(D^{-1}E, v),$$

$$\begin{aligned} q_+\left(\lambda^{-1/3}|D + Ev|^{1/2}, x, \delta\right) = & q_+\left(\lambda^{-1/3} D^{1/2}, x, \delta\right) \\ & + \tfrac{1}{2} q_+'(\lambda^{-1/3} D^{1/2}, x, \delta) \lambda^{-1/3} D^{-1/2} Ev \\ & + \lambda^{-1/3} D^{1/2} (D^{-1}E)^2 r_2(\lambda^{-1/3} D^{1/2}, D^{-1}E, v, x, \delta), \end{aligned}$$

where q'_+ denotes the partial derivative of q_+ with respect to the first variable and where r_1 and r_2 are smooth, real-valued functions. This implies that

$$\begin{aligned}|D+Ev|^{3/2}q_+(\lambda^{-1/3}|D+Ev|^{1/2},x,\delta) &= D^{3/2}q_+(\lambda^{-1/3}D^{1/2},x,\delta)\\ &+v\left[\tfrac{1}{2}q'_+(\lambda^{-1/3}D^{1/2},x,\delta)\lambda^{-1/3}DE+\tfrac{3}{2}q'_+(\lambda^{-1/3}D^{1/2},x,\delta)D^{1/2}E\right]\\ &+r(\lambda^{-1/3}D^{1/2},D^{-1}E,v,x,\delta),\end{aligned} \tag{8.121}$$

again with a smooth, real-valued function r. In combination with (8.112), we then find that the complete phase in the oscillatory integral defining $\mu^1_{l,\lambda}(x)$ is of the form

$$s_3[A'+B'v+r],$$

with $A'=\lambda Q_{A'}(x,\delta)$, $B'=2^{l/3}\lambda^{2/3}Q_{B'}(x,\delta)$, where

$$\begin{aligned}Q_{A'}(x,\delta) &:= Q_A(x,\delta)+Q_D(x,\delta)^{3/2}q_+\big(Q_D(x,\delta)^{1/2},x,\delta\big),\\ Q_{B'}(x,\delta) &:= Q_B(x,\delta)+\tfrac{1}{2}q'_+\big(Q_D(x,\delta)^{1/2},x,\delta\big)Q_D(x,\delta)\delta_0\\ &\quad+\tfrac{3}{2}q'_+\big(Q_D(x,\delta)^{1/2},x,\delta\big)Q_D(x,\delta)^{1/2}\delta_0.\end{aligned}$$

Thus, if $|B'|\gg 1$, an integration by parts in v shows that

$$|\mu^1_{l,\lambda}(x)|\lesssim |D|^{-1/4}|B'|^{-1}.$$

This estimate allows to control the sum over all l such that $|B'|\gg 1$ and, subsequently, the sum over all dyadic λ such that $|D|\gtrsim 1$, and we arrive at the desired uniform estimate in x and δ.

Next, if $|B'|\lesssim 1$, then we can argue in a similar way as before and apply Lemma 2.7 to the summation in l by letting

$$\begin{aligned}H_{\lambda,x}(u_1,\dots,u_7) := \int & e^{-is_3[A'+u_1v+r(Q_D(x)^{1/2},D^{-1}u_3,v,x,\delta)]}|D+u_2v|^{-1/4}\chi_1(s_3)\,\chi_1(v)\\ &\times a_+\big(|Q_D(x,\delta)+u_3v|^{1/2},v,u_4,u_5,\delta\big)\\ &\times \chi_0(|u_6+u_7v|^{1/2})\,dv\,ds_3,\end{aligned}$$

and choosing the cuboid Q in the obvious way. Then we easily see that $\|H_{\lambda,x}\|_{C^1(Q)}\lesssim |D|^{-1/4}$, and so after summation in those l for which $|B'|\ll 1$, we can also sum (absolutely) in the λs for which $|D|\gtrsim 1$. Observe that this requires again that $\gamma(\zeta)$ contains the factor $\gamma_1(\zeta)$.

This concludes the proof of the uniform estimate of $\nu^1_{II,1+it}(x)$ in x and δ and thus also of estimate (8.109).

8.7.4 Contribution by the $\nu^\lambda_{l,I}$

Let us next consider the contribution of the terms $\nu^\lambda_{l,I}$ in (8.106), that is, let us look at

$$\nu_{I,1+it} := \gamma(1+it)\sum_{2^M\le\lambda\le 2^M\rho^{-1}}\ \sum_{\{l:M_0\le 2^l\le\lambda/M_1\}} 2^{-itl/2}\lambda^{-2it}\,\mu_{l,\lambda,I}(x),$$

where we have set $\mu_{l,\lambda,I} := 2^{-l/3}\lambda^{-4/3}\,\nu^{\lambda}_{l,I}$. We want to prove that

$$|\nu_{I,\,1+it}(x)| \le C \qquad \forall t \in \mathbb{R}, x \in \mathbb{R}^3. \tag{8.122}$$

Our discussion in Section 8.3 shows that

$$\begin{aligned}\mu_{l,\lambda,I}(x) = \lambda^{1/3}\,2^l \int & e^{-is_3\tilde{\Psi}(u,z,s_2,x,\delta,\lambda,l)}\tilde{a}(\sigma_{2^l\lambda^{-1}}u,\,(2^l\lambda^{-1})^{2/3}z,\,s_2,\,\delta)\\ & \times \chi_1(s_2,s_3)\,(1-\chi_0)(\varepsilon(2^l\lambda^{-1})^{-1/3}(x_1 - s_2^{1/(n-2)}G_1(s_2,\delta)))\\ & \times \chi_0(u)\chi_1(u_1)\,\chi_1(z)\,du_1\,du_2\,dz\,ds_2\,ds_3,\end{aligned}$$

where $\varepsilon > 0$ is small. Moreover,

$$\begin{aligned}\tilde{\Psi}(u,z,s_2,x,\delta,\lambda,l) = {} & \lambda^{1/3}2^{2l/3}(x_1 - s_2^{1/(n-2)}G_1(s_2,\delta) - (2^l\lambda^{-1})^{1/3}u_1)\,z\\ & + \lambda(s_2^{n/(n-2)}G_5(s_2,\delta) - x_1 s_2^{(n-1)/(n-2)}G_3(s_2,\delta) - s_2x_2 - x_3)\\ & + 2^l(u_1^3\,B_3(s_2,\delta_1,(2^{-l}\lambda)^{-1/3}u_1) + \phi^{\sharp}_{2^{-l}\lambda}(u_1,u_2,\tilde{\delta}^{2^{-l}\lambda},s_2)),\end{aligned}$$

with $\phi^{\sharp}$ given by (8.13). Observe that the first term is here much bigger than 2^l, so that it dominates the third term.

We change variables from s_2 to $v := x_1 - s_2^{1/(n-2)}G_1(s_2,\delta)$. Then s_2 is a smooth function $s_2\big(v,x_1,\delta\big)$, and we may rewrite

$$\begin{aligned}\mu_{l,\lambda,I}(x) = \lambda^{1/3}\,2^l \int & e^{-is_3\tilde{\Psi}_1(u,z,v,x,\delta,\lambda,l)}\,a_1\big(\sigma_{2^l\lambda^{-1}}u,\,(2^l\lambda^{-1})^{1/3},\,z,\,v,\,x,\,\delta\big)\\ & \times \chi_0(v)\chi_1(s_3)\,(1-\chi_0)\big(\varepsilon(2^l\lambda^{-1})^{-1/3}v\big)\,\chi_0(u)\chi_1(u_1)\\ & \times \chi_1(z)\,du_1\,du_2\,dz\,dv\,ds_3,\end{aligned}$$

where $\tilde{\Psi}_1$ is of the form

$$\begin{aligned}\tilde{\Psi}_1(u,z,v,x,\delta,\lambda,l) = {} & \big(\lambda^{1/3}2^{2l/3}\,v - 2^l u_1\big)z + 2^l g_1\big(u,v,x,(2^l\lambda^{-1})^{1/3},\tilde{\delta}^{2^{-l}\lambda},\delta\big)\\ & + \lambda g_2\big(u,v,x,\delta\big),\end{aligned}$$

with smooth, real-valued functions g_1 and g_2. Integrating N times by parts in z, and subsequently changing coordinates from v to $w := (2^l\lambda^{-1})^{-1/3}v$, we arrive at the following expression for $\mu_{l,\lambda,I}(x)$:

$$\begin{aligned}\mu_{l,\lambda,I}(x) = 2^{(4/3-N)l} \int & e^{-is_3\Psi(y,z,w,x,\delta,\lambda,l)}\\ & \times a\big(\sigma_{2^l\lambda^{-1}}y,\,(2^l\lambda^{-1})^{1/3},\,z,\,(2^l\lambda^{-1})^{1/3}w,\,\delta\big)\,\chi_1(s_3)\\ & \times \chi_0((2^l\lambda^{-1})^{1/3}w)\,(1-\chi_0)(\varepsilon w)\,\chi_0(y)\chi_1(y_1)\\ & \times \chi_1(z)\frac{1}{(w-y_1)^N}\,dy_1\,dy_2\,dz\,ds_3\,dw,\end{aligned} \tag{8.123}$$

with phase Ψ of the form

$$\begin{aligned}\Psi(y,z,w,x,\delta,\lambda,l) = {} & 2^l(w-y_1)z + 2^l g_1\big(y,(2^l\lambda^{-1})^{1/3}w,x,(2^l\lambda^{-1})^{1/3},\tilde{\delta}^{2^{-l}\lambda},\delta\big)\\ & + \lambda g_2\big(y,(2^l\lambda^{-1})^{1/3}w,x,\delta\big).\end{aligned} \tag{8.124}$$

Notice that we have also changed the names of variables u to y in order to avoid possible confusion in the later application of Lemma 2.7. Recall also that in this integral, $|w| \gg 1 \sim u_1$.

A Taylor expansion of g_2 in the "variable" $(2^l\lambda^{-1})^{1/3}w$ then shows that we may rewrite the phase in the form

$$\begin{aligned}\Psi(y, z, w, x, \delta, \lambda, l) = 2^l\big[(w - y_1)z + g_1\big(y, (2^l\lambda^{-1})^{1/3}w, x, (2^l\lambda^{-1})^{1/3}, \tilde{\delta}^{2^{-l}\lambda}, \delta\big)\\ + h_0(y, (2^l\lambda^{-1})^{1/3}w, x, \delta)w^3\big] + \lambda h_3\big(y, x, \delta\big)\\ + \lambda^{2/3}2^{l/3}\, h_2\,(y, x, \delta)\, w + \lambda^{1/3}\, 2^{2l/3}h_1\,(y, x, \delta)\, w^2,\end{aligned} \tag{8.125}$$

where $h_0, \ldots, h_3$ are again smooth, real-valued functions of their (bounded) variables.

The following lemma will be useful.

Lemma 8.14. *Let $\beta_1, \ldots, \beta_n \in]0, \infty[$ be given pairwise distinct positive numbers. For any complex numbers $\alpha_1, \ldots, \alpha_n \in \mathbb{C}$, denote by Λ the set of all dyadic numbers $\lambda = 2^j$ such that $\max_{k=1,\ldots,n} \lambda^{\beta_k}|\alpha_k| \geq 1$. Then there exists an exceptional set $\Lambda_e \subset \Lambda$ depending on the α_k and β_k whose cardinality is bounded by a constant $C_1(\beta_1, \ldots, \beta_n)$ depending only on the $\beta_1, \ldots, \beta_n$ such that for all $\lambda \in \Lambda \setminus \Lambda_e$ we have that $|\sum_{k=1}^{n} \lambda^{\beta_k}\alpha_k| \geq \frac{2}{3}$, and moreover*

$$\sum_{\lambda\in\Lambda\setminus\Lambda_e} \Big|\sum_{k=1}^{n} \lambda^{\beta_k}\alpha_k\Big|^{-1} \leq C_2(\beta_1, \ldots, \beta_n), \tag{8.126}$$

where the constant $C_2(\beta_1, \ldots, \beta_n)$ depends only on the numbers β_k.

Proof. We may assume without loss of generality that $\alpha_k \neq 0$ for every k. Observe that if $k \neq l$ and, say, $\beta_k > \beta_l$, then we have that $\frac{1}{4} \leq (2^{\beta_k j}|\alpha_k|)/(2^{\beta_l j}|\alpha_l|) \leq 4$ if and only if

$$\left| j + \frac{\log_2 |\frac{\alpha_k}{\alpha_l}|}{\beta_k - \beta_l} \right| \leq \frac{2}{\beta_k - \beta_l}.$$

We therefore define the set Λ_e to be the set of all dyadic numbers $\lambda = 2^j \in \Lambda$ satisfying this conditions for at least one pair $k \neq l$. The cardinality of Λ_e is then clearly bounded by $\binom{n}{2}\, 4 \max_{k\neq l} |\beta_k - \beta_l|^{-1}$. Moreover, if $\lambda \in \Lambda \setminus \Lambda_e$ and if we choose a permutation $(k(1), \ldots, k(n))$ of $(1, \ldots, n)$ so that

$$\lambda^{\beta_{k(1)}}|\alpha_{k(1)}| > \lambda^{\beta_{k(2)}}|\alpha_{k(2)}| > \cdots > \lambda^{\beta_{k(n)}}|\alpha_{k(n)}|,$$

then we have

$$\lambda^{\beta_{k(1)}}|\alpha_{k(1)}| > 4\lambda^{\beta_{k(2)}}|\alpha_{k(2)}| > \cdots > 4^{n-1}\lambda^{\beta_{k(n)}}|\alpha_{k(n)}|.$$

This implies that

$$\Big|\sum_{k=1}^{n} \lambda^{\beta_k}\alpha_k\Big| \geq \lambda^{\beta_{k(1)}}|\alpha_{k(1)}|\Big(1 - \sum_{l=1}^{n-1} 4^{-l}\Big) \geq \tfrac{2}{3}\lambda^{\beta_{k(1)}}|\alpha_{k(1)}| \geq \tfrac{2}{3}.$$

And, since $\sum_{j\in\mathbb{Z}:2^{\beta_l j}|\alpha_l|\geq 1}(2^{\beta_l j}|\alpha_l|)^{-1}\leq(1-2^{-\beta_l})^{-1}$, we obtain (8.126), with

$$C_2(\beta_1,\ldots,\beta_n):=\frac{3}{2}n!\max_k\frac{1}{1-2^{-\beta_k}}.$$

Q.E.D.

In order to prove (8.122), let us consider

$$F(t,y,z,w,x,\delta,l):=\sum_{2^M\leq\lambda\leq 2^M\rho^{-1}}\lambda^{-2it}\int e^{-is_3\Psi(y,z,w,x,\delta,\lambda,l)}\,\chi_1(s_3)\,ds_3.$$

We shall prove that

$$|F(t,y,z,w,x,\delta,\lambda,l)|\leq C\frac{2^l(1+|w|^3)}{|2^{-i2t}-1|},\tag{8.127}$$

with a constant C not depending on t, y, z, x, δ and l. By choosing N in (8.123) sufficiently large, we see that this estimate will imply (8.122), provided $\gamma(\zeta)$ contains a factor

$$\gamma_4(\zeta):=\frac{2^{2(1-\zeta)}-1}{3}.$$

Let us put $\beta_3:=1$, $\beta_2:=\frac{2}{3}$, $\beta_1:=\frac{1}{3}$ and, given y, w, x, δ and l, $\alpha_3:=h_3(y,x,\delta)$, $\alpha_2:=2^{l/3}h_2\big(y,x,\delta\big)w$, $\alpha_1:=2^{2l/3}h_1\big(y,x,\delta\big)w^2$. Accordingly, we set

$$\begin{aligned}\Lambda_1=\Lambda_1(y,w,x,\delta,l)&:=\big\{\lambda=2^j:2^M\leq\lambda\leq 2^M\rho^{-1}\text{ and}\\&\quad\max\{\lambda|h_3\big(y,x,\delta\big)|,\,\lambda^{2/3}|2^{l/3}h_2\big(y,x,\delta\big)w|,\,\lambda^{1/3}2^{2l/3}|h_1\big(y,x,\delta\big)w^2|\}\geq 1\big\}\\&=\big\{\lambda=2^j:2^M\leq\lambda\leq 2^M\rho^{-1}\text{ and }\max_{k=1,\ldots,3}\lambda^{\beta_k}|\alpha_k|\geq 1\big\}\end{aligned}$$

and

$$\begin{aligned}\Lambda_2=\Lambda_2(y,w,x,\delta,l)&:=\big\{\lambda=2^j:2^M\leq\lambda\leq 2^M\rho^{-1}\text{ and}\\&\quad\max\{\lambda|h_3\big(y,x,\delta\big)|,\,\lambda^{2/3}|2^{l/3}h_2\big(y,x,\delta\big)w|,\,\lambda^{1/3}2^{2l/3}|h_1\big(y,x,\delta\big)w^2|\}<1\big\}.\end{aligned}$$

We also denote by $\Lambda_e\subset\Lambda$ the set of exceptional λs given by Lemma 8.14 for this choice of β_k and α_k. Correspondingly, we decompose $F=F_e+F_1+F_2$, where F_e, F_1 and F_2 are defined as F, only with summation over the dyadic λs restricted to the subsets Λ_e, $\Lambda_1\setminus\Lambda_e$ and Λ_2, respectively.

For F_e, we then trivially get the estimate $|F(t,y,z,w,x,\delta,\lambda,l)|\leq C$, since the cardinality of Λ_e is bounded by a constant not depending on the arguments of F_e.

Next, in order to estimate F_1, let us rewrite

$$\int e^{-is_3\Psi(y,z,w,x,\delta,\lambda,l)}\,\chi_1(s_3)\,ds_3=\int e^{-is_3(\lambda^{\beta_1}\alpha_1+\lambda^{\beta_2}\alpha_2+\lambda^{\beta_3}\alpha_3)}\times\big(e^{-is_3\Psi_0(y,z,w,x,\delta,\lambda,l)}\,\chi_1(s_3)\big)\,ds_3,$$

where Ψ_0 denotes the phase

$$\Psi_0(y,z,w,x,\delta,\lambda,l):=2^l\Big[(w-y_1)z+g_1\big(y,(2^l\lambda^{-1})^{1/3}w,x,(2^l\lambda^{-1})^{1/3},\tilde{\delta}^{2^{-l}\lambda},\delta\big)\\ +h_0\big(y,(2^l\lambda^{-1})^{1/3}w,x,\delta\big)\,w^3\Big].$$

Observe that $|\Psi_0(y,z,w,x,\delta,\lambda,l)|\le C2^l(1+|w|)$. Integrating by parts in s_3 therefore shows that

$$\Big|\int e^{-is_3\Psi(y,z,w,x,\delta,\lambda,l)}\,\chi_1(s_3)\,ds_3\Big|\le C\,\frac{2^l(1+|w|)}{|\lambda^{\beta_1}\alpha_1+\lambda^{\beta_2}\alpha_2+\lambda^{\beta_3}\alpha_3|}.$$

We may thus control the sum over all $\lambda\in\Lambda_1$ by means of Lemma 8.14 and obtain the estimate

$$|F_1(t,y,z,w,x,\delta,\lambda,l)|\le C2^l(1+|w|).$$

Finally, F_2 can again be estimated by means of Lemma 2.7. Indeed, observe that in the sum defining $F_2(t,y,z,w,x,\delta,\lambda,l)$, the expressions

$$(2^l\lambda^{-1})^{1/3}w,\ (2^l\lambda^{-1})^{1/3},\ \tilde{\delta}^{2^{-l}\lambda},\ \lambda h_3\big(y,x,\delta\big),\\ \lambda^{2/3}2^{l/3}\,h_2\big(y,x,\delta\big)w,\ \lambda^{1/3}\,2^{2l/3}h_1\big(y,x,\delta\big)w^2$$

are all uniformly bounded. Therefore, we may let

$$H(u_1,\dots,u_6):=\int e^{-is_3\left(2^l[(w-y_1)z+g_1(y,u_1,x,u_2,u_3,\delta)+h_0(y,u_1,x,\delta)w^3]+u_4+u_5+u_6\right)}\\ \times\chi_1(s_3)\,ds_3,$$

with the a_k in Lemma 2.7 given by

$$a_1:=2^{l/3}2^w,\ a_2:=2^{l/3},\dots,a_4:=h_3(y,x,\delta),\\ a_5:=2^{l/3}\,h_2(y,x,\delta)w,\ a_6:=2^{2l/3}h_1(y,x,\delta)w^2,$$

and the obvious corresponding cuboid Q. Then clearly $\|H\|_{C^1(Q)}\le C2^l(1+|w|^3)$, and thus Lemma 2.7 yields the estimate

$$|F_2(t,y,z,w,x,\delta,\lambda,l)|\le C\frac{2^l(1+|w|^3)}{|2^{-i2t}-1|}.$$

This concludes the proof of estimate (8.127) and, thus, also of (8.122).

8.7.5 Contribution by the $\nu^\lambda_{l,III}$

The contribution of the terms $\nu^\lambda_{l,III}$ in (8.106) can be treated in a very similar way as the one by the terms $\nu^\lambda_{l,I}$. Indeed, arguing as before, we here arrive at the following expression for $\mu_{l,\lambda,III} := 2^{-l/3}\lambda^{-4/3}\,\nu^{\lambda}_{l,III}$:

$$\mu_{l,\lambda,III}(x) = 2^{(4/3-N)l} \int e^{-is_3\Psi(y,z,w,x,\delta,\lambda,l)}$$
$$\times\, a\big(\sigma_{2^l\lambda^{-1}}y, (2^l\lambda^{-1})^{1/3}, z, (2^l\lambda^{-1})^{1/3}w, \delta\big)\,\chi_1(s_3)$$
$$\times\, \chi_0\big((2^l\lambda^{-1})^{1/3}w\big)\,\chi_0\Big(\frac{w}{\varepsilon}\Big)\,\chi_0(y)\chi_1(y_1)$$
$$\times\, \chi_1(z)\frac{1}{(w-y_1)^N}\,dy_1\,dy_2\,dz\,ds_3\,dw. \tag{8.128}$$

The phase Ψ is still given by (8.124). Notice that now $|w| \ll 1 \sim u_1$. The arguments used in the preceding subsection therefore carry over to this case, with minor modifications (even simplifications).

8.7.6 Contribution by the $\nu^{\lambda}_{l,\infty}$

Let us next look at

$$\nu_{\infty,1+it} := \gamma(1+it) \sum_{2^M\le\lambda\le 2^M\rho^{-1}} \;\sum_{\{l: M_0\le 2^l\le\lambda/M_1\}} 2^{-itl/2}\lambda^{-2it}\,\mu_{l,\lambda,\infty}(x), \tag{8.129}$$

where we have set $\mu_{l,\lambda,\infty} := 2^{-l/3}\lambda^{-4/3}\,\nu^{\lambda}_{l,\infty}$. We want to prove that

$$|\nu_{\infty,\,1+it}(x)| \le C \qquad \forall t\in\mathbb{R}, x\in\mathbb{R}^3. \tag{8.130}$$

To this end, recall formulas (8.33) and (8.34) for $\nu^{\lambda}_{l,\infty}(x)$. From these formulas, it is easy to see that $|\nu^{\lambda}_{l,\infty}(x)| \lesssim 2^{-lN}\lambda^{-N}$ if $|x| \gg 1$, and thus summation in l and λ is no problem in this case. So, assume that $|x| \lesssim 1$. Then the second term in the phase $\Psi(z, s_2, \delta)$ in (8.34) can be absorbed into the amplitude $a_{N,l}$, and we arrive at an expression of the following form for $\mu_{l,\lambda,\infty}(x)$:

$$\mu_{l,\lambda,\infty}(x) = 2^{-lN}\,\lambda^{1/3} \int e^{-is_3\lambda(s_2^{n/(n-2)}G_5(s_2,\delta) - x_1 s_2^{(n-1)/(n-2)}G_3(s_2,\delta) - s_2x_2 - x_3)}$$
$$\times\, a_{N,l}(z, s_2, s_3, \delta_0^{\mathfrak{r}}, \tilde\delta^{2^{-l}\lambda}, (2^{-l}\lambda)^{-1/3}, \lambda^{-1/9})$$
$$\times\, \chi_1(z)\,\chi_1(s_2)\chi_1(s_3)\,dz\,ds_2\,ds_3,$$

where $a_{N,l}$ is a smooth function of all its (bounded) variables such that $\|a_{N,l}\|_{C^k}$ is uniformly bounded in l. Denote by $\Psi(s_2) = \Psi(s_2, x, \delta) = s_2^{n/(n-2)}G_5(s_2,\delta) - x_1 s_2^{(n-1)/(n-2)}G_3(s_2,\delta) - s_2x_2 - x_3$ the phase appearing in this integral. We can now argue in a similar way as in Subsection 8.7.1.

Since $s_2 \sim 1$ in the integral, we see that if $|x_1| \ll 1$, then $|\partial^2_{s_2}\Psi(s_2)| \sim 1$, and van der Corput's estimate implies that $|\mu_{l,\lambda,\infty}(x)| \lesssim 2^{-lN}\lambda^{1/3-1/2}$. We can then sum the series (8.129) absolutely and arrive at (8.130). Let us therefore assume from now on that $|x_1| \sim 1$ and that the sign of x_1 is such that there is a point $s_2^c(x,\delta) \sim 1$ such that $\partial^2_{s_2}\Psi(s_2^c(x,\delta), x, \delta) = 0$. This point is then unique, by the implicit function theorem, since $|\partial_{s_2}^3 \Psi(s_2^c(x,\delta),x,\delta)| \sim 1$. Changing coordinates from s_2 to $v := s_2 - s_2^c(x,\delta)$ and applying a Taylor expansion in v, we see that the phase can be written in the form

$$Q_3(v,x,\delta)\,v^3 - Q_1(x,\delta)\,v + Q_0(x,\delta),$$

with smooth functions Q_j. Scaling in v by the factor $\lambda^{-1/3}$ then leads to an expression of the following form for $\mu_{l,\lambda,\infty}(x)$:

$$\begin{aligned}\mu_{l,\lambda,\infty}(x) = 2^{-lN} \int & e^{-is_3\left(Q_3(\lambda^{-1/3}w,x,\delta)\,w^3+\lambda^{2/3}Q_1(x,\delta)\,w+\lambda Q_0(x,\delta)\right)} \\ & \times a_{N,l}\left(z,\lambda^{-1/3}w,s_3,\delta_0^{\mathfrak{r}},\tilde{\delta}^{2^{-l}\lambda},(2^{-l}\lambda)^{-1/3},\lambda^{-1/9}\right)\chi_1(z)\,\chi_0(\lambda^{-1/3}w) \\ & \times \chi_1(s_3)\,ds_3\,dw\,dz. \end{aligned} \tag{8.131}$$

Performing first the integration in s_3, this easily implies the following estimate:

$$\begin{aligned}|\mu_{l,\lambda,\infty}(x)| \lesssim 2^{-lN} \int\int_{-\lambda^{1/3}}^{\lambda^{1/3}} & \Big(1+\big|Q_3(\lambda^{-1/3}w,x,\delta)\,w^3 \\ & + \lambda^{2/3}Q_1(x,\delta)\,w + \lambda Q_0(x,\delta)\big|\Big)^{-N}\chi_1(z)\,\chi_0(\lambda^{-1/3}w)\,dw\,dz.\end{aligned}$$

Putting $A := \lambda Q_0(x,\delta)$, $B := \lambda^{2/3}Q_1(x,\delta)$, and $T := \lambda^{1/3}$ in Lemma 2.4, we then find that

$$|\mu_{l,\lambda,\infty}(x)| \lesssim 2^{-lN}\max\{|A|^{1/3},|B|^{1/2}\}^{\epsilon-1/2}.$$

This estimate allows us to sum over all λ such that $\max\{\lambda|Q_0(x,\delta)|, \lambda^{2/3}|Q_1(x,\delta)|\} > 1$, even absolutely.

There remains the summation over all λ such that $\lambda|Q_0(x,\delta)| \le 1$ and $\lambda^{2/3}|Q_1(x,\delta)| \le 1$. However, in view of (8.131), this sum can easily be controlled by means of Lemma 2.7, as we have done in many similar cases before, and we shall therefore skip the details. Altogether, we arrive at (8.130).

8.7.7 Contribution by the $\nu_{l,00}^{\lambda}$

We finally come to the contribution of the terms $\nu_{l,00}^{\lambda}$ in (8.106). Recall from Subsection 8.3.2 that

$$\begin{aligned}\widehat{\nu_{l,00}^{\lambda}}(\xi) := {} & (2^{-l}\lambda)^{-2/3}\chi_1(s,s_3)\,\chi_1((2^{-l}\lambda)^{2/3}B_1(s,\delta_1))\,e^{-is_3\lambda B_0(s,\delta_1)} \\ & \times \iint e^{-is_3 2^l\Phi(u_1,u_2,s,\delta,\lambda,l)}a(\sigma_{2^l\lambda^{-1}}u,\delta,s)\,\chi_0(u)\chi_0\left(\frac{u_1}{\varepsilon}\right)du_1\,du_2,\end{aligned}$$

where we assume $\varepsilon > 0$ to be sufficiently small. Moreover, the phase Φ is given by (8.28) and (8.29), with $B = 3$. Since

$$|\tilde{\delta}^{2^{-l}\lambda}| \ll 1 \quad \text{and} \quad (2^{-l}\lambda)^{2/3}|B_1(s,\delta_1)| \sim 1,$$

we see that we can again integrate by parts in u_1 in order to gain factors 2^{-lN}, and then the same type of argument that led to expression (8.30) for $\widehat{\nu_{l,\infty}^{\lambda}}(\xi)$ can be applied in order to see that an analogous expression,

$$\widehat{\nu_{l,00}^{\lambda}}(\xi) = 2^{-lN}\,\lambda^{-2/3}\chi_1(s,s_3)\,\chi_1((2^{-l}\lambda)^{2/3}B_1(s,\,\delta_1))\,e^{-is_3\lambda B_0(s,\delta_1)}$$
$$\times\,\tilde{a}_{N,l}((2^{-l}\lambda)^{2/3}B_1(s,\,\delta_1),\,s,\,s_3,\,\tilde{\delta}^{2^{-l}\lambda},\,\delta_0^{\mathfrak{r}},\,(2^{-l}\lambda)^{-1/3},\,\lambda^{-1/3B}),$$

can be obtained for $\widehat{\nu_{l,00}^{\lambda}}(\xi)$ too, where $\tilde{a}_{N,l}$ is again a smooth function of all its (bounded) variables such that $\|a_{N,l}\|_{C^k}$ is uniformly bounded in l. From here on, we can argue exactly as for $\nu_{l,\infty}^{\lambda}$.

This concludes the proof of the second estimate in (8.101) and, thus, the proof of part (a) of Proposition 8.12 as well.

8.8 PROOF OF PROPOSITION 8.12(b): COMPLEX INTERPOLATION

In this section, we assume that $B=3$ and $A=0$ in (7.89) and that $\lambda\rho \gg 1$.

8.8.1 Estimation of $T_{\delta,Ai}^{III}$

As usual, we embed $\nu_{\delta,Ai}^{III}$ into an analytic family of measures

$$\nu_{\delta,\zeta}^{III} := \gamma(\zeta) \sum_{2^M\rho^{-1}<\lambda\le 2^{-M}\delta_0^{-3}} \big(\rho^{-4/5}(\lambda\rho)\big)^{5(1-3\zeta)/6}\nu_{\delta,0}^{\lambda},$$

where ζ lies in the complex strip Σ given by $0\le \operatorname{Re}\zeta\le 1$. Since the supports of the $\widehat{\nu_{\delta,0}^{\lambda}}$ are almost disjoint and since, according to (8.90), $\|\widehat{\nu_{\delta,0}^{\lambda}}\|_\infty \lesssim \rho^{2/3}(\lambda\rho)^{-5/6}$, we see that

$$\|\widehat{\nu_{\delta,it}^{III}}\|_\infty \lesssim 1 \qquad \forall t\in\mathbb{R}.$$

Again, by Stein's interpolation theorem, it will therefore suffice to prove the following estimate:

$$|\nu_{\delta,1+it}^{III}(x)| \le C \qquad \forall t\in\mathbb{R},\, x\in\mathbb{R}^3. \tag{8.132}$$

Now, if $|x|\gg 1$, arguing in a similar way as for the case $B=4$ in Subsection 8.5.1, we see that $|\nu_{\delta,0}^{\lambda}(x)| \lesssim \rho^{2/3+2/3}\lambda^3(\lambda\rho^{2/3})^{-N}$ for every $N\in\mathbb{N}$, which allows to sum absolutely in λ and obtain (8.132).

From now on, we shall therefore assume that $|x|\lesssim 1$. We then rewrite

$$\nu_{\delta,1+it}^{III}(x) = \gamma(1+it) \sum_{2^M\rho^{-1}<\lambda\le 2^{-M}\delta_0^{-3}} \big(\rho^{-4/5}(\lambda\rho)\big)^{-5it/2}\mu_\lambda(x), \tag{8.133}$$

where $\mu_\lambda := \rho^{4/3}(\lambda\rho)^{-5/3}\nu_{\delta,0}^{\lambda}$. Recall also from (8.93) that $\|\nu_{\delta,0}^{\lambda}\|_\infty \lesssim \rho^{-4/3}$ $\times\,(\lambda\rho)^{5/3}$, which barely fails to be sufficient to obtain (8.132).

We therefore again need a more refined reasoning. Following our discussion for the case $B=4$ in Subsection 8.5.1, we decompose $\nu_{\delta,0}^{\lambda} = \nu_{0,I}^{\lambda} + \nu_{0,II}^{\lambda}$ as in (8.72). In the same way in which we derived (8.77), we find here that

$$\|\nu_{0,I}^{\lambda}(x)\|_\infty \le C_N\rho^{2/3}\lambda^2\,(\lambda\rho)^{-N}. \tag{8.134}$$

If we set $\mu_{\lambda,I} := \rho^{4/3}(\lambda\rho)^{-5/3}\nu_{0,I}^{\lambda}$, then this estimate shows that we can sum the corresponding series in (8.133), with μ_λ replaced by $\mu_{\lambda,I}$, absolutely, and obtain the desired uniform estimate in x and δ.

What remains are the contributions by the $\nu_{0,II}^{\lambda}$. In order to keep the notation simple, we therefore shall assume from now on that $\mu_\lambda = \rho^{4/3}(\lambda\rho)^{-5/3}\nu_{0,II}^{\lambda}$.

In analogy with our formulas (8.78) and (8.79), we then find the following expressions for μ_λ:

$$\mu_\lambda(x) = (\lambda\rho)^{1/3}\int e^{-i\lambda s_3\Phi_4(y_1,u_2,w,x,\delta)}\,\widehat{\chi_0}(s_3y_1)\,\chi_0(w)\,\chi_0(w-(\lambda\rho)^{-1}y_1)$$
$$\times\,\chi_0(u_2)\,a_4\left((\lambda\rho^{2/3})^{-1}y_1,\rho^{1/3}u_2,w,s_1,\rho^{1/3},x,\delta\right)\tilde\chi_1(s_3)\,dy_1\,du_2\,dw\,ds_3,$$

with phase Φ_4 of the form

$$\begin{aligned}\Phi_4(y_1,u_2,w,x,\delta) = \Psi_3\left(\rho^{1/3}w,x,\delta\right)&\\
+\rho\left(w-(\lambda\rho)^{-1}y_1\right)^3\tilde B_3\left(\rho^{1/3}w,(\lambda\rho^{2/3})^{-1}y_1,x,\delta\right)&\\
+\rho\left(u_2^3\,b\left(\rho^{1/3}w,(\lambda\rho^{2/3})^{-1}y_1,\rho^{1/3}u_2,x,\delta_0^{\mathfrak r}\right)\right.&\\
+\delta'_{3,0}u_2\,\tilde\alpha_1\left(\rho^{1/3}w,x,\delta_0^{\mathfrak r}\right)&\\
\left.+\delta'_0u_2\left(w-(\lambda\rho)^{-1}y_1\right)\alpha_{1,1}\left(\rho^{1/3}w,(\lambda\rho^{2/3})^{-1}y_1,x,\delta_0^{\mathfrak r}\right)\right).&\end{aligned}$$

Recall that in this integral, $|u_2|+|w|\lesssim 1$ and $|y_1|\lesssim\lambda\rho$. Moreover, the factor $\widehat{\chi_0}(s_3y_1)$ guarantees the absolute convergence of this integral with respect to the variable y_1. We also recall that $\delta'_0+\delta'_{3,0}\sim 1$ and that the coefficient δ'_0 does not appear in Case ND, where $\alpha_{1,1}=0$. Finally, we perform the change of variables $u_2=(\lambda\rho)^{-1/3}y_2$ and obtain

$$\begin{aligned}\mu_\lambda(x) = \int e^{-is_3\Phi_5(y,w,x,\delta;\lambda)}\,&\widehat{\chi_0}(s_3y_1)\,\chi_0(w)\,\chi_0(w-(\lambda\rho)^{-1}y_1)\\
\times\,&\chi_0((\lambda\rho)^{-1/3}y_2)\,a_5((\lambda\rho^{2/3})^{-1}y_1,\lambda^{-1/3}y_2,w,s_1,\rho^{1/3},x,\delta)\\
\times\,&\tilde\chi_1(s_3)\,ds_3\,dy_1\,dy_2\,dw,\end{aligned}\tag{8.135}$$

with phase Φ_5 of the form

$$\begin{aligned}\Phi_5(y,w,x,\delta;\lambda) = \lambda\Psi_3\left(\rho^{1/3}w,x,\delta\right)&\\
+\lambda\rho\left(w-(\lambda\rho)^{-1}y_1\right)^3\tilde B_3\left(\rho^{1/3}w,(\lambda\rho^{2/3})^{-1}y_1,x,\delta\right)&\\
+y_2^3\,\tilde b\left(\rho^{1/3}w,(\lambda\rho^{2/3})^{-1}y_1,\lambda^{-1/3}y_2,x,\delta_0^{\mathfrak r}\right)&\\
+y_2(\lambda\rho)^{2/3}\left(\delta'_{3,0}\tilde\alpha_1\left(\rho^{1/3}w,x,\delta_0^{\mathfrak r}\right)\right.&\\
\left.+\delta'_0\left(w-(\lambda\rho)^{-1}y_1\right)\tilde\alpha_{1,1}\left(\rho^{1/3}w,(\lambda\rho^{2/3})^{-1}y_1,x,\delta_0^{\mathfrak r}\right)\right).&\end{aligned}\tag{8.136}$$

Performing a Taylor expansion with respect to the bounded quantities $(\lambda\rho^{2/3})^{-1}y_1$ and $(\lambda\rho)^{-1}y_1$, we see that we may rewrite the phase as

$$\begin{aligned}\Phi_5(y,w,x,\delta;\lambda) = A+By_2+\tilde b\big(\rho^{1/3}w,(\lambda\rho^{2/3})^{-1}y_1,\lambda^{-1/3}y_2,x,\delta_0^{\mathfrak r}\big)\,y_2^3&\\
+r_1(y_1)+(\lambda\rho)^{-1/3}y_2r_2(y_1),&\end{aligned}\tag{8.137}$$

where

$$A := \lambda \left[\Psi_3 \left(\rho^{1/3} w, x, \delta \right) + \rho w^3 \tilde{B}_3 \left(\rho^{1/3} w, 0, x, \delta \right) \right] =: \lambda\, Q_A(\rho^{1/3} w, x, \delta);$$

$$B := (\lambda\rho)^{2/3} \left(\delta'_{3,0} \tilde{\alpha}_1 \left(\rho^{1/3} w, x, \delta_0^{\mathfrak{r}} \right) + \delta'_0 w \tilde{\alpha}_{1,1} \left(\rho^{1/3} w, 0, x, \delta_0^{\mathfrak{r}} \right) \right)$$

$$=: \lambda^{2/3}\, Q_B(\rho^{1/3} w, x, \delta),$$

and where r_1 and r_2 are of the form

$$r_1(y_1) = R_1\big(w, (\lambda\rho)^{-1} y_1, \rho^{1/3}, x, \delta\big)\, y_1,$$
$$r_2(y_1) = R_2\big(w, (\lambda\rho)^{-1} y_1, \rho^{1/3}, x, \delta\big)\, y_1,$$

with smooth functions R_1, R_2 of their bounded entries w, $(\lambda\rho)^{-1} y_1$, $\rho^{1/3}$, x, and δ.

Let us put, for w fixed such that $|w| \lesssim 1$,

$$\begin{aligned} \mu_\lambda(w, x) := \int & e^{-is_3 \Phi_5(y,w,x,\delta;\lambda)}\, \widehat{\chi_0}(s_3 y_1)\, \chi_0(w - (\lambda\rho)^{-1} y_1) \\ & \times \chi_0((\lambda\rho)^{-1/3} y_2)\, a_5\left((\lambda\rho^{2/3})^{-1} y_1, \lambda^{-1/3} y_2, w, s_1, \rho^{1/3}, x, \delta \right) \\ & \times \tilde{\chi}_1(s_3)\, ds_3\, dy_1\, dy_2. \end{aligned}$$

By means of integrations by parts in s_3 and exploiting the rapid decay of $\widehat{\chi_0}(s_3 y_1)$ in y_1, we may estimate

$$\begin{aligned} |\mu_\lambda(w, x)| \le C_N \int \int_{|y_2| \lesssim (\lambda\rho)^{1/3}} \int \Big(1 + \Big| A + B y_2 & \\ + \tilde{b}\big(\rho^{1/3} w, (\lambda\rho^{2/3})^{-1} y_1, \lambda^{-1/3} y_2, x, \delta_0^{\mathfrak{r}}\big)\, y_2^3 & \\ + r_1(y_1) + (\lambda\rho)^{-1/3} y_2 r_2(y_1) \Big| \Big)^{-N} & (1 + |y_1|)^{-N}\, dy_1\, dy_2. \end{aligned}$$

Observe first that $|B| \lesssim (\lambda\rho)^{2/3}$. Thus, if $|A| \gg \lambda\rho$, then the term A becomes dominant, and we can clearly estimate $|\mu_\lambda(x)| \le C|A|^{-N}$ for every $N \in \mathbb{N}$. Otherwise, if $|A| \lesssim \lambda\rho$, then by choosing $T := c(\lambda\rho)^{1/3}$ in Lemma 2.4, with a suitable constant $c > 0$, we see that all assumptions of this lemma are satisfied, and we obtain the estimate

$$|\mu_\lambda(w, x)| \le C \max\{|A|^{1/3}, |B|^{1/2}\}^{-1/2}. \tag{8.138}$$

This estimate thus holds true no matter how large $|A|$ is. Consider the function

$$F(t, w, x, \delta) := \sum_{2^M \rho^{-1} < \lambda \le 2^{-M} \delta_0^{-3}} \left(\rho^{-4/5} (\lambda\rho) \right)^{-5it/2} \mu_\lambda(w, x),$$

for $|w| \lesssim 1$. We shall prove that

$$|F(t, w, x, \delta)| \le C \frac{1}{|2^{-5it/2} - 1|}, \tag{8.139}$$

with a constant C not depending on t, w, x, and δ. This estimate will immediately yield the desired estimate for the contributions of the $\nu_{0,II}^\lambda$ and thus complete the proof of (8.132), provided we choose

$$\gamma(\zeta) := \frac{2^{5(1-\zeta)/2} - 1}{2^{5/3} - 1}.$$

Given w, x, δ, denote by $\Lambda(w, x, \delta)$ the set of all dyadic λ from our summation range $\Lambda := \{\lambda = 2^j : 2^M \rho^{-1} < \lambda \leq 2^{-M} \delta_0^{-3}\}$, for which either $|A| = \lambda\, |Q_A(\rho^{1/3} w, x, \delta)| > 1$ or $|B| = \lambda^{2/3}\, |Q_B(\rho^{1/3} w, x, \delta)| > 1$. We then decompose $F(t, w, x, \delta) = F_1(t, w, x, \delta) + F_2(t, w, x, \delta)$, where $F_1(t, w, x, \delta)$ and $F_2(t, w, x, \delta)$ are defined like F, only with summation restricted to the subsets $\Lambda(w, x, \delta)$ and $\Lambda \setminus \Lambda(w, x, \delta)$, respectively. Then, by (8.138), we clearly have that

$$|F_1(t, w, x, \delta)| \leq \sum_{\lambda \in \Lambda(w, x, \delta)} |\mu_\lambda(w, x)| \leq C,$$

and we are thus left with $F_2(t, w, x, \delta)$.

In the corresponding sum, we have $\lambda\, |Q_A(\rho^{1/3} w, x, \delta)| \leq 1$ and $\lambda^{2/3} \times |Q_B(\rho^{1/3} w, x, \delta)| \leq 1$, and therefore F_2 can again be estimated by means of Lemma 2.7. Indeed, we may here put

$$\begin{aligned}
&H(u_1, \dots, u_6) \\
&\quad := \int e^{-is_3\left(u_1 + u_2 y_2 + \tilde{b}(\rho^{1/3} w, u_3 y_1, u_4 y_2, x, \delta_0^{\mathfrak{r}})\, y_2^3 + R_1(w, u_5 y_1, \rho^{1/3}, x, \delta)\, y_1 + u_6 y_2 R_2(w, u_5 y_1, \rho^{1/3}, x, \delta)\right)} \\
&\qquad \times \widehat{\chi_0}(s_3 y_1)\, \chi_0(w - u_5 y_1) \chi_0(u_6 y_2)\, a_5\left(u_3 y_1, u_4 y_2, w, s_1, \rho^{1/3}, x, \delta\right) \\
&\qquad \times \tilde{\chi}_1(s_3)\, ds_3\, dy_1\, dy_2\,,
\end{aligned}$$

where the variables $u_1, \dots, u_6$ correspond to the bounded expressions $\lambda Q_A(\rho^{1/3} w, x, \delta)$, $\lambda^{2/3} Q_B(\rho^{1/3} w, x, \delta)$, $(\lambda \rho^{2/3})^{-1}$, $\lambda^{-1/3}$, $(\lambda\rho)^{-1}$, and $(\lambda\rho)^{-1/3}$, respectively. By means of integrations by parts in the variable y_2 for $|y_2| \gg 1$ (or, alternatively, in s_3), it is then easily verified that $\|H\|_{C^1(Q)} \leq C$, where Q denotes the obvious cuboid Q appearing in this situation. Thus, estimate (8.139) follows from Lemma 2.7.

8.8.2 Estimation of T_δ^{IV}

The estimation of the operator T_δ^{IV} will follow similar ideas as the one for T_δ^{II}. Nevertheless, for the convenience of the reader, we will give some details.

As usually, we embed ν_δ^{IV} into an analytic family of measures

$$\nu_{\delta,\zeta}^{IV} := \gamma(\zeta) \sum_{\{l : M_0 \leq 2^l \leq \rho^{-1}/M_1\}} \sum_{2^M \rho^{-1} < \lambda \leq 2^{-M} \delta_0^{-3}} \left((2^l \rho)^{-4/5} (\lambda 2^l \rho)\right)^{5(1-3\zeta)/6} \nu_{l,0}^\lambda,$$

where ζ lies in the complex strip Σ given by $0 \leq \operatorname{Re} \zeta \leq 1$. Since the supports of the $\widehat{\nu_{\delta,0}^\lambda}$ are almost disjoint, estimate (8.97) shows that

$$\|\widehat{\nu_{\delta,it}^{IV}}\|_\infty \lesssim 1 \qquad \forall t \in \mathbb{R}.$$

Again, by Stein's interpolation theorem, it will therefore suffice to prove the following estimate:

$$|\nu_{\delta,1+it}^{IV}(x)| \leq C \qquad \forall t \in \mathbb{R}, x \in \mathbb{R}^3, \tag{8.140}$$

where we write

$$\nu^{IV}_{\delta,1+it}(x)=\gamma(1+it)\sum_{\{l:M_0\le 2^l\le\rho^{-1}/M_1\}}\ \sum_{2^M\rho^{-1}<\lambda\le 2^{-M}\delta_0^{-3}}(\lambda(2^l\rho)^{1/5})^{-5it/2}\mu_{l,\lambda}(x), \tag{8.141}$$

with $\mu_{l,\lambda}:=\lambda^{-5/3}(2^l\rho)^{-1/3}\nu^\lambda_{l,0}$.

Regretfully, it seems that the approach in the previous subsection cannot be applied in the present situation, and a more refined analysis is needed, similar to our discussion in Subsection 8.7.2. From (8.98) to (8.100) we see that

$$\mu_{l,\lambda}=(2^l\rho)\lambda^{4/3}\int e^{-i\lambda s_3\Phi_2(u,z,s_2,x,\delta)}\chi_1(z)\chi_0(u)a(\sigma_{2^l\rho}u,(2^l\rho)^{2/3}z,s,\delta)\tilde\chi_1(s_2,s_3)$$
$$\times\,du\,dz\,ds_2\,ds_3,$$

where

$$\begin{aligned}\Phi_2(u,z,s_2,x,\delta)=&\,s_2^{n/(n-2)}G_5(s_2,\delta)-x_1s_2^{(n-1)/(n-2)}G_3(s_2,\delta)-s_2x_2-x_3\\&+z(2^l\rho)((2^l\rho)^{-1/3}(x_1-s_2^{1/(n-2)}G_1(s_2,\delta))-u_1)\\&+(2^l\rho)u_1^3B_3\left(s_2,\delta_1,(2^l\rho)^{1/3}u_1\right)\\&+(2^l\rho)\left(u_2^3b(\sigma_{2^l\rho}u,\delta_0^\mathfrak{r},s_2)+\delta'_{3,0}u_2\tilde\alpha_1(\delta_0^\mathfrak{r},s_2)\right.\\&\left.+\delta_0'u_1u_2\alpha_{1,1}\left((2^l\rho)^{1/3}u_1,\delta_0^\mathfrak{r},s_2\right)\right).\end{aligned}\tag{8.142}$$

Now, if $|x|\gg 1$, we see that $|\mu_{l,\lambda}(x)|\lesssim 2^l\rho\lambda^{4/3}(\lambda(2^l\rho)^{2/3})^{-N}$ for every $N\in\mathbb{N}$, which is stronger than what is needed for (8.140).

From now on we shall therefore assume that $|x|\lesssim 1$. For such x fixed, we again decompose

$$\nu^\lambda_{l,0}=\nu^\lambda_{l,I}+\nu^\lambda_{l,II},$$

where $\nu^\lambda_{l,I}$ and $\nu^\lambda_{l,II}$ denote the contributions to the preceding integral by the region L_I, where $|x_1-s_2^{1/(n-2)}G_1(s_2,\delta)|\gg(2^l\rho)^{1/3}$, and the region L_{II}, where $|x_1-s_2^{1/(n-2)}G_1(s_2,\delta)|\lesssim(2^l\rho)^{1/3}$, respectively. Then, in analogy with (8.134), by means of integrations by parts in z we obtain

$$\|\nu^\lambda_{l,I}(x)\|_\infty\le C_N(2^l\rho)^{2/3}\lambda^2(\lambda 2^l\rho)^{-N}.$$

If we denote by $\mu_{l,\lambda,I}:=\lambda^{-5/3}(2^l\rho)^{-1/3}\nu^\lambda_{l,I}$, then this estimate shows that we can sum the corresponding series in (8.141), with $\mu_{l,\lambda}$ replaced by $\mu_{l,\lambda,I}$, absolutely, and obtain the desired uniform estimate in x and δ.

What remains are the contributions by the $\nu^\lambda_{l,II}$. In order to keep the notation simple, we shall assume from now on that $\mu_{l,\lambda}=\lambda^{-5/3}(2^l\rho)^{-1/3}\nu^\lambda_{l,I}$, that is, that

$$\mu_{l,\lambda}(x)=(2^l\rho)\lambda^{4/3}\int e^{-i\lambda s_3\Phi_2(u,z,s_2,x,\delta)}a\left((2^l\rho)^{1/3}u,(2^l\rho)^{2/3}z,s,\delta\right)\chi_1(z)\chi_0(u)$$
$$\times\,\chi_0\left((2^l\rho)^{-1/3}(x_1-s_2^{1/(n-2)}G_1(s_2,\delta))\right)\tilde\chi_1(s_2,s_3)\,du\,dz\,ds_2\,ds_3. \tag{8.143}$$

Given a point u^0, s_2^0, z^0 at which the amplitude in this integral does not vanish, we want to understand the contribution of a small neighborhood of this point to the integral. Assume first that $\partial_{u_1}\Phi_2(u^0, z^0, s_2^0, x, \delta) \neq 0$. Then, integrations by parts in u_1 allow us to gain factors $(\lambda 2^l\rho)^{-N}$, and so we can again sum the corresponding contributions to $\nu^{IV}_{\delta, 1+it}(x)$ absolutely.

Let us next assume that $\partial_{u_1}\Phi_2(u^0, z^0, s_2^0, x, \delta) = 0$. For a short while, it will then be helpful to change coordinates from s_2 first to $v := x_1 - s_2^{1/(n-2)}G_1(s_2, \delta)$ and then to $w := (2^l\rho)^{-1/3}v = (2^l\rho)^{-1/3}\big(x_1 - s_2^{1/(n-2)}G_1(s_2, \delta)\big)$ in a similar way as in Subsection 8.5.1 and rewrite

$$\mu_{l,\lambda}(x) = (\lambda 2^l\rho)^{4/3}\int e^{-i\lambda s_3\tilde{\Phi}_2(u,z,w,s_2,x,\delta)}a\big((2^l\rho)^{1/3}u, (2^l\rho)^{2/3}z, (2^l\rho)^{1/3}w, x, \delta\big)$$
$$\times\, \chi_1(z)\chi_0(u)\,\chi_0(w))\,\chi_1(s_3)\,du\,dz\,dw\,ds_3,$$

where

$$\tilde{\Phi}_2 = z(2^l\rho)(w - u_1) + \Psi_3(\rho^{1/3}w, x, \delta) + (2^l\rho)u_1^3\, B_3((2^l\rho)^{1/3}w, (2^l\rho)^{1/3}u_1, x, \delta)$$
$$+ (2^l\rho)(u_2^3\, b((2^l\rho)^{1/3}u, (2^l\rho)^{1/3}w, x, \delta_0^{\mathfrak{r}}) + \delta'_{3,0}u_2\,\tilde{\alpha}_1(2^l\rho)^{1/3}w, x, \delta_0^{\mathfrak{r}})$$
$$+ \delta'_0 u_1 u_2\,\alpha_{1,1}(2^l\rho)^{1/3}u_1, 2^l\rho)^{1/3}w, x, \delta_0^{\mathfrak{r}})).$$

By w^0 we denote the value of w corresponding to s_2^0. We now can see that there is also a unique critical point of the phase with respect to the variable z, at z^0, provided $w = u_1^0$. Thus, if $w^0 = u_1^0$, then the phase has a critical point with respect to the pairs of variables (u_1, s_2), respectively, (u_1, w); otherwise, we can again integrate by parts in z, which allows us to gain factors $(\lambda 2^l\rho)^{-N}$ as before, and we are done. So, assume that $w^0 = u_1^0$. Since the phase is linear in z and since $|\partial_z\partial_{u_1}\tilde{\Phi}_2| \sim 2^l\rho$ at the critical point, we see that we may apply the method of stationary phase to the double integration with respect to the pair of variables (u_1, z) and gain, in particular, a factor $(\lambda 2^l\rho)^{-1}$. Having realized this, we may come back to our previous formula (8.143), and knowing that we may apply the method of stationary phase to the integration with respect to the pair of variables (u_1, z) as well, we see that we may essentially write

$$\mu_{l,\lambda}(x) = \lambda^{1/3}\int e^{-i\lambda s_3\Psi_2(u_2,s_2,x,\delta,l)}a_2\left((2^l\rho)^{1/3}u_2, (2^l\rho)^{1/3}, s, \delta\right)\,\chi_0(u_2)$$
$$\times\, \chi_0((2^l\rho)^{-1/3}(x_1 - s_2^{1/(n-2)}G_1(s_2, \delta))\,\tilde{\chi}_1(s_2, s_3)\,du_2\,ds_2\,ds_3, \tag{8.144}$$

where the phase Ψ_2 arises from Φ_2 by replacing (u_1, z) by the critical point (u_1^0, z^0).

Now, arguing exactly as in Section 8.5.1, by means of Lemma 5.6 we find that the phase Ψ_2 is given by the expression in (8.42), with $B = 3$, that is,

$$\Psi_2 = s_2x_1^2\omega(\delta_1x_1) + x_1^n\alpha(\delta_1x_1) + s_2\delta_0y_2 + y_2^3\,b(x_1, y_2, \delta)$$
$$+ r(x_1, y_2, \delta) - s_2x_2 - x_3.$$

Moreover, since we here have changed coordinates from y_2 to u_2 so that $y_2 = (2^l\rho)^{1/3}u_2$, this means that

$$\begin{aligned}\Psi_2(u_2, s_2, x, \delta, l) = {} & s_2 x_1^2 \omega(\delta_1 x_1) + x_1^n \alpha(\delta_1 x_1) - s_2 x_2 - x_3 \\ & + (2^l\rho)\, u_2^3\, b\big(x_1, (2^l\rho)^{1/3}u_2, \delta\big) \\ & + (2^l\rho)^{1/3} u_2 \big(\delta_0 s_2 + \delta_3 x_1^{n_1} \alpha_1(\delta_1 x_1)\big) \end{aligned} \tag{8.145}$$

(compare (8.2)). Note that $\partial_{s_2}(s_2^{1/(n-2)} G_1(s_2, \delta)) \sim 1$ because $s_2 \sim 1$ and $G_1(s_2, 0) = 1$. Therefore, the relation $|x_1 - s_2^{1/(n-2)} G_1(s_2, \delta)| \lesssim (2^l\rho)^{1/3}$ can be rewritten as $|s_2 - \tilde{G}_1(x_1, \delta)| \lesssim (2^l\rho)^{1/3}$, where $\tilde{G}_1$ is again a smooth function such that $|\tilde{G}_1| \sim 1$. If we write

$$s_2 = (2^l\rho)^{1/3} v + \tilde{G}_1(x_1, \delta),$$

then this means that $|v| \lesssim 1$. We shall therefore change variables from s_2 to v, which leads to the following expression for $\mu_{l,\lambda}(x)$:

$$\begin{aligned}\mu_{l,\lambda}(x) = {} & (\lambda 2^l \rho)^{1/3} \int e^{-i\lambda s_3 \Psi_3(u_2, v, x, \delta, l)} a_3\big((2^l\rho)^{1/3}u_2, (2^l\rho)^{1/3}v, x, \delta\big) \\ & \times \chi_0(u_2)\chi_0(v)\, \chi_1(s_3)\, du_2\, dv\, ds_3, \end{aligned}$$

with a smooth amplitude a_3 and the new phase function

$$\begin{aligned}\Psi_3(u_2, v, x, \delta, l) = {} & v\,(2^l\rho)^{1/3}\big(x_1^2\omega(\delta_1 x_1) - x_2\big) + (2^l\rho)^{2/3}\delta_0\, v u_2 + Q_A(x, \delta) \\ & + (2^l\rho)\, u_2^3\, b\big(x_1, (2^l\rho)^{1/3}u_2, \delta\big) + (2^l\rho)^{1/3} u_2 Q_D(x, \delta) \end{aligned}$$

(compare with the corresponding expressions in (8.110)–(8.112)). Finally, putting $y_2 := (\lambda 2^l\rho)^{1/3}u_2$, we find that

$$\begin{aligned}\mu_{l,\lambda}(x) = {} & \int e^{-is_3\Psi_4(y_2, v, x, \delta, \lambda, l)} a_4\left(\lambda^{-1/3}y_2, (2^l\rho)^{1/3}v, x, \delta\right) \\ & \times \chi_0\left((\lambda 2^l\rho)^{-1/3}y_2\right)\, \chi_1(s_3)\, \chi_0(v)\, dv\, ds_3\, dy_2 \\ = {} & \int \widehat{\chi_1}\,(\Psi_4(y_2, v, x, \delta, \lambda, l))\; a_4\left(\lambda^{-1/3}y_2, (2^l\rho)^{1/3}v, x, \delta\right)\, \chi_0(v) \\ & \times \chi_0\left((\lambda 2^l\rho)^{-1/3}y_2\right)\, dy_2\, dv, \end{aligned} \tag{8.146}$$

with a smooth amplitude a_4 and phase function

$$\begin{aligned}\Psi_4(y_2, v, x, \delta, \lambda, l) = {} & v\,\lambda(2^l\rho)^{1/3}\big(x_1^2\omega(\delta_1 x_1) - x_2\big) + \lambda^{2/3}(2^l\rho)^{1/3}\delta_0\, v y_2 \\ & + \lambda Q_A(x, \delta) + y_2^3\, b\big(x_1, \lambda^{-1/3}y_2, \delta\big) + \lambda^{2/3} y_2 Q_D(x, \delta). \end{aligned}$$

We shall write this as

$$\Psi_4(y_2, v, x, \delta, \lambda, l) = A + Bv + y_2^3\, b\big(x_1, \lambda^{-1/3}y_2, \delta\big) + y_2(D + Ev), \tag{8.147}$$

with

$$\begin{aligned} A &:= \lambda Q_A(x, \delta), & B &:= \lambda 2^{l/3}\rho^{1/3} Q_B(x, \delta), \\ D &:= \lambda^{2/3} Q_D(x, \delta), & E &:= \lambda^{2/3} 2^{l/3}(\rho^{1/3}\delta_0)\,. \end{aligned} \tag{8.148}$$

Here, $Q_A(x, \delta)$, $Q_B(x, \delta)$, and $Q_D(x, \delta)$ are as in (8.112).

Applying Lemma 2.5, with $T := (\lambda 2^l \rho)^{1/3} = (\lambda\rho)^{1/3} 2^{l/3} \gg 1$ and $\delta := \lambda^{-1/3}$ so that $\delta T = (2^l\rho)^{1/3} \ll 1$, and subsequently Lemma 8.13 in a similar way as in Subsection 8.7.2, we find that in analogy with (8.114) we have that

$$|\mu_{l,\lambda}(x)| \lesssim \big(\max\{|A|, |B|, |D|, |E|\}\big)^{-1/6}, \tag{8.149}$$

now with A, B, D and E given by (8.148).

In order to estimate $\nu^{IV}_{\delta,1+it}(x)$, we shall again distinguish various cases.

As in the discussion of $\nu^{II}_{\delta,1+it}(x)$ in Subsection 8.7.2, the contributions by those terms in (8.141) for which either $|D| \gtrsim 1$ and $|E| \gtrsim 1$, or $|A| \gtrsim 1$ and $|B| \gtrsim 1$, can easily handled by means of estimate (8.149) (compare with (8.148)).

Consider next the terms for which $|D| \ll 1$ and $|E| \ll 1$. For these terms, it will be useful to rewrite $\mu_{l,\lambda}(x)$ in analogy with (8.115) as

$$\mu_{l,\lambda}(x) = \int e^{-is_3(A+Bv)} J(v, s_3)\, \chi_1(s_3)\, \chi_0(v)\, \chi_0(D+Ev)\, dv\, ds_3,$$

with

$$\begin{aligned} J(v, s_3) := \int & e^{-is_3(y_2^3 b(x_1, \lambda^{-1/3} y_2, \delta) + y_2(D+Ev))} a_4(\lambda^{-1/3} y_2, (2^l\rho)^{1/3} v, x, \delta) \\ & \times \chi_0\big((\lambda 2^l \rho)^{-1/3} y_2\big)\, dy_2 \,. \end{aligned} \tag{8.150}$$

From here we arrive without loss of generality at the following analogue of (8.117):

$$\mu_{l,\lambda}(x) = \int g(D+Ev, x, \lambda^{-1/3}, (2^l\rho)^{1/3}, \delta)\, \widehat{\chi_1}(A+Bv)\, \chi_0(v)\, dv,$$

where g is a smooth function of its bounded arguments.

Recall that we assume that either $|A| \gtrsim 1$ and $|B| \ll 1$, or $|A| \ll 1$ and $|B| \gtrsim 1$, or $|A| \ll 1$ and $|B| \ll 1$.

If $|A| \gtrsim 1$ and $|B| \ll 1$, then we can treat the summation in l by means of Lemma 2.7, where we choose, for λ fixed,

$$H_{\lambda,x}(u_1, u_2, u_3, u_4) := \int g(D + u_1 v, x, u_3, u_4, \delta)\, \widehat{\chi_1}(A + u_2 v)\, \chi_0(v)\, dv.$$

Then clearly $\|H_{\lambda,x}\|_{C^1(Q)} \lesssim |A|^{-1}$, and so after summation in those l for which $|E| \ll 1$ and $|B| \ll 1$, we can also sum (absolutely) in the λs for which $|A| \gtrsim 1$. Observe that this requires that $\gamma(\zeta)$ contain a factor

$$\gamma_1(\zeta) := \frac{2^{(1-\zeta)/2} - 1}{2^{1/3} - 1}.$$

Consider next the case where $|B| \gtrsim 1$ and $|A| \ll 1$. If we write $\lambda = 2^j$, then $\lambda 2^{l/3} = 2^{k/3}$ if we let $k := l + 3j$. We therefore pass from the summation variables j and l to the variables j and k, which allows us to write $B = 2^{k/3}\rho^{1/3} Q_B(x)$. For k fixed, we then sum first in j by means of Lemma 2.7, which gives an estimate of order $O(|B|^{-1})$, which then in turn allows to sum (absolutely) in those k for which

$|B| \gtrsim 1$. The application of Lemma 2.7 requires in this case that $\gamma(\zeta)$ contains a factor

$$\gamma_2(\zeta) := \frac{2^{1-\zeta}-1}{2^{2/3}-1}.$$

There remains the subcase where $|A|+|B| \ll 1$ and $|D|+|E| \lesssim 1$. The summation over all ls and λs for which these conditions are satisfied can easily be treated by means of the double summation Lemma 2.9, in a very similar way as in Subsection 8.7.2.

What remains are the contributions by those l and λ for which either $|D| \gtrsim 1$ and $|E| \ll 1$ or $|D| \ll 1$ and $|E| \gtrsim 1$.

We begin with the case where $|E| \gtrsim 1$ and $|D| \ll 1$. Then we may assume in addition that $|B| \ll 1$, for otherwise by (8.149) we have $|\mu_{l,\lambda}(x)| \lesssim |E|^{-1/12}|B|^{-1/12}$, which allows us to sum absolutely in j and l. In a very similar way, we may also assume that $|A| \ll 1$.

Recall next from (8.146) and (8.147) that

$$\mu_{l,\lambda}(x) = \int e^{-is_3\left(A+y_2^3\, b(x_1,\lambda^{-1/3}y_2,\delta)+Dy_2+v(B+Ey_2)\right)} a_4\left(\lambda^{-1/3}y_2, (2^l\rho)^{1/3}v, x, \delta\right)$$
$$\times\, \chi_1(s_3)\chi_0(v)\,\chi_0\left((\lambda 2^l\rho)^{-1/3}y_2\right)\, dy_2\, dv\, ds_3.$$

Again, by our usual argument, we may assume without loss of generality that a_4 is independent of v. Then we find that, in analogy with (8.118),

$$\mu_{l,\lambda}(x) = \int e^{-is_3\left(A+y_2^3\, b(x_1,\lambda^{-1/3}y_2,\delta)+Dy_2\right)}\,\widehat{\chi_0}\left(s_3(B+Ey_2)\right)$$
$$\times\, a_4\left(\lambda^{-1/3}y_2, (2^l\rho)^{1/3}, x, \delta\right)\chi_1(s_3)\,\chi_0\left((\lambda 2^l\rho)^{-1/3}y_2\right)\, dy_2\, ds_3.$$

We then change the summation variables from j, l to j, k, where $k := 2j + l$, so that $E = 2^{k/3}\rho^{1/3}\delta_0$. Then, for k fixed, we can treat the summation in j by means of Lemma 2.7, where we choose

$$H_{k,x}(u_1, u_2, u_3, u_4, u_5, u_6) := \int e^{-is_3\left(u_1+y_2^3\, b(x_1,u_2y_2,\delta)+u_6y_2\right)}\,\widehat{\chi_0}\left(s_3(u_3+Ey_2)\right)$$
$$\times\, a_4(u_2y_2, u_4, x, \delta)\chi_1(s_3)\,\chi_1(v)\,\chi_0(u_5y_2)\, dy_2\, ds_3.$$

Arguing in the same way as in the corresponding case of Subsection 8.7.2, we find by means of the change of variables $y_2 \mapsto y_2/E$ that $\|H_{\lambda,x}\|_{C^1(Q)} \lesssim |E|^{-1}$, and thus after the summation over the $\lambda = 2^j$ this allows to subsequently also sum over the k for which $|E| \gtrsim 1$.

There remains the contribution by those l and λ for which $|D| \gtrsim 1$ and $|E| \ll 1$. Observe that here we have $|D + Ev| \gtrsim 1$ in (8.150).

Applying the change of variables $y_2 = \lambda^{1/3}t$ in the integral defining $J(v, s_3)$, we obtain

$$J(v, s_3) = \lambda^{1/3}\int e^{-is_3\lambda\left(t^3 b(x_1,t,\delta)+t(\lambda^{-2/3}(D+Ev))\right)} a_4\left(t, (2^l\rho)^{1/3}v, x, \delta\right)$$
$$\times\, \chi_0\left((2^l\rho)^{-1/3}t\right)\,\chi_0(v)\,\chi_1(s_3)\, dt.$$

Observe also that by (8.148)

$$\lambda^{-2/3} D = Q_D(x,\delta), \quad \lambda^{-2/3} E = \delta_0 \ll 1,$$

so that, in particular, $\lambda^{-2/3}|D + vE| \lesssim 1$.

Arguing in a similar way as in Subsection 8.7.2, we find that for every $N \in \mathbb{N}$,

$$\begin{aligned} J(v,s_3) = {} & |D+Ev|^{-1/4} a_+ \left(\lambda^{-1/3}|D+Ev|^{1/2}, v, x, (2^l\rho)^{1/3}, \delta\right) \\ & \times \chi_0\left((\lambda 2^l \rho)^{-1/3}|D+Ev|^{1/2}\right) e^{-is_3|D+Ev|^{3/2} q_+(\lambda^{-1/3}|D+Ev|^{1/2}, x, \delta)} \\ & + |D+Ev|^{-1/4} a_- \left(\lambda^{-1/3}|D+Ev|^{1/2}, v, x, (2^l\rho)^{1/3}, \delta\right) \\ & \times \chi_0\left((\lambda 2^l \rho)^{-1/3}|D+Ev|^{1/2}\right) e^{-is_3|D+Ev|^{3/2} q_-(\lambda^{-1/3}|D+Ev|^{1/2}, x, \delta)} \\ & + (D+Ev)^{-N} F_N \left(|D+vE|^{3/2}, \lambda^{-1/3}|D+Ev|^{1/2}, v, x, 2^{-l/3}, \right. \\ & \left. \qquad (2^l\lambda^{-1})^{1/3}, \delta\right), \end{aligned}$$

where $a_\pm$, $q_\pm$ and F_N are smooth functions of their (bounded) variables. Moreover, $|q_\pm\left(0, x, (2^l\lambda^{-1})^{1/3}, \delta\right)| \sim 1$.

We shall concentrate on the first term only. The second term can be treated in the same way as the first one, and the last term can be handled in an even easier way by a similar method, since it is of order $= O(|D|^{-N})$ and, unlike the first term, carries no oscillatory factor.

We denote by

$$\begin{aligned} \mu^1_{l,\lambda}(x) := \int & e^{-is_3\left[(A+Bv) + |D+Ev|^{3/2} q_+(\lambda^{-1/3}|D+Ev|^{1/2}, x, \delta)\right]} |D+Ev|^{-1/4} \chi_1(s_3)\, \chi_0(v) \\ & \times a_+\left(\lambda^{-1/3}|D+Ev|^{1/2}, v, x, (2^l\rho)^{1/3}, \delta\right) \chi_0 \\ & \times \left((\lambda 2^l\rho)^{-1/3}|D+Ev|^{1/2}\right) dv\, ds_3 \end{aligned}$$

the contribution by the first term in (8.151) to $\mu_{l,\lambda}(x)$, and by $\nu^1_{\delta,1+it}(x)$ the contribution of the $\mu^1_{l,\lambda}(x)$ to the sum defining $\nu^{IV}_{\delta,1+it}(x)$.

Assuming, for instance, that $D > 0$ and making use of (8.121), we here find that the complete phase in the oscillatory integral defining $\mu^1_{l,\lambda}(x)$ is of the form

$$s_3[A' + B'v + r],$$

with $A' = \lambda Q_{A'}(x,\delta)$, $B' = \lambda 2^{l/3} Q_{B'}(x,\delta)$, where

$$\begin{aligned} Q_{A'}(x,\delta) &:= Q_A(x,\delta) + Q_D(x,\delta)^{3/2} q_+\left(Q_D(x,\delta)^{1/2}, x, \delta\right), \\ Q_{B'}(x,\delta) &:= \rho^{1/3} Q_B(x,\delta) + \tfrac{1}{2} q_+'\left(Q_D(x,\delta)^{1/2}, x, \delta\right)\rho^{1/3} Q_D(x,\delta)\delta_0 \\ & \quad + \tfrac{3}{2} q_+'\left(Q_D(x,\delta)^{1/2}, x, \delta\right)\rho^{1/3} Q_D(x,\delta)^{1/2}\delta_0, \end{aligned}$$

and where r is again a bounded error term.

Thus, if $|B'| \gg 1$, then an integration by parts in v shows that

$$|\mu^1_{l,\lambda}(x)| \lesssim |D|^{-1/4}|B'|^{-1}.$$

This estimate allows to control the sum over all l such that $|B'| \gg 1$ and, subsequently, the sum over all dyadic λ such that $|D| \gtrsim 1$, and we arrive at the desired uniform estimate in x and δ.

Next, if $|B'| \lesssim 1$, then we can argue in a similar way as before and apply Lemma 2.7 to the summation in l by letting

$$H_{\lambda,x}(u_1,\ldots,u_7) := \int e^{-is_3[A'+u_1v+r(Q_D(x)^{1/2},D^{-1}u_3,v,x,\delta)]}|D+u_2v|^{-1/4}\chi_1(s_3)\,\chi_0(v)$$
$$\times\, a_+\big(|Q_D(x,\delta)+u_3v|^{1/2},v,x,u_4,\delta\big)$$
$$\times\, \chi_0\left(|u_6+u_7v|^{1/2}\right)\,dv\,ds_3$$

and choosing the cuboid Q in the obvious way. Then we easily see that $\|H_{\lambda,x}\|_{C^1(Q)} \lesssim |D|^{-1/4}$, and so after summation in those l for which $|B'| \ll 1$, we can also sum (absolutely) in the λs for which $|D| \gtrsim 1$. Observe that this requires again that $\gamma(\zeta)$ contains the factor $\gamma_1(\zeta)$.

This concludes the proof of the uniform estimate of $\nu^1_{\delta,1+it}(x)$ in x and δ and, thus, also of estimate (8.140).

The proof of Proposition 8.12 is now complete and, thus, also ultimately the proof of our main result, Theorem 1.14.

Chapter Nine

Proofs of Propositions 1.7 and 1.17

We conclude this monograph with the remaining proofs of two results from the introduction.

9.1 APPENDIX A: PROOF OF PROPOSITION 1.7 ON THE CHARACTERIZATION OF LINEARLY ADAPTED COORDINATES

In order to prove that condition (a) implies condition (b) in Proposition 1.7, assume that $d_x := d(\phi) = h_{\text{lin}}(\phi)$. By interchanging the coordinates x_1 and x_2, if necessary, we may assume that $\kappa_2/\kappa_1 \geq 1$, where we recall that $\kappa_2/\kappa_1 \in \mathbb{N}$. Now, if we had $\kappa_2/\kappa_1 = 1$, then, by Varchenko's algorithm, there would exist a linear change of coordinates of the form $y_1 = x_1, y_2 = x_2 - cx_1$ so that $d_y > d_x = d$, which would contradict the maximality of d_x. Thus, necessarily, $\kappa_2/\kappa_1 \geq 2$.

Conversely, assume without loss of generality that $\kappa_2/\kappa_1 \geq 2$. Consider any matrix $T = \begin{pmatrix} a & b \\ c & d \end{pmatrix} \in GL(2, \mathbb{R})$ and the corresponding linear coordinates y given by

$$x_1 = ay_1 + by_2, \quad x_2 = cy_1 + dy_2.$$

To prove (a), we have to show that $d_y \leq d_x$ for all such matrices T

Case 1. $a \neq 0$. Then we may factor $T = T_1T_2$, where

$$T_1 := \begin{pmatrix} a & 0 \\ c & \dfrac{ad - bc}{a} \end{pmatrix}, \quad T_2 := \begin{pmatrix} 1 & \dfrac{b}{a} \\ 0 & 1 \end{pmatrix}.$$

We first consider T_2. Since $\phi_{\text{pr}}(T_2y) = \phi_\kappa(y_1 + \frac{b}{a}y_2, y_2)$ and since y_2 is κ-homogenous of degree $\kappa_2 > \kappa_1$, whereas y_1 is κ-homogenous of degree κ_1, we see that the κ-principal part of $\phi \circ T_2$ is given by $(\phi \circ T_2)_\kappa = \phi_\kappa$, so that $\phi \circ T_2$ and ϕ have the same principal face and, in particular, the same Newton distance. This shows that we may assume without loss of generality that $b = 0$. Then, necessarily, $d \neq 0$. But then our change of coordinates is of the type $x_1 = ay_1, x_2 = cy_1 + dy_2$ considered in Lemma 3.2 of [IM11a], so that this lemma implies that $d_y \leq d_x$. Indeed, one finds more precisely that $d_y < d_x$, if $c \neq 0$, and $d_y = d_x$ otherwise.

Case 2. $a = 0, d = 0$. Since separate scalings of the coordinates have no effect on the Newton polyhedra, T then essentially interchanges the roles of x_1 and x_2, that is, the Newton polyhedron is reflected at the bisectrix under this coordinate change. This shows that here $d_y = d_x$.

Case 3. $a = 0, d \neq 0$. Then we may factor $T = \begin{pmatrix} 0 & b \\ c & d \end{pmatrix} = T_1 T_2$, where

$$T_1 := \begin{pmatrix} 0 & 1 \\ 1 & 0 \end{pmatrix}, \quad T_2 := \begin{pmatrix} c & d \\ 0 & b \end{pmatrix}.$$

We have seen in the previous cases that both T_1 and T_2 do not change the Newton distance, and thus here $d_y = d_x$. This concludes the proof of the first part of Proposition 1.7.

Assume finally that x and y are two linearly adapted coordinate systems for ϕ for which the corresponding principal weights κ and κ' satisfy $\kappa_2/\kappa_1 > 1$ and $\kappa'_2/\kappa'_1 > 1$, respectively. Choose $T \in GL(2, \mathbb{R})$ such that $x = Ty$.

Inspecting the three cases from the previous argument, we see that in Case 1 the mapping T_2 does not change the principal face and that necessarily $c = 0$, since otherwise we had $d_y < d_x$. But then T_1 also does not change the principal face. Case 2 cannot arise here, since we assume that both $\kappa_2/\kappa_1 > 1$ and $\kappa'_2/\kappa'_1 > 1$, and similarly Case 3 cannot apply. This proves also the second statement in the proposition. Q.E.D.

9.2 APPENDIX B: A DIRECT PROOF OF PROPOSITION 1.17 ON AN INVARIANT DESCRIPTION OF THE NOTION OF r-HEIGHT

We shall prove both parts (a) and (b) of Proposition 1.17 at the same time. So, let us assume that our coordinates (x_1, x_2) are linearly adapted to ϕ, and let $f(x_1)$ be any nonflat fractionally smooth, real-valued function of x_1, say for $x_1 > 0$, with corresponding fractional shear

$$y_1 := x_1, \quad y_2 := x_2 - f(x_1)$$

on the half plane H^+. In order to prove Proposition 1.17, what still remains to be shown is that

$$h^f(\phi) \le \begin{cases} d(\phi), & \text{if the coordinates } (x_1, x_2) \text{ are adapted to } \phi, \\ h^r(\phi), & \text{if the coordinates } (x_1, x_2) \text{ are not adapted to } \phi, \end{cases} \tag{9.1}$$

without making recourse to the Theorems 1.5 and 1.14, as we did in our "indirect" proof of this proposition in Section 1.5. Indeed, the reverse inequalities had already been shown directly in Section 1.5.

Recall from (1.19) and (1.9) that $c_0 x_1^{m_0}$ denotes the leading term in the Puiseux series expansion of $f(x_1)$, whereas $b_1 x_1^m$ denotes the leading term of the principal root jet $\psi(x_1)$; m_0 is rational, and m is an integer.

In analogy with the definition of the supporting line L^f to the Newton polyhedron $\mathcal{N}(\phi^f)$, we choose the weight $\kappa^0 = (\kappa_1^0, \kappa_2^0)$ so that $\kappa_2^0/\kappa_1^0 = m_0$ and so that the line

$$L^0 := \{(t_1, t_2) \in \mathbb{R}^2 : \kappa_1^0 t_1 + \kappa_2^0 t_2 = 1\}$$

is a supporting line to the Newton polyhedron $\mathcal{N}(\phi)$ of ϕ. Then $\mathcal{N}(\phi) \cap L^0$ is an interval of the form $[T_1, T_2]$, whose left endpoint we shall denote by $T_1 = (A, B)$. Note that if $T_1 = T_2$, then $[T_1, T_2]$ is a vertex of $\mathcal{N}(\phi)$ and otherwise is a compact edge. We shall distinguish three cases, which will depend on the relative position of $[T_1, T_2]$ to the bisectrix $\Delta = \{t_1 = t_2\}$.

The case where $[T_1, T_2] \subset \{t_1 \leq t_2\}$. In this case, we will prove that

$$h^f(\phi) = d^f(\phi) \leq d(\phi), \tag{9.2}$$

where $d^f(\phi) := d^f$. This clearly will imply (9.1).

Indeed, we shall show that the Newton polyhedron $\mathcal{N}(\phi^f)$ has a compact edge of the form $[T_1, \tilde{T}_2] \subset L^0$, where T_1 lies on or above the bisectrix, whereas $\tilde{T}_2$ lies on or below the bisectrix. In particular, this will imply that $L^f = L^0$ and that the augmented Newton polyhedron $\mathcal{N}^f(\phi^f)$ will have only one edge within the half space bounded from below by the bisectrix, and this edge lies again on the line L^f. By the definition (1.20) of $h^f(\phi)$ and its geometric interpretation in terms of the augmented Newton polyhedron $\mathcal{N}^f(\phi^f)$, this will mean that $h^f(\phi) = d^f$. It will be useful to recall here that the second coordinate of the point of intersection of the line $\Delta^{(m_0)}$ with the line L^f is given by $d^f + 1$. On the other hand, by the definition of d^f, (d^f, d^f) is the point of intersection of the line L^f with the bisectrix Δ, and thus it obvious from the geometry of the Newton polyhedron $\mathcal{N}(\phi^f)$ that $d^f \leq d$, and we arrive at (9.2).

We begin with the case where $\mathcal{N}(\phi) \cap L^0$ is a vertex, that is, where $T_1 = T_2 = (A, B)$. Since the point (A, B) lies on or above the bisectrix, we have $B \geq A$. Moreover, the κ^0-principal part of ϕ is of the form $\phi_{\kappa^0}(x_1, x_2) = cx_1^A x_2^B$, and it is then easily seen that the κ^0-principal part of $\phi^f(y_1, y_2) = \phi(y_1, y_2 + f(y_1))$ is given by the fractionally smooth function $cy_1^A(y_2 + c_0 y_1^{m_0})^B$ (compare the proof of Lemma 3.2 in [IM11a]).

So, the associated edge of the Newton polyhedron $\mathcal{N}(\phi^f)$ is the closed interval $[T_1, \tilde{T}_2] \subset L^0$, with $T_1 = (A, B)$ lying above the bisectrix and $\tilde{T}_2 := (A + m_0 B, 0)$ lying below the bisectrix. Hence, the interval $[T_1, \tilde{T}_2]$ is, in fact, the principal face of the Newton polyhedron $\mathcal{N}(\phi^f)$, which proves our claim in this case.

Let us next consider the case where $\mathcal{N}(\phi) \cap L^0$ is a compact edge γ of $\mathcal{N}(\phi)$. By Proposition 2.2 in [IM11a], we may then assume that

$$\kappa_1^0 := \frac{q}{k}, \quad \kappa_2^0 := \frac{p}{k}, \quad (p, q, k) = 1, \quad (p, q) = 1,$$

and that the corresponding polynomial ϕ_{κ^0} can be factored as

$$\phi_{\kappa^0} := cx_1^{\nu_1} x_2^{\nu_2} \prod_{l=1}^{M} (x_2^q - \lambda_l x_1^p)^{n_l},$$

with $M \geq 1$, distinct $\lambda_l \in \mathbb{C} \setminus \{0\}$, and $n_l \in \mathbb{N} \setminus \{0\}$ and with $\nu_1, \nu_2 \in \mathbb{N} \setminus \{0\}$. We let $n := \sum_{l=1}^{M} n_l$. Then $m_0 = \kappa_2^0/\kappa_1^0 = p/q$, and the edge γ is given by the interval $\gamma = [(\nu_1, \nu_2 + nq), (\nu_1 + np, \nu_2)]$. Since the edge γ is contained in the half space where $t_2 \geq t_1$, we have $\nu_2 \geq \nu_1 + np$.

If $c_0 x_1^{m_0} = c_0 x_1^{p/q}$ does not coincide with any root of the polynomial ϕ_{κ^0}, then we have

$$\begin{aligned}\phi^f_{\kappa^0}(y_1, y_2) &= ((\phi_{\kappa^0})^f)_{\kappa^0}(y_1, y_2)\\ &= c y_1^{\nu_1}(y_2 + c_0 y_1^{p/q})^{\nu_2} \prod_{l=1}^{M}\Big((y_2 + c_0 y_1^{p/q})^q - \lambda_l y_1^p\Big)^{n_l}.\end{aligned}$$

So, as before, the principal face of the Newton polyhedron $\mathcal{N}(\phi^f)$ will be an interval $[T_1, \tilde{T}_2] \subset L^0$, with $T_1 = (\nu_1, \nu_2 + nq) = (A, B)$ and $\tilde{T}_2 := (\nu_1 + \nu_2 p/q + np, 0)$ lying on the first coordinate axis, so that again this interval is the principal face of the Newton polyhedron $\mathcal{N}(\phi^f)$.

Assume next that $c_0 x_1^{m_0}$ does coincide with some real root of the polynomial function ϕ_{κ^0}, say, $c_0 = \lambda_{l_0}^{1/q}$. Then we have

$$((\phi_{\kappa^0})^f)_{\kappa^0}(y_1, y_2) = c y_1^{\nu_1}(y_2 + c_0 y_1^{p/q})^{\nu_2} y_2^{n_{l_0}} \prod_{j=1}^{q-1}\Big(y_2 - c_0(\varepsilon_j - 1) y_1^{p/q}\Big)^{n_{l_0}}$$

$$\times \prod_{l \neq l_0}^{M}\Big((y_2 + c_0 y_1^{p/q})^q - \lambda_l y_1^p\Big)^{n_l},$$

where $\{\varepsilon_j\}_{j=1,\dots,q-1}$ denotes the set of qth roots of unity that are different from 1. This shows that the principal edge of the Newton polyhedron $\mathcal{N}(\phi^f)$ is now given by the interval $[T_1, \tilde{T}_2] \subset L^0$, with $T_1 = (\nu_1, \nu_2 + nq) = (A, B)$ as before and now with

$$\tilde{T}_2 := \Big(\nu_1 + \frac{\nu_2 p}{q} + \frac{(q-1)pn_{l_0}}{q} + (n - n_{l_0})p,\ n_{l_0}\Big).$$

Observe that T_1 lies above the bisectrix, whereas $\tilde{T}_2$ lies on or below the bisectrix, which again proves our claim. Indeed, since $n \geq n_{l_0}$ and $\nu_2 \geq \nu_1 + np$, we have

$$\frac{\nu_2 p}{q} + \frac{(q-1)pn_{l_0}}{q} \geq pn_{l_0}\frac{(q-1+p)}{q} \geq n_{l_0}.$$

The case where $[T_1, T_2] \subset \{t_1 \geq t_2\}$. Then $A \geq B$, and so the principal face of $\mathcal{N}(\phi^f)$ agrees with the principal face of $\mathcal{N}(\phi)$. Arguing in a similar way as in the preceding case, we see that again $L^0 = L^f$, but this time L^f will touch the Newton polyhedron of ϕ^f only in points lying on or below the bisectrix. This shows that the line $\Delta^{(m_0)}$ will intersect the boundary of the augmented Newton polyhedron $\mathcal{N}^f(\phi^f)$ at some point of the line L^f, so that again $h^f(\phi) = d^f(\phi)$. Moreover, from the geometry of the lines L and L^f, it is again clear that $d^f(\phi) \leq d(\phi)$; that is, (9.1) does hold true also in this case.

There remains the case where T_1 lies strictly above the bisectrix and T_2 strictly below it. Then clearly the compact edge $[T_1, T_2]$ is the principal face $\pi(\phi) \subset L$ of $\mathcal{N}(\phi)$, and necessarily we have $L^0 = L$ and $m_0 = m$. What remains is the following case.

The case where $\pi(\phi)$ is a compact edge given by $[T_1, T_2]$. Let us first assume that the coordinates (x_1, x_2) are adapted to ϕ and denote by κ the principal weight associated to the principal face $\pi(\phi)$ of the Newton polyhedron of ϕ, so that $\pi(\phi)$ is contained in the principal line

$$L = \{(t_1, t_2) \in \mathbb{R}^2 : \kappa t_1 + \kappa_2 t_2 = 1\}.$$

Since the coordinates of ϕ are adapted to ϕ, Proposition 1.2 shows that the κ-principal part ϕ_κ has only real roots of multiplicity $\leq d(\phi)$. By a similar reasoning as in the case where $[T_1, T_2] \subset \{t_1 \leq t_2\}$, we then see that the Newton polyhedron $\mathcal{N}(\phi^f)$ contains a compact edge of the form $[T_1, \tilde{T}_2] \subset L$, where T_1 is the left endpoint of the principal edge $\pi(\phi)$ and thus lies above the bisectrix and where the second coordinate of the point $\tilde{T}_2$ is still $\leq d(\phi)$, so that $\tilde{T}_2$ again lies on or below the bisectrix. This implies that we indeed have $L^f = L^0 = L$ and $d^f(\phi) = d(\phi)$.

Let us finally assume that the coordinates (x_1, x_2) are not adapted to ϕ and that in adapted coordinates, ϕ is represented by ϕ^a. Recall also that $m_0 = m$. We then write

$$\phi^f(y_1, y_2) = \phi(y_1, y_2 + \psi(y_1) + (f(y_1) - \psi(y_1))) = \phi^a(y_1, y_2 + f_1(y_1)),$$

where $f_1(y_1) := f(y_1) - \psi(y_1)$, that is, $\phi^f = (\phi^a)^{f_1}$.

If f_1 is flat, then the Newton polyhedra of ϕ^f and ϕ^a are the same, which clearly implies that $h^r(\phi) = h^f(\phi)$. Let us therefore assume that f_1 is nonflat. Then f_1 has a formal Puiseux series expansion

$$f_1(y_1) \sim \sum_{j \geq 1} c_j y_1^{m_j}$$

with leading term $c_1 y_1^{m_1}$, where $c_1 \neq 0$. Then $m = m_0 \leq m_1 < m_2 < \cdots$ are rational numbers with a fixed common denominator.

(i) Consider first the case where the principal face of the Newton polyhedron $\mathcal{N}(\phi^a)$ is a compact edge whose slope has modulus $1/a$.

If $m_1 \geq a$, then we have $m_1 \geq a > m_0 = m$. It is then easily seen that $L^f = L$ and that the augmented Newton polyhedra $\mathcal{N}^r(\phi^a)$ and $\mathcal{N}^f(\phi^f)$ do agree in the half space bounded from below by the bisectrix, since the "perturbation" of ϕ^a by $f_1(y_1)$ will have no effect on the Newton polyhedron of ϕ^a within this half space. Consequently, we find that $h^f(\phi) = h^r(\phi)$.

Indeed, if $m_1 > a$, then the claim about the effect of the "perturbation" of ϕ^a by $f_1(y_1)$ is obvious.

And, if $m_1 = a$, then let us denote by κ^{pr} the principal weight associated to the principal face $\pi(\phi^a)$ of the Newton polyhedron of ϕ^a, so that $\pi(\phi^a)$ is contained in the line

$$L^{\text{pr}} := \{(t_1, t_2) \in \mathbb{R}^2 : \kappa_1^{\text{pr}} t_1 + \kappa_2^{\text{pr}} t_2 = 1\}.$$

Since the coordinates of ϕ^a are adapted to ϕ^a, arguing in a similar way as in the case where the coordinates (x_1, x_2) were adapted to ϕ, we then see that the Newton polyhedron $\mathcal{N}((\phi^a)^{f_1})$ contains a compact edge of the form $[T_1, \tilde{T}_2] \subset L^{\text{pr}}$, where the left endpoint T_1 lies above the bisectrix and where the second coordinate of

$\tilde{T}_2$ is still $\leq h$, so that $\tilde{T}_2$ lies on or below the bisectrix. This implies again that $h^f(\phi) = h^r(\phi)$.

Assume finally that $m_1 < a$. Then $m = m_0 \leq m_1 < a$. We shall then prove that, within the closed half space lying above the bisectrix, the augmented Newton polyhedron of ϕ^a is contained in the one of ϕ^f. To this end, observe first that again $L = L^f$. We shall argue in a somewhat similar way as in the case where $[T_1, T_2] \subset \{t_1 \leq t_2\}$, but with ϕ^a taking over the role of ϕ, f_1 the role of f, and m_1 the role of m_0. The basis for our arguments will be the identity $\phi^f = (\phi^a)^{f_1}$.

Indeed, denote by L^{f_1} the supporting line to $\mathcal{N}(\phi^a)$ with slope $1/m_1$, and denote by $S_1 = (a_1, b_1)$ the vertex of $\mathcal{N}(\phi^a) \cap L^{f_1}$ with largest second coordinate. The same kind of reasoning that we had employed in the case where $[T_1, T_2] \subset \{t_1 \leq t_2\}$ (distinguishing between the cases where $\mathcal{N}(\phi^a) \cap L^{f_1}$ is a single point or a compact interval) then shows that indeed the principal face of $(\phi^a)^{f_1}$ will be an interval of the form $[S_1, \tilde{S}_2]$ contained in the line L^{f_1}. Moreover, within the half space where $t_1 \geq a_1$, the augmented Newton polyhedra $\mathcal{N}^r(\phi^a)$ and $\mathcal{N}^f(\phi^f)$ will agree since $\phi^f = (\phi^a)^{f_1}$. And, since L^{f_1} is a supporting line to the convex set $\mathcal{N}(\phi^a)$, which is less steep than the line L, all this together shows that we have

$$\mathcal{N}^r(\phi^a) \cap \{t_2 \geq t_1\} \subset \mathcal{N}^f(\phi^f) \cap \{t_2 \geq t_1\}.$$

But this clearly implies that the line $\Delta^{(m)} = \Delta^{(m_0)}$ will intersect the boundary of the augmented Newton polyhedron $\mathcal{N}^f(\phi^f)$ at some point whose second coordinate is less than or equal to the second coordinate of the point of intersection with the boundary of $\mathcal{N}^r(\phi^a)$, and thus $h^f(\phi) \leq h^r(\phi)$.

(ii) Consider next the case where the principal face of $\mathcal{N}(\phi^a)$ is a noncompact horizontal edge. This case can be viewed as the limiting case as $a \to \infty$ of the previous case (i) when $m_1 < a$ and can thus be treated in a very similar way. We leave the details to the interested reader.

(iii) Finally, we consider the case where the principal face $\mathcal{N}(\phi^a)$ is a vertex (h, h). We then chose the index $l_{\rm pr}$ as in the beginning of Chapter 6, so that (h, h) is the right endpoint $(A_{l_{\rm pr}-1}, B_{l_{\rm pr}-1})$ of the compact edge $\gamma_{l_{\rm pr}-1} = [(A_{l_{\rm pr}-2}, B_{l_{\rm pr}-2}), (A_{l_{\rm pr}-1}, B_{l_{\rm pr}-1})]$, whose (modulus) of slope is given by $1/a_{l_{\rm pr}-1}$.

We may then argue essentially in the same way as in case (i), by replacing the edge given by the principal face in (i) by the edge $\gamma_{l_{\rm pr}-1}$ and the exponent a by $a_{l_{\rm pr}-1}$. Observe that if $m_1 = a_{l_{\rm pr}-1}$, then the change of coordinates given by the nonlinear shear defined by f_1 will transform the edge $\gamma_{l_{\rm pr}-1}$ into another edge $[(A_{l_{\rm pr}-2}, B_{l_{\rm pr}-2}), (\tilde{A}_2, \tilde{B}_2)]$ lying on the same line as $\gamma_{l_{\rm pr}-1}$, and the right endpoint $(\tilde{A}_2, \tilde{B}_2)$ will still lie on or below the bisectrix. This follows from the same kind of reasoning that we applied in the case where $[T_1, T_2] \subset \{t_1 \leq t_2\}$. Again we leave the details to the interested reader.

This concludes the proof of the inequality (9.1) and, hence, also our direct proof of Proposition 1.17.

Bibliography

[AKC79] Arhipov, G. I., Karacuba, A. A., Čubarikov, V. N., Trigonometric integrals. *Izv. Akad. Nauk SSSR Ser. Mat.*, 43 (1979), 971–1003, 1197 (Russian); English translation in *Math. USSR-Izv.*, 15 (1980), 211–239.

[Arn73] Arnol'd, V. I., Remarks on the method of stationary phase and on the Coxeter numbers. *Uspekhi Mat. Nauk*, 28 (1973), 17–44 (Russsian); English translation in *Russian Math. Surveys*, 28 (1973), 19–48.

[AGV88] Arnol'd, V. I., Gusein-Zade, S. M., Varchenko, A. N., Singularities of differentiable maps. Vol. II, Monodromy and asymptotics of integrals, *Monographs in Mathematics, 83*. Birkhäuser, Boston Inc., Boston, MA, 1988.

[At70] Atiyah, M., Resolution of singularities and division of distributions. *Comm. Pure Appl. Math.*, 23 (1970) 145–150.

[BS11] Bak, J.-G., Seeger, A., Extensions of the Stein-Tomas theorem. *Math. Res. Lett..*, 18 (2011), no. 4, 767–81.

[Ba85] Barcelo Taberner, B., On the restriction of the Fourier transform to a conical surface. *Trans. Amer. Math. Soc.*, 292 (1985), no. 1, 321–33.

[Ba86] ———. The restriction of the Fourier transform to some curves and surfaces. *Studia Math.*, 84 (1986), no. 1, 39–69.

[BCT06] Bennet, J., Carbery, A,, Tao, T., On the multilinear restriction and Kakeya conjectures, *Acta Math.*, 196 (2006), 261–302.

[BG69] Bernstein, I. N., Gelfand, S. I., Meromorphy of the function P^λ. *Funktsional. Anal. i Priložen.*, 3 (1) (1969) 84–85.

[Bo95] Borel, E., Sur quelques points de la théorie des fonctions. *Ann. Sci. École Norm. Sup.*, 12 no. 3 (1895), 9–55.

[Bou85] Bourgain, J., Estimations de certaines fonctions maximales. *C. R. Acad. Sci. Paris Sér. I Math.*, 301 (1985) no. 10, 499–502.

[Bou91] ———. Besicovitch-type maximal operators and applications to Fourier analysis. *Geom. Funct. Anal. 1*, no. 2 (1991), 147–87.

[Bou95] ———. Some new estimates on oscillatory integrals. *Essays in Fourier Analysis in honor of E. M. Stein*. Princeton Math. Ser. 42, Princeton University Press, Princeton, NJ, 1995, 83–112.

[BG11] Bourgain, J., Guth, L., Bounds on oscillatory integral operators. *C. R. Acad. Sci. Paris, Ser. I*, 349 (2011) 137–41.

[BNW88] Bruna, J., Nagel, A., Wainger, S., Convex hypersurfaces and Fourier transforms, *Ann. of Math. (2)*, 127 (1988), 333–365.

[Bu12] Buschenhenke, S., A sharp L^p-L^q Fourier restriction theorem for a conical surface of finite type. Mathematische Zeitschrift, Vol. (1) (2015), 367–99.

[BMV15] Buschenhenke, S., Müller, D., Vargas, A., restriction theorem for a two-dimensional surface of finite type, arXiv:1508.00791.

[CS72] Carleson, L., P. Sjölin, P., Oscillatory integrals and a multiplier problem for the disc. *Studia Math.*, 44 (1972), 287–99.

[CaCW99] Carbery, C., Christ, M., Wright, J., Multidimensional van der Corput and sublevel set estimates. *J. Amer. Math. Soc.*, 12 (1999), no. 4, 981–1015.

[vdC21] van der Corput, J. G., Zahlentheoretische Abschätzungen. *Math. Ann.*, 84 (1921), 53–79.

[D77] Domar, Y., On the Banach algebra $A(G)$ for smooth sets $\Gamma \subset \mathbb{R}^n$. *Comment. Math. Helv.*, 52, no. 3 (1977), 357–71.

[Dru85] Drury, S. W., Restrictions of Fourier transforms to curves. *Ann. Inst. Fourier*, 35 (1985), 117–23.

[Dui74] Duistermaat, J. J., Oscillatory integrals, Lagrange immersions and unfolding of singularities. *Comm. Pure Appl. Math.*, 27 (1974), 207–81.

[F70] Fefferman, C., Inequalities for strongly singular convolution operators. *Acta Math.*, (1970), 9–36.

[FU04] Ferreyra, E., Urciuolo, M., Restriction theorems for the Fourier transform to homogeneous polynomial surfaces in $\mathbb{R}^3$. *Studia Math.*, 160 (3) (2004), 249–65.

[FU08] ———. Restriction theorems for anisotropically homogeneous hypersurfaces of $\mathbb{R}^{n+1}$ *Georgian Math. J.*, 15 (4) (2008), 643–51.

[FU09] ———. Fourier restriction estimates to mixed homogeneous surfaces *JIPAM. J. Inequal. Pure Appl. Math.*, 10 (2009), no. 2, Article 35, 11 pp.

[GV92] Ginibre, J., Velo, G., Smoothing properties and retarded estimates for some dispersive evolution equations *Commun. Math. Phys.*, 144 (1992), 163–88.

[G09] Grafakos, L., Modern Fourier analysis. *Graduate Texts in Mathematics* 250. Springer, New York, 2009.

[Gb09] Greenblatt, M., The asymptotic behavior of degenerate oscillatory integrals in two dimensions. *J. Funct. Anal.*, 257 (2009), no. 6, 1759–98.

[Gl81] Greenleaf, A., Principal curvature and harmonic analysis. *Indiana Univ. Math. J.*, 30(4) (1981), 519–37.

[GS99] Greenleaf, A., Seeger, A., On oscillatory integrals with folding canonical relations. *Studia Mathematica*, 132 (1999), 125–39.

[H73] Hörmander, L., Oscillatory integrals and multipliers on FLp. *Ark. Mat.*, 11 (1973), 1–11.

[H90] ———. The analysis of linear partial differential operators. I. Distribution theory and Fourier analysis. Reprint of the second (1990) edition [Springer, Berlin; MR1065993 (91m:35001a)]. *Classics in Mathematics,* Springer-Verlag, Berlin, 2003. x+440 pp. ISBN: 3-540-00662-1 35-02

[IKM10] Ikromov, I. A., Kempe, M., Müller, D., *Estimates for maximal functions associated to hypersurfaces in* $\mathbb{R}^3$ *and related problems of harmonic analysis.* Acta Math. 204 (2010), 151–271.

[IM11a] Ikromov, I. A., Müller, D., On adapted coordinate systems. *Trans. Amer. Math. Soc.,* 363 (2011), no. 6, 2821–48.

[IM11b] ———. Uniform estimates for the Fourier transform of surface carried measures in $\mathbb{R}^3$ and an application to Fourier restriction. *J. Fourier Anal. Appl.,* 17 (2011), no. 6, 1292–1332.

[I99] Iosevich, A., Fourier transform, L^2 restriction theorem, and scaling. *Boll. Unione Mat. Ital. Sez. B Artic. Ric. Mat.* (8) 2 (1999), 383–87.

[K84] Karpushkin, V. N., A theorem on uniform estimates for oscillatory integrals with a phase depending on two variables. *Trudy Sem. Petrovsk.* 10 (1984), 150–69, 238 (Russian); English translation in *J. Soviet Math.*, 35 (1986), 2809–26.

[KT98] Keel, M., Tao, T., Endpoint Strichartz estimates. *Amer. Jour. of Math.*, 120 (1998), no. 5, 955–80.

[LB09] Łaba, I. and Pramanik, M., Arithmetic progressions in sets of fractional Hausdorff dimension. *Geom. Funct. Anal.*, 19 (2009), no. 2, 429–56.

[LV10] Lee, S., Vargas, A., Restriction estimates for some surfaces with vanishing curvatures. *J. Funct. Anal.*, 258 (2010), no. 9, 2884–2909.

[M09] Magyar, A., On Fourier restriction and the Newton polygon. *Proceedings Amer. Math. Soc.* 137 (2009), 615–25.

[M00] Mockenhaupt, G., Salem sets and restriction properties of Fourier transforms. *Geom. Funct. Anal.,* 10 (2000), 1579–87.

[MVV96] Moyua, A., Vargas, A., Vega, L., Schrödinger maximal function and restriction properties of the Fourier transform. *Internat. Math. Res. Notices* 16 (1996), 793–815.

[M14] Müller, D., Problems in harmonic analysis related to finite-type hypersurfacs in $\mathbb{R}^3$, and Newton polyhedra. *Advances in Analysis: The Legacy of Elias M. Stein,* Princeton University Press, Princeton, NJ, 2014, 303–45.

[PS97] Phong, D. H., Stein, E. M., The Newton polyhedron and oscillatory integral operators. *Acta Math.*, 179 (1997), no. 1, 105–52.

[PSS99] Phong, D. H., Stein, E. M., Sturm, J. A., On the growth and stability of real-analytic functions. *Amer. J. Math.*, 121 (1999), no. 3, 519–54.

[R69] Randol, B., On the Fourier transform of the indicator function of a planar set. *Trans. Amer. Math. Soc.,* 139 (1969), 271–78.

[RS87] Ricci, F., Stein, E. M., Harmonic analysis on nilpotent groups and singular integrals. I. Oscillatory integrals. *J. Funct. Anal.*, 73 (1987), 179–94.

[Si74] Siersma, D., Classification and deformation of singularities. *Doctoral dissertation, University of Amsterdam*, (1974), 1–115.

[So93] Sogge, C. D., Fourier integrals in classical analysis. *Cambridge Tracts in Mathematics.* Cambridge University Press, Cambridge, 1993.

[Sc90] Schulz, H., On the decay of the Fourier transform of measures on hypersurfaces, generated by radial functions, and related restriction theorems. *unpublished preprint, 1990.*

[S93] Stein, E. M., Harmonic analysis: Real-variable methods, orthogonality, and oscillatory integrals. *Princeton Mathematical Series* 43. Princeton University Press, Princeton, NJ, 1993.

[SW71] Stein, E. M., Weiss, G,, Introduction to Fourier analysis on Euclidean spaces. *Princeton Mathematical Series* No. 32. Princeton University Press, Princeton, NJ, 1971.

[Str77] Strichartz, R. S., Restrictions of Fourier transforms to quadratic surfaces and decay of solutions of wave equations. *Duke Math. J.*, 44 (1977), 705–714.

[T03] Tao, T., A sharp bilinear restriction estimate for paraboloids. *Geom. Funct. Anal.* 13 (2003) no. 6, 1359–84,

[T04] ———. Some recent progress on the restriction conjecture. *Fourier analysis and convexity*, 217–243, *Appl. Numer. Harmon. Anal.*, Birkhäuser Boston, Boston, MA, 2004

[TV00] Tao, T., Vargas, A., A bilinear approach to cone multipliers. I. Restriction estimates. *Geom. Funct. Anal.* 10 (2000) no.1, 185–215.

[TVV98] Tao, T., Vargas, A., Vega, L., A bilinear approach to the restriction and Kakeya conjectures. *J. Amer. Math. Soc.* 11 (1998) no. 4, 967–1000.

[To75] Tomas, P. A., A restriction theorem for the Fourier transform. *Bull. Amer. Math. Soc.* 81 (1975), 477–78.

[V76] Varchenko, A. N., Newton polyhedra and estimates of oscillating integrals. *Funkcional. Anal. i Priložen*, 10 (1976), 13–38 (Russian); English translation in *Functional Anal. Appl.*, 18 (1976), 175–96.

[W95] Wolff, T., An improved bound for Kakeya type maximal functions. *Revista Mat. Iberoamericana*, 11 (1995), 651–74.

[W00] ———. Local smoothing type estimates on L^p for large p. *Geom. Funct. Anal.*, 10 (2000) no. 3, 1237–88.

[W01] ———. A sharp bilinear cone restriction estimate. *Ann. of Math.* (2), 153 (2001) no. 5, 661–98.

[Z74] Zygmund, A., On Fourier coefficients and transforms of functions of two variables. *Studia Math.*, 50 (1974), 189–201.

Index